In the Arms of Saguaros

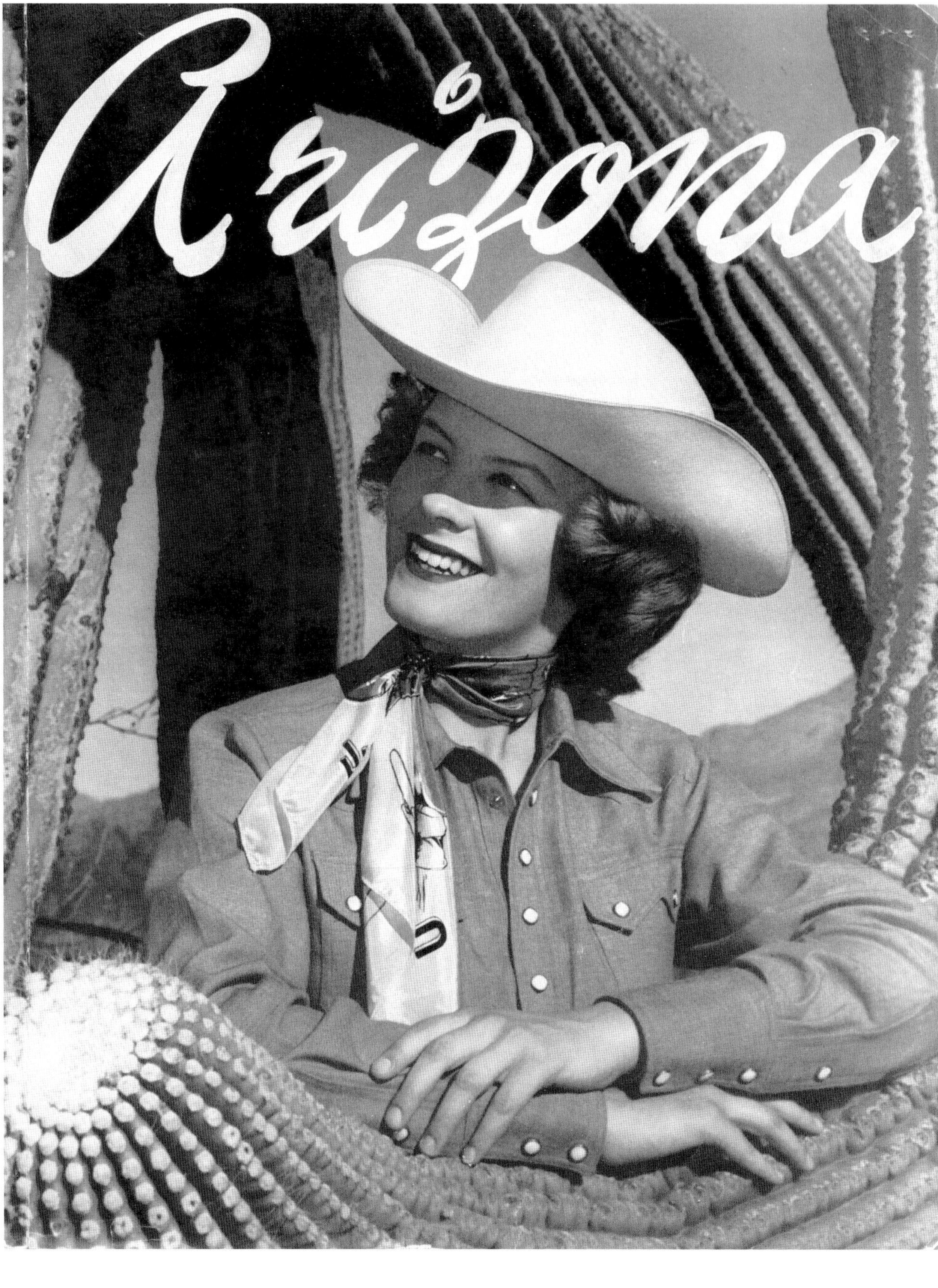
Arizona

In the Arms of Saguaros

Iconography of the Giant Cactus

William L. Bird Jr.

THE UNIVERSITY OF
ARIZONA PRESS
TUCSON

The University of Arizona Press
www.uapress.arizona.edu

We respectfully acknowledge the University of Arizona is on the land and territories of Indigenous peoples. Today, Arizona is home to twenty-two federally recognized tribes, with Tucson being home to the O'odham and the Yaqui. Committed to diversity and inclusion, the University strives to build sustainable relationships with sovereign Native Nations and Indigenous communities through education offerings, partnerships, and community service.

Originally published in 2021 as volume 3 of the Proceedings of the Desert Laboratory on Tumamoc Hill, University of Arizona

ISBN-13: 978-0-8165-5283-2 (paperback)
ISBN-13: 978-0-8165-5284-9 (ebook)

Cover design by Leigh McDonald
Cover art: Saguaro and background adapted from brochure cover. Hotels and Guest Ranches in Arizona. Rock Island Railroad, 1933. Postcard inset, Double U Ranch, Tucson. Kodachrome original by Ray Manley. Published and distributed by Bob Petley, Phoenix, Arizona. Postmarked 1953. Back cover images: "Tall tale postcard," 1913; Road map. "Golden Anniversary Statehood of Arizona." Arizona Highway Department, 1962; Brochure cover (detail). Tucson Arizona Sunshine Climate Club, ca. 1952. Collection of the author. Inside front cover: Stamp sheet. Tucson, Arizona, The Sunshine City. Inside back cover: Stamp sheet. Phoenix, Arizona. Fun under the sun in the Valley of the Sun. Frontispiece: Marcia Louise Luglan posed on a saguaro arm. 1950. Western Ways, Pictorial America–Arizona Edition. Photograph by Ray Manley.

Designed by Amanda González Moreno
Editor: Benjamin T. Wilder
Managing Editor: Anna Seiferle-Valencia

Library of Congress Cataloging-in-Publication Data is on the last page of the book.

Printed in the United States of America
♾ This paper meets the requirements of ANSI/NISO Z39.48-1992 (Permanence of Paper).

CONTENTS

In the Arms of Saguaros

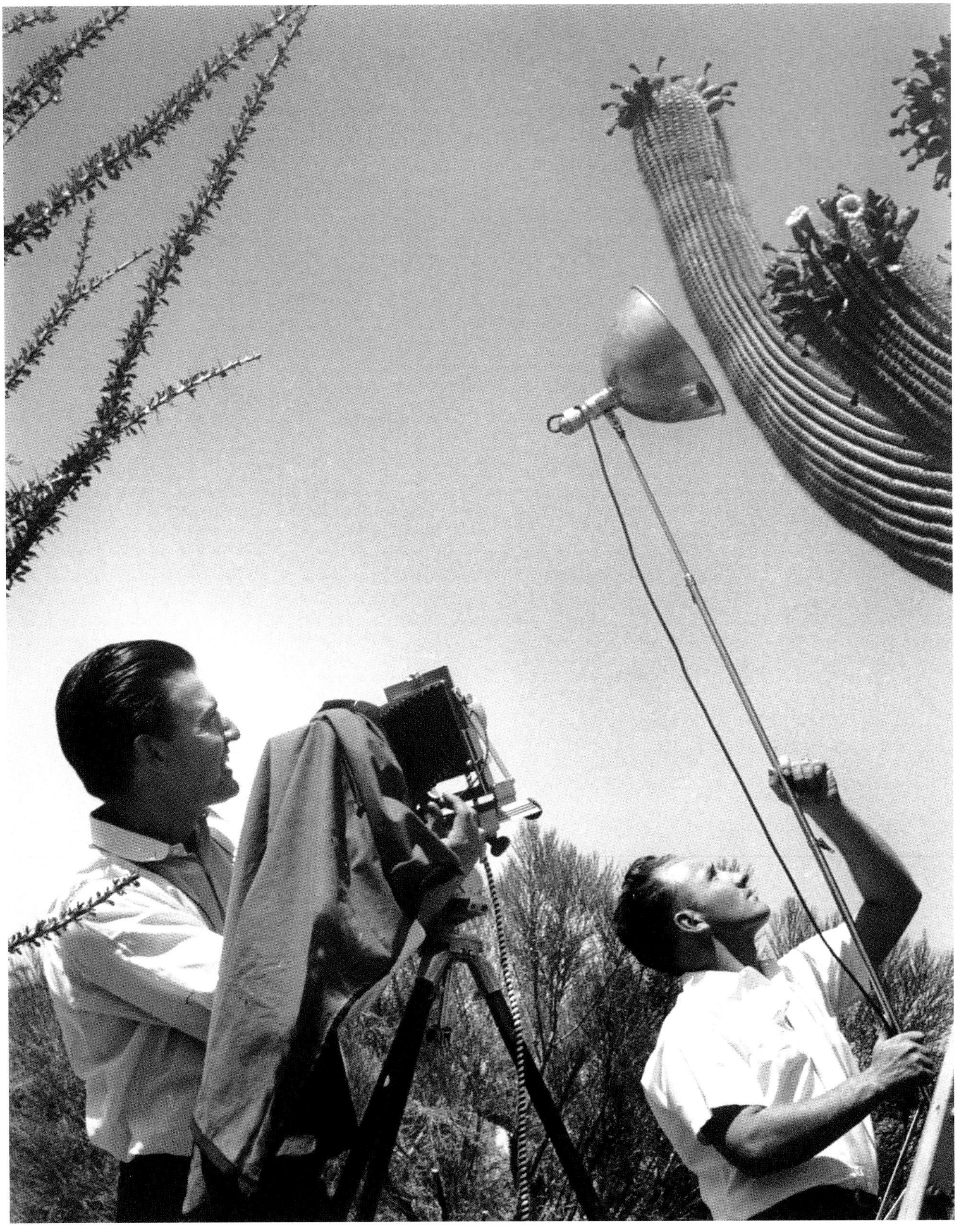

PREFACE: THE SOCIAL SAGUARO

In the Arms of Saguaros explores the historical development of saguaro cactus iconography from the botanical investigators of the nineteenth century to the leaders of the travel and tourism industry of the twentieth century. In this time the arms of the giant cactus came to symbolize the American Southwest to the world. In its guise as a welcoming metaphor, its outstretched arms appeared in travel literature often accompanied by freshly starched western wear, horseback rides, and swimming pools. The art direction of the saguaro's active social life idealized an arboreal desert of mesquite trees and creosote scrub with distant mountain views. Whether pictured in a magazine, a theatrical film, or a postcard, audiences responded. They saw, they came, and kept coming. Transfixed, many built desert homes, framed by saguaro views, that provided contemplative space for creative writing, painting, and photography that rounded the promotional circle in deeply personal ways (Figures I and II)—reflecting what desert publicists described as the anthropomorphic moods of man (Figure III).

Saguaro iconography illuminated the trail of scientific discovery through the Sonoran Desert in the heyday of American botanical exploration. Those encountering saguaros for the first time typically drew up short and marked the moment. This was no less true after the railroad's penetration of the Arizona territory for travelers who had become familiar with the desert giant through illustrated news and magazine features (Figure IV). Many who followed the railroad track and the irrigation ditch found little use for the wonders of botanical imagery outside of its application to crops. The packaging of Arizona as an agricultural paradise relegated the saguaro to a curiosity, if pictured at all. But what a curiosity. Botanists and amateur plant collectors alike built reputations on the study and acquisition of columnar specimens of the plant, furthering its fame as a fully armed and flowering wonder in the wild.

Highly imaginative and ever expansive, the saguaro's iconic range grew from its ethnobotanical significance as the species around which life revolved in the Sonoran Desert (Figures V and VI). Alternately described as a vegetable standpipe, a chandelier, and a sentinel, the projection of the saguaro as relatable, flexible, and

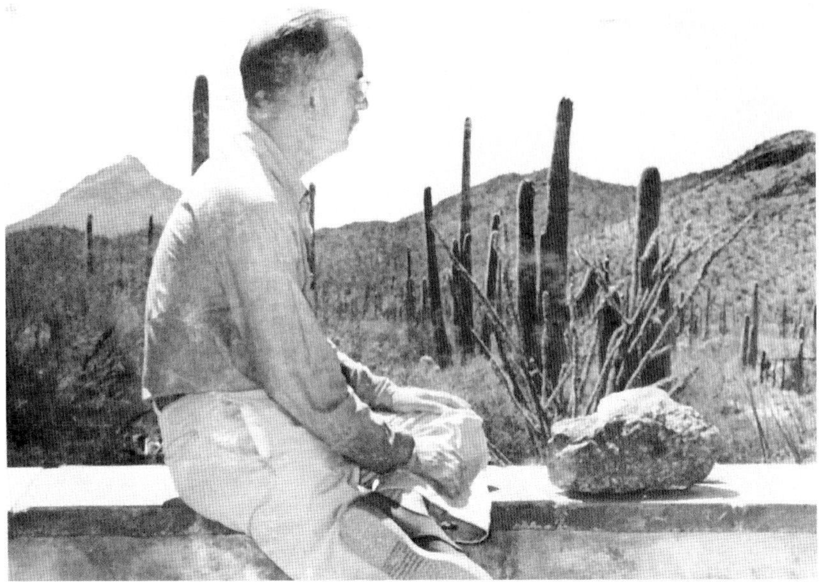

FIGURE I. Joseph Wood Krutch dustjacket photo. 1954. *Voice of the Desert*. Photograph by Ray Manley.

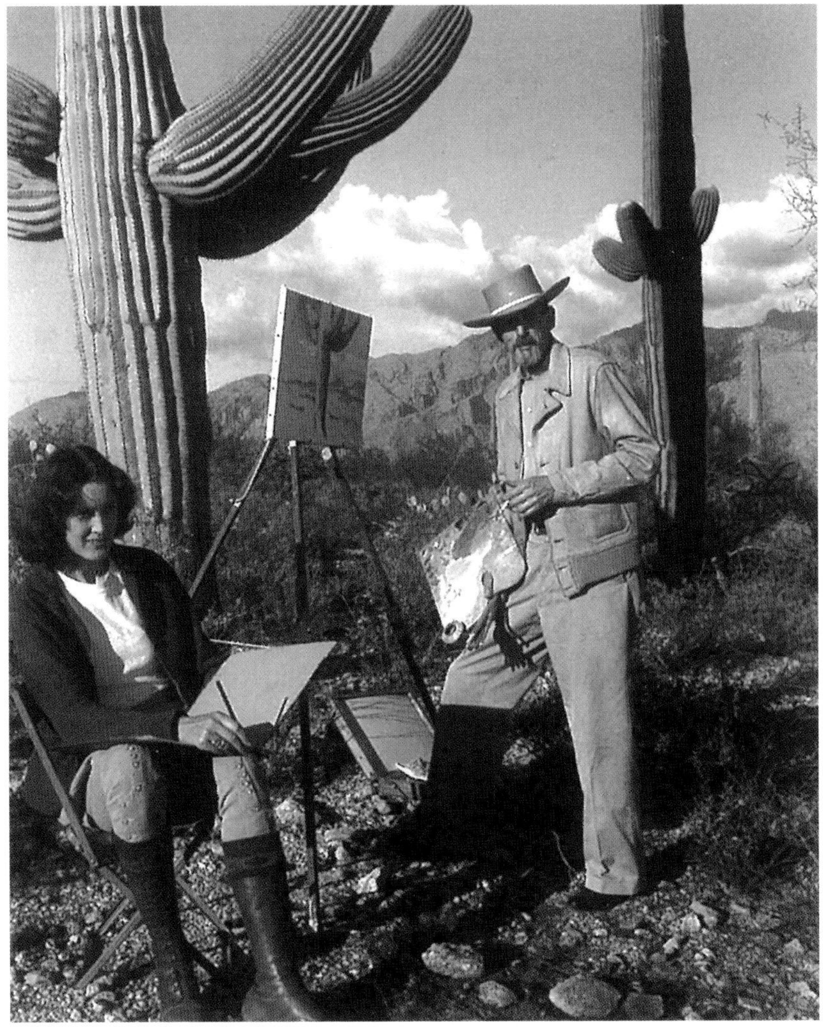

FIGURE II. Maynard Dixon and Edith Hamlin, Tucson, Arizona. 1942. Photograph by Charles Abbott.

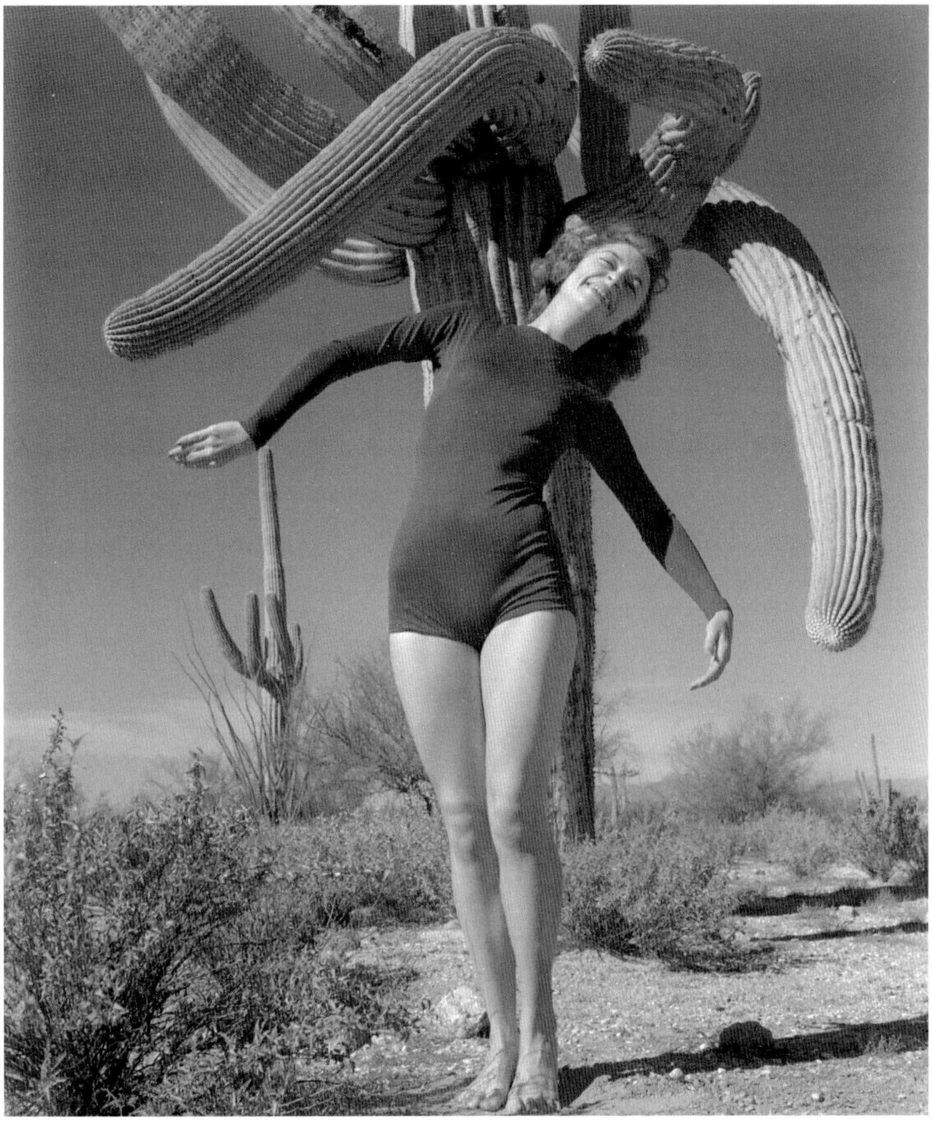

FIGURE III. Cactus ballet. 1950. Sarah Keil, director of the Tucson studio of the dance, demonstrates how the lonely sentinels of the desert take form from the moods of man. Photograph by Charles Herbert.

durable—particularly its ability to withstand hardship—suggests how, in fits and starts, the saguaro grew its iconic range.

Though not the largest columnar cactus—the *cardón* of Mexico is taller—the saguaro became a beloved icon and meme for the American West. Yet, oddly enough, the pioneering hospitality industries of Phoenix and Tucson initially regarded it with indifference. In the 1920s, for example, Tucson's Sunshine Climate Club launched a national advertising campaign to promote the city's winter climate. The saguaro played a minor role in club booklets mailed to midwestern and eastern audiences that focused upon trees, lawns, impressive homes, and winter weather for golf. Later, when the Sunshine Climate Club reduced its commitment to paid advertising, it sent its own photo publicity to newspapers and magazines. The typical

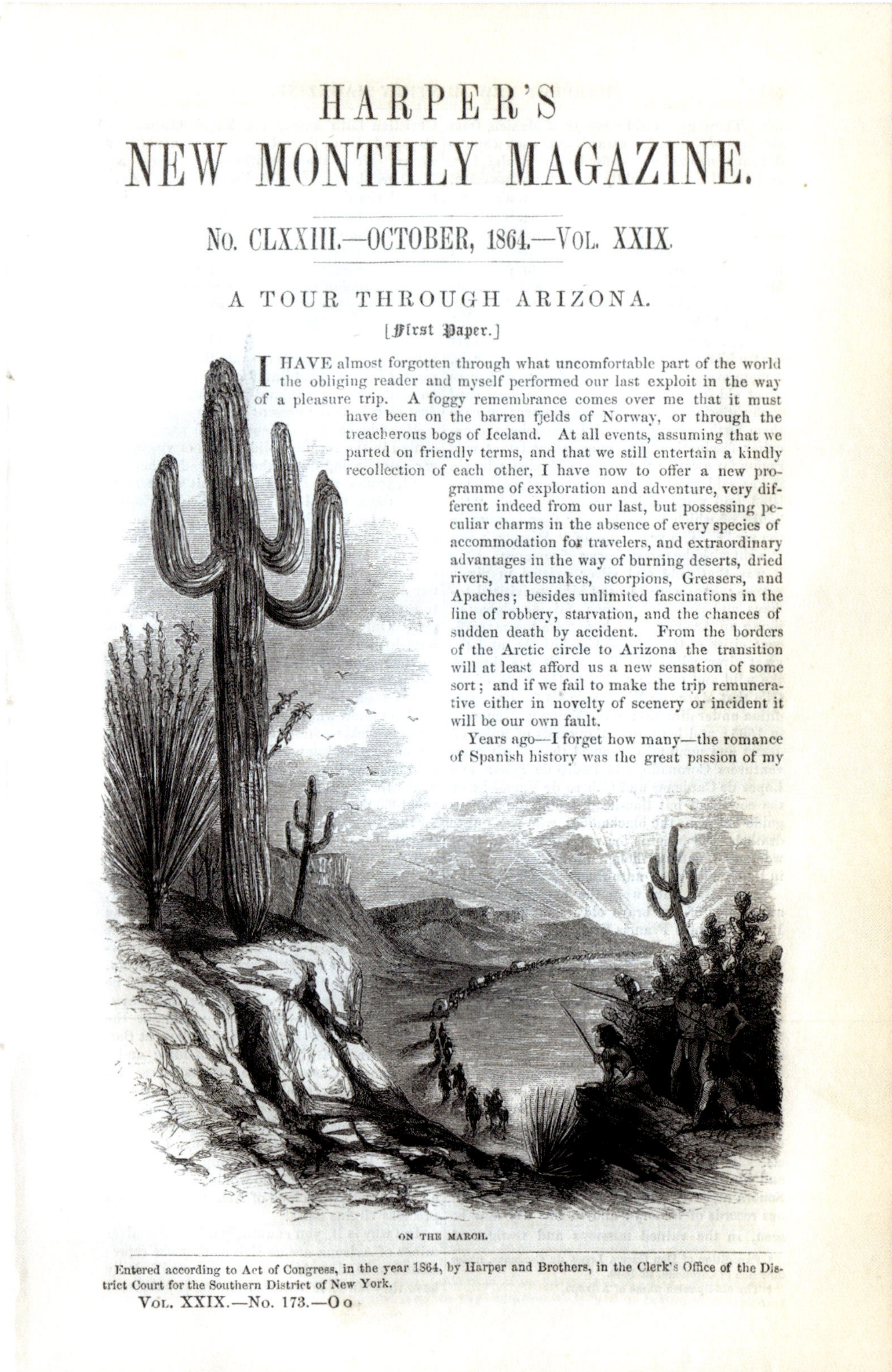

HARPER'S
NEW MONTHLY MAGAZINE.

No. CLXXIII.—OCTOBER, 1864.—Vol. XXIX.

A TOUR THROUGH ARIZONA.

[First Paper.]

I HAVE almost forgotten through what uncomfortable part of the world the obliging reader and myself performed our last exploit in the way of a pleasure trip. A foggy remembrance comes over me that it must have been on the barren fjelds of Norway, or through the treacherous bogs of Iceland. At all events, assuming that we parted on friendly terms, and that we still entertain a kindly recollection of each other, I have now to offer a new programme of exploration and adventure, very different indeed from our last, but possessing peculiar charms in the absence of every species of accommodation for travelers, and extraordinary advantages in the way of burning deserts, dried rivers, rattlesnakes, scorpions, Greasers, and Apaches; besides unlimited fascinations in the line of robbery, starvation, and the chances of sudden death by accident. From the borders of the Arctic circle to Arizona the transition will at least afford us a new sensation of some sort; and if we fail to make the trip remunerative either in novelty of scenery or incident it will be our own fault.

Years ago—I forget how many—the romance of Spanish history was the great passion of my

ON THE MARCH.

Entered according to Act of Congress, in the year 1864, by Harper and Brothers, in the Clerk's Office of the District Court for the Southern District of New York.

Vol. XXIX.—No. 173.—Oo

FIGURE IV. This *Harper's* woodcut illustrating "A tour through Arizona" is one example of the dissemination of saguaro cactus imagery dating to the mid-nineteenth century.

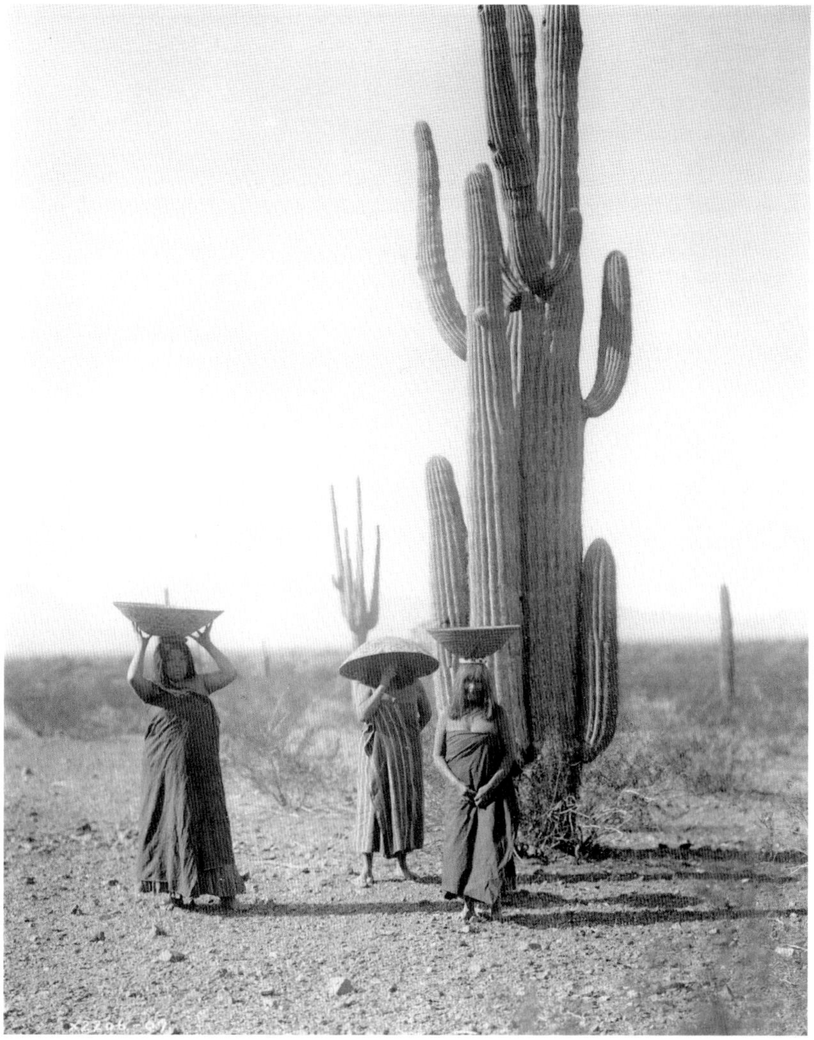

FIGURE V. Maricopa women with baskets for gathering fruit from saguaro cacti. 1907. In his pursuit of North American Indian portraiture, Edward Curtis staged the activities of the fruit harvest out of season with the noticeable absence of fruit. Photograph by Edward S. Curtis.

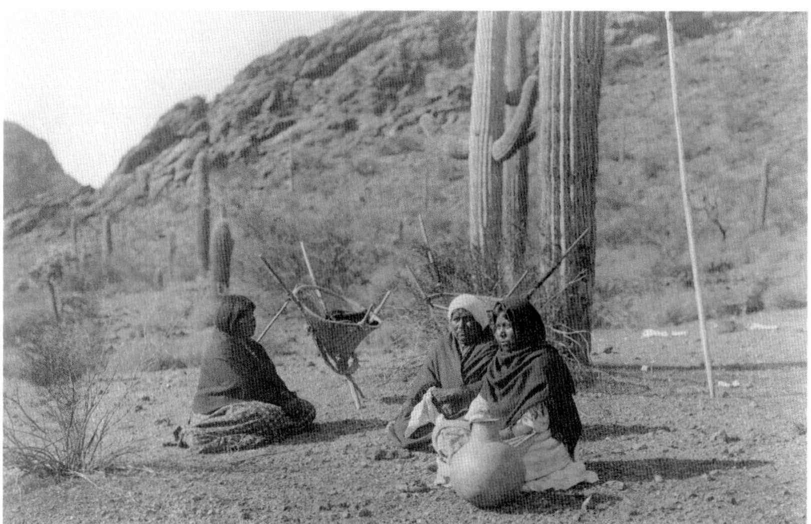

FIGURE VI. Qahatika women resting in Harvest Field—Qahatika. 1907. The pole leaning against the plant, right, was used to dislodge the fruit from the plants' tall columns. Photograph by Edward S. Curtis.

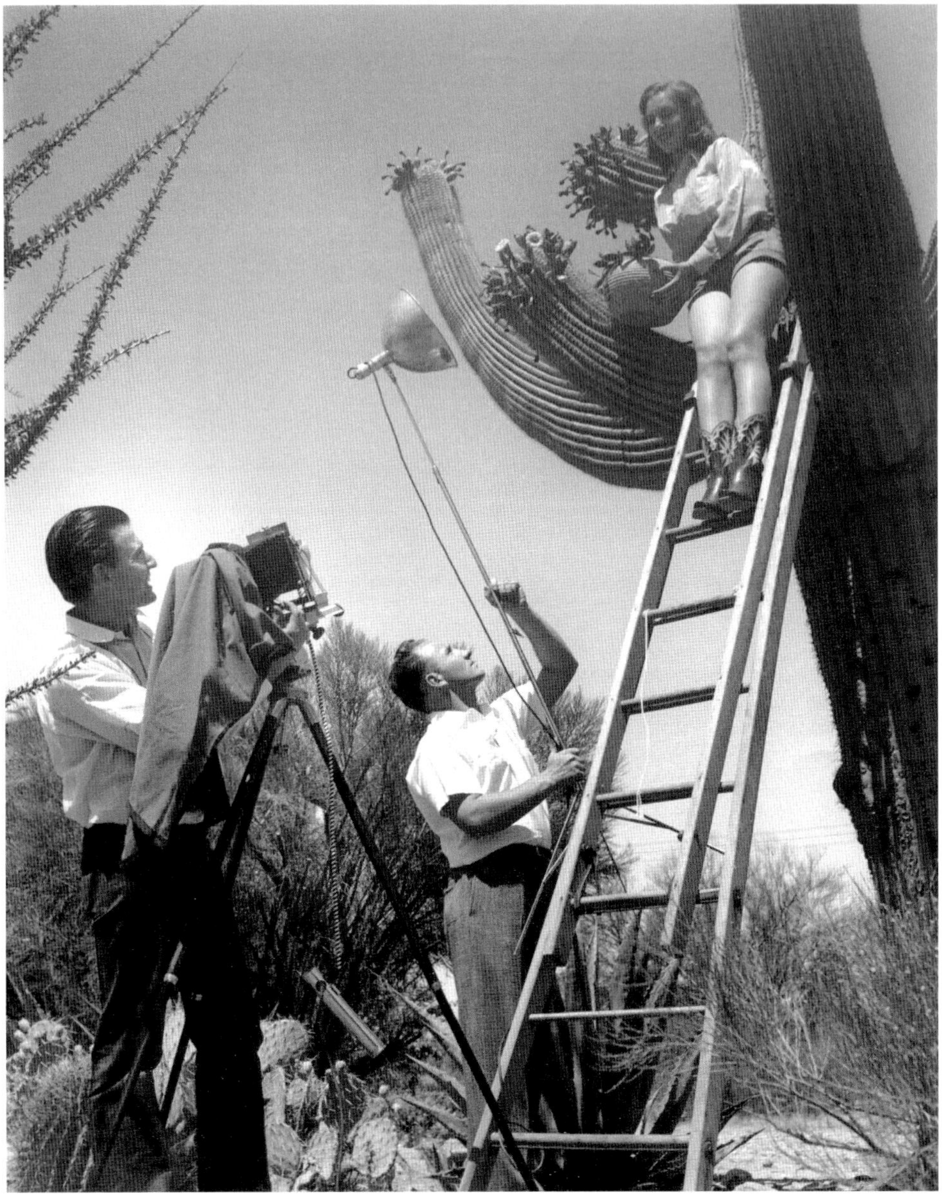

FIGURE VII. Photographer Ray Manley, with Howard Kinney, pose Jane McIntosh for a Tucson Sunshine Climate Club publicity photo. 1954. The photo's caption sheet explains that poses known as "cactus gags," became a staple of southern Arizona travel and tourism publicity: "Spines are removed to make gag shot possible." In this case, the photographers padded the topmost rungs of the ladder.

photo featured a young woman seated on the arm of a saguaro whose spines had been discretely cut or padded for the purpose of the pose (Figure VII).

In the Arms of Saguaros explores the role of advertising and creative publicity in the pell-mell development of the American Southwest. As desert publicists went about building a regional identity, they discovered that there was nothing like cozying up to the saguaro cactus for human interest, wit, and appeal. They packaged an audience for a *social saguaro* whose outstretched arms gathered resources, beckoned visitors, and welcomed new residents to a compelling new land in which anyone might find her or himself in the picture (Figure VIII).

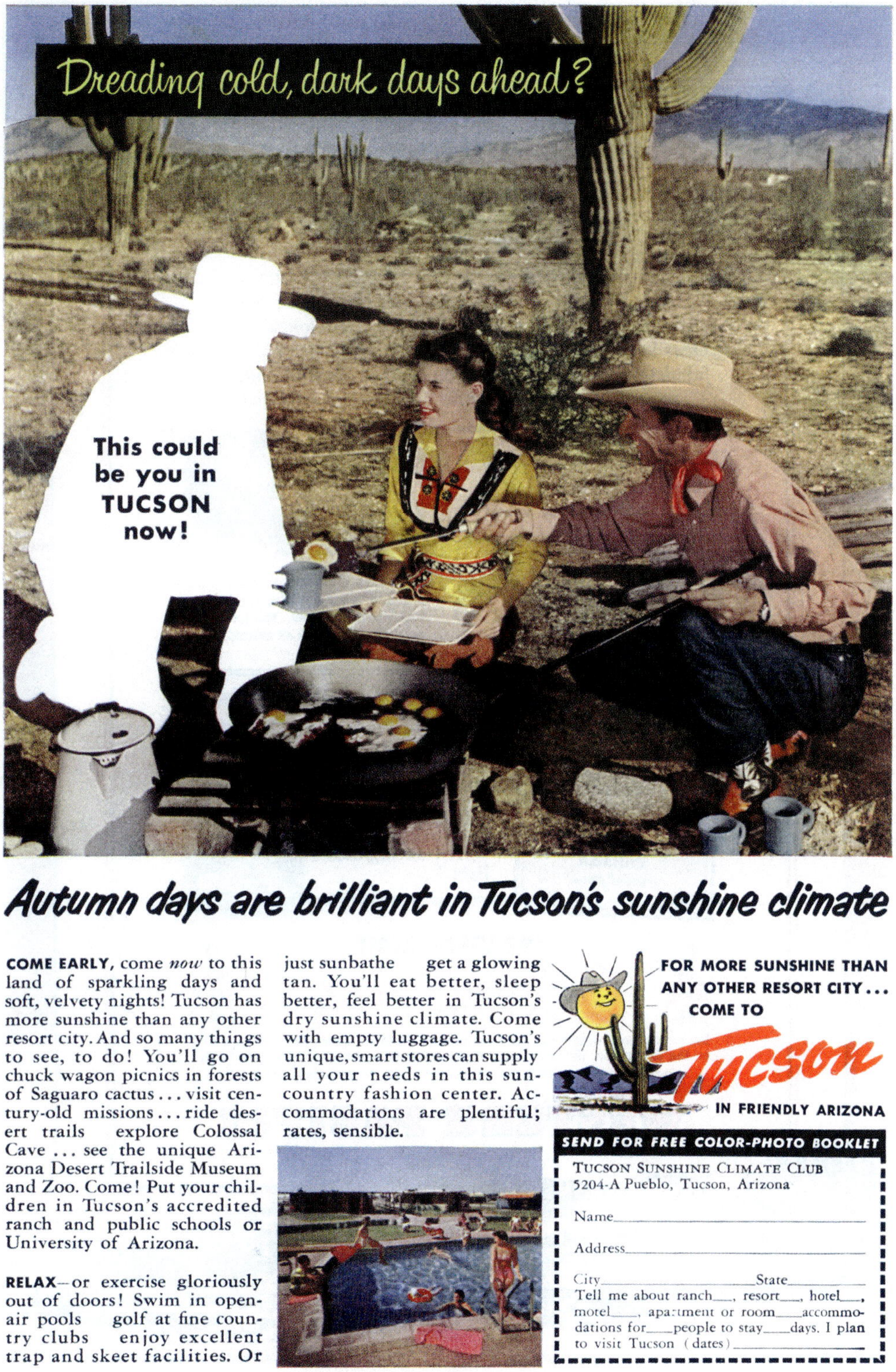

FIGURE VIII. Advertisement, Sunshine Climate Club. 1952. Dreading cold, dark, days ahead? Completing the picture, like the coupon, required little imagination. *National Geographic*.

ze feet in circumference

60 feet, sometimes a

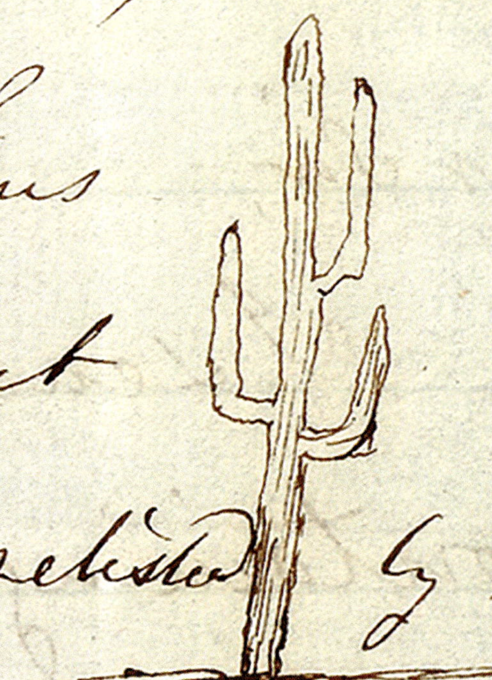

es thus it ha

ecollect correct

uch relished by Indians &

treat of my journal

, In the first

troduce your name

CHAPTER ONE
CACTUS WORDS AND PICTURES

The geographical range of the saguaro cactus fans south from Phoenix, Arizona across the Gila River Valley into Sonora, Mexico. Its western range crosses the Colorado River, touching southern California (Figure 1.1; Britton and Rose 1919–1920:vol. 2, pp. 164–167; Dimmit 2000; Yetman 2007:51–58; Drezner 2014). In the realm of the imagination, the saguaro enjoys an infinite range (Francaviglia 1994; Towne 2000).

With the exception of *Rhipsalis*, an epiphytic cactus native to Africa, cacti are exclusively native to the Western Hemisphere (Griffiths and Thompson 1929:2). The published record of the saguaro began in the mid-1500s. The earliest published accounts of southwestern explorers rendered all cacti as fruity and strange. Francisco Vasquez de Coronado, Juan de Oñate, and Father Eusebio Francisco Kino, for example, drank the wine that Indians distilled from the fruit of "pitahaya" and "great thistle" (Winship 1896:516; MacDougal 1905c). The wine's cactus source may have been a prickly pear, an organ pipe, or a saguaro, as each bears fruit. Little changed in the mid-eighteenth and early nineteenth-century accounts published by Don Antonio de Alcedo, Alexander de Humboldt, and Robert William Hale Hardy. Venegas (1759:82, 98, 100, 253, 324) and Alcedo (Thompson 1815:80) described the fruitful gathering of the pitahayas. Not until the 1820s did specific references to tall, saguaro-like columns appear in the published accounts of Humboldt (Black 1822:290–281) and Hardy (1829:212–213), who respectively noted "cylindrical cacti shooting up to extraordinary heights" and the tall poles that the Indians improvised to harvest their fruit.

Columnar *Cereus*, ball-shaped *Mammillaria*, and prickly pear *Opuntia* soon leapt from the pages of travelers' descriptive accounts to hothouses in England and continental Europe (Miller 1768; de Candolle 1829; Pfeiffer 1843; Brown 1998:i–iv). The botanists at London's Royal Botanical Gardens at Kew, for example, had long experience, if limited success, with large specimens collected in Mexico. In 1845 the garden acquired a large *biznaga* (barrel cactus) from San Luis Potosí. Collecting the cactus involved the logistical challenges of mule team cartage and a long ocean voyage from the port of Veracruz, the access point for cactus specimens

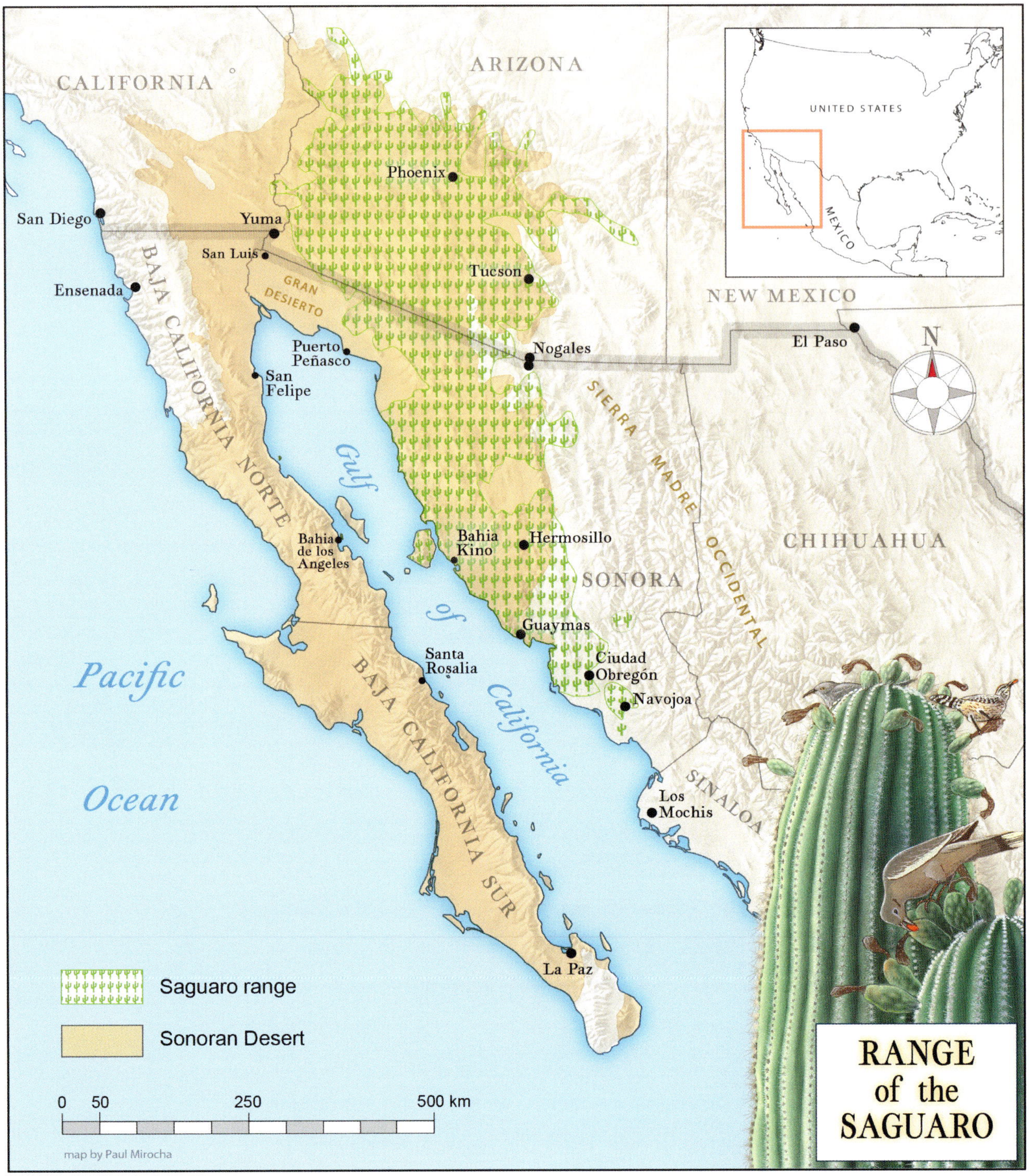

FIGURE 1.1. Range of the saguaro. Map by Paul Mirocha. Saguaro distribution based on Yetman et al. (2020).

dominated by large barrels, impossible to move *cardones* and columnar specimens that would be seen long before encountering saguaros (Yetman et al. 2020:9). The public exhibition of the seven-foot, 700-pound barrel at Kew took on the proportion of a living legend. The cactus, however, soon died. A second and even larger barrel specimen received the following year topped out at nine feet and weighed one ton. It, too, enjoyed a brief exhibit life (Figure 1.2). A columnar *Cereus senilis* next appeared at the garden. Tall, unbranched *Cereus senilis* specimens collected in Mexico became popular garden attractions (Figure 1.3; Illustrated London News 1845, 1846, 1847; Hooker 1847:43, 1849:92–94; Allnut 1877:48–54; Bean 1908:147–148; Torre 2017).

Encountering the saguaro cactus for the first time, desert travelers typically stopped. Traversing the Gila River Valley near the present-day city of Phoenix, Arizona, in 1824, the fur trader Frank Ohio Pattie (1962 [1831]:vol. 2, p. 193; MacDougal 1905a) recalled the moment:

> A species of tree, which I had never seen before, here arrested my attention. It grows to a height of forty or fifty feet. The top is cone-shaped, and almost without foliage. The bark resembles that of the prickly pear; and the body is covered with thorns. I have seen some three feet in diameter at the root, and throwing up twelve distinctive shafts.

Approaching the same area from the south in 1830, Asa Bement Clarke (1852:94) and a traveling companion encountered a "gigantic cactus" with branches "springing from the sides" and "fruit more delicious than a fig." Clarke observed "several specimens, whose branches six in number, grew regularly in the form of a chandelier." Clarke's characterization of the saguaro as a chandelier became a familiar metaphor among travelers who hid from the sun during the day and took advantage of the phases of the moon to light the way at night: "Their naked trunks seen by moonlight scattered over the ground reminded us of the monuments of the dead, and a friend remarked that it seemed as though we were passing a magnificent grave yard."

The Gila River Valley became highly studied terrain in the era of the Mexican War. A reconnaissance survey made by Lieutenant William Hemsley Emory of the U.S. Topographical Engineers confirmed what many Americans wished to believe: that the valley offered the most direct and flattest route for a railroad to California.

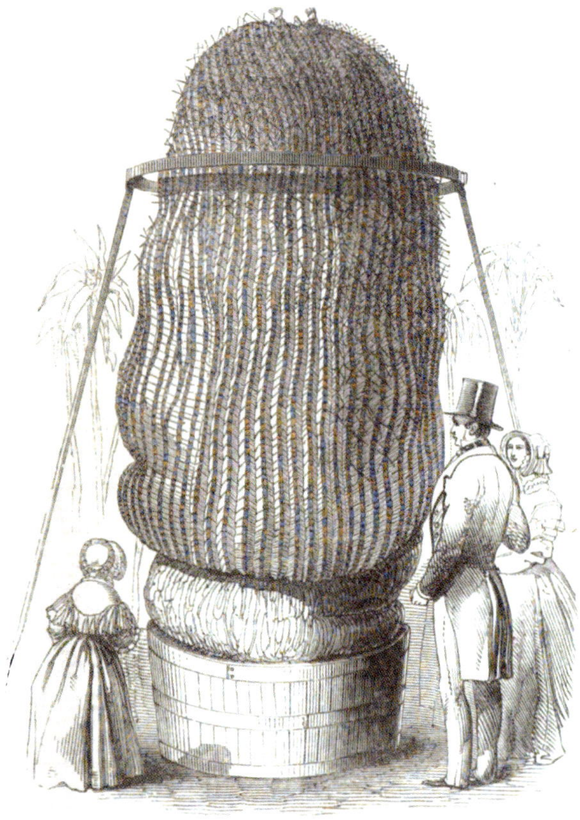

FIGURE 1.2. Woodcut. 1846. The exhibition of unusual cactus specimens from Mexico and their promotion in the press helped build audiences for the Royal Botanic Gardens at Kew. *Illustrated London News.*

It is hard to overstate the influence of Emory's and other publications from this period that documented the valley's natural resources, including the memoir of artist and U.S. Boundary Survey Commissioner John Russell Bartlett. The hard-scrabble work of survey botanists and artists who rendered what they saw in the field created a new and far-flung audience for the Sonoran Desert's plant life (Kearney et al. 1942:5–7; McKelvey 1955:990–1018; Hine 1968).

Topographical engineers who multitasked as artists found ready employment in boundary survey parties. The description of new plant specimens encountered in the line of march ranked in importance just below the imperative of mapmaking. Like every West Point cadet, the leadership of the Topographical Engineers received rudimentary training in composition and drawing, tasks readily applied to describing field specimens (Tyler 1999:158; Kelsey 2007:37). Even before Emory began his journey from Fort Leavenworth to San Diego by way of the Gila River Valley, two artists assigned to him dropped out at St. Louis when they became sick. Reaching Santa Fe, Emory encountered artist John Mix Stanley, who joined the reconnaissance as a plant collector and sketch artist, in addition to his formal assignment as a topographical engineer (Clarke 1966; Hassrick and Besaw 2015). Field sketches of large saguaro specimens and natural features that could not be collected informed scientific description and classification as contributions to knowledge (Pope 2020).

Before departing for the west in 1846, Emory made the acquaintance of George Engelmann, a St. Louis physician compiling a herbarium of cactus specimens collected from the Texas and New Mexico territory by his friend and future medical partner Dr. George Wislizenus, whose travel account, like Emory's, was published by the U.S. Senate in 1848 (Emory 1848c:155; Wislizenus 1848; McKelvey 1955:941–960).

Engelmann was one of a handful of American botanists who consulted a library of reference works to catalog plant specimens (Soule 1970; Shaw 1986). Botanists engaged in western study included Professor John Torrey of Princeton and his former pupil, Professor Asa Gray of Harvard. Engelmann, Torrey, and Gray exchanged plant specimens. In 1847, Engelmann prepared instructions for plant collectors, though it is not known whether he communicated those instructions to Emory or perhaps to the collector Augustus Fendler, whom Engelmann and Gray recruited and who independently followed Emory to Santa Fe by some eight weeks in 1847. Engelmann (1848, 1849, 1986) stressed the importance of collecting flowers, fruit, and seeds and of pressing each specimen in paper for study and preservation in a herbarium (McKelvey 1955:1019–1030; Shaw 1982:1–70; Stieber and Lange 1986).

Emory later acknowledged that he halted his party's botanical collecting before reaching Santa Fe, and none of it included cacti. "The infinite variety of cacti could not be brought home for analysis," he explained, "and this department of the Flora must be left to the enterprise of some traveler, with greater means of transportation than we possessed" (Emory 1848c:54). Emory did have the illustrative services of Stanley, who drew cactus specimens, including the massive saguaros the party encountered in the Gila River Valley.

Returning east from San Diego, Emory took up correspondence with Torrey to ask him to classify the dried plant specimens collected by the party and, with help from Engelmann, the cactus field drawings worked into small oil paintings by Stanley. Torrey (1847) accepted Emory's offer, including the services of an artist whom Torrey would engage, not to exceed $20, to draw up additional selected plant specimens at $2.50 per drawing. Torrey's participation gave heft to Emory's (1848c) published *Notes of a Military Reconnoissance*. Emory (1847) explained, "a very superficial determination [would] answer the purposes of the

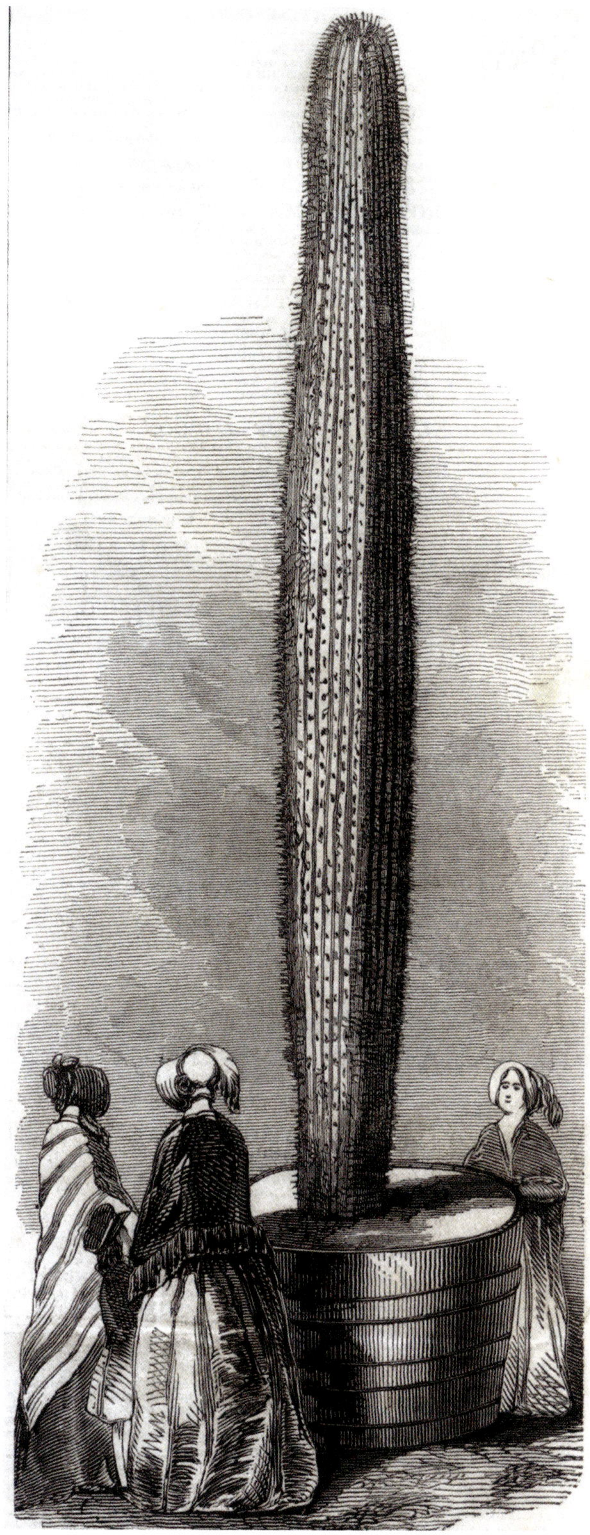

FIGURE 1.3. Woodcut. 1847. The exhibition of a *Cereus senilis* from Real Del Monte at Kew anticipated the exhibition of columnar saguaro specimens in botanical gardens throughout Europe and the United States. *Illustrated London News*.

government better than the more elaborate ones of other botanists whose services I might be enabled at once to obtain." Thus, Emory commenced a working relationship with Torrey, the leading American botanist, who became a clearinghouse for the classification of plants published in the expedition's report, with a nod to Engelmann, the leading expert on cacti (Reveal 2010).

The publication by the U.S. Congress of Emory's *Notes of a Military Reconnoissance* (1848c) and two volume *Report on the United States and Mexican Boundary Survey* (Emory 1857, 1859) exemplified the close coordination of lithographs and engravings with text, a recent development in publishing led by government reports (Seavey 1990).

Ideally, the picture production process worked this way: the artist, in this case Stanley, made a field sketch. At the conclusion of the trip, the artist worked the sketches into small oil paintings (Bryant 1849; Hassrick and Besaw 2015). Stanley gave his paintings and drawings to Emory, who sent them to Engelmann for botanical identification. Engelmann consulted his herbarium of flowers and seeds. Engelmann identified each plant so pictured and returned the art to Emory so named. Emory forwarded the artwork as a package to a contract lithographer or engraver in Washington, D.C., to produce stone lithographic or metal engraved plates capable of serially producing upwards of 10,000 fine prints for publication (U.S. Senate 1850; Metropolitan Museum of Art 2020). An example of each page-sized print would be sent to its original artist for approval pending publication.

At any point the production process might break down and, in the case of the saguaros, it did. The package of artwork sent to Engelmann did not include Stanley's saguaro figure that Emory described as a drawing. Somehow it bypassed Engelmann and went straight to the lithographer. But Emory needed its botanical name. In the body of a letter to Engelmann, Emory (1848a) apologetically sketched a saguaro from memory (Figure 1.4).

From Emory's crude outline and seeds and cactus fragments contributed by others, Engelmann (1848) named the saguaro *Cereus giganteus* (Emory 1848c:158). The new name distinguished the saguaro from the "pitaya," "pitahaya," or "petahaya," a name previously applied to all columnar cacti. *Cereus giganteus* appeared for the first time in Emory's *Notes*, including the caption for an impossibly large saguaro by lithographer C.D. Graham rendered from Stanley's errant artwork. It is not known if Stanley saw Graham's lithograph before publication (Figure 1.5). Stanley's original field sketches and paintings cannot be found today.

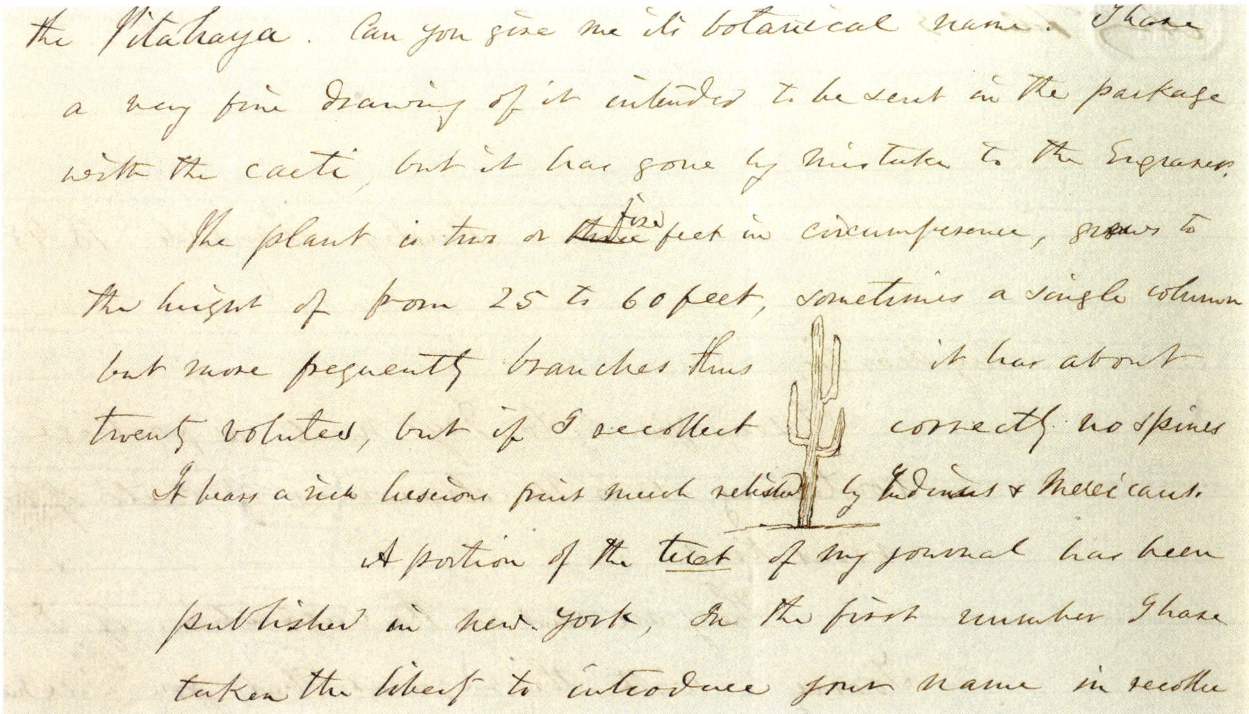

the Pitahaya. Can you give me its botanical name. I have
a very fine drawing of it intended to be sent in the package
with the cacti, but it has gone by mistake to the Engraver.
The plant is two or ~~three~~ five feet in circumference, grows to
the height of from 25 to 60 feet, sometimes a single column
but more frequently branches thus it has about
twenty volutes, but if I recollect correctly no spines
It bears a rich luscious fruit much relished by Indians & Mexicans.
A portion of the text of my journal has been
published in New York, In the first number I have
taken the liberty to introduce your name in recollec

FIGURE 1.4. William H. Emory's saguaro sketch in his letter to George Engelmann requesting a botanical name for the new plant. 1848.

Upon the publication of *Notes,* botanists at London's Kew Gardens duly noted Engelmann's botanical identification of the saguaro. Reviewer William Jackson Hooker (1849:92–93) in the *Journal of Botany* stated, "Assuredly the most remarkable plants met with were the *Cacti*, and of them the most striking is called by Dr. Engelmann *Cereus giganteus.*" However striking, Hooker found fault with the figures' lack of detail, which wanted for "botanical character." Concurring with Hooker, Engelmann wrote Torrey that Stanley's figures were beautifully rendered. But as botanical illustrations they were, in fact, useless. In the future, Engelmann (1848) would redouble his efforts to provide instructions for survey artists and plant collectors.

With the conclusion of the Mexican War, a new and much enlarged American survey party returned to document the Gila River Valley in detail. The Treaty of Guadalupe Hidalgo provided instructions for marking the border of the United States and Mexico in a rough line from San Diego to the Rio Grande River above El Paso del Norte, and then down the river to the Gulf of Mexico. To fulfill their mutual treaty obligations, each nation organized cooperative boundary survey parties to place stone pylons and markers in the field, confirmed by astronomical observation. The United States Congress created a Boundary Survey Commission administered by the Department of the Interior. John Russell Bartlett, a politically connected

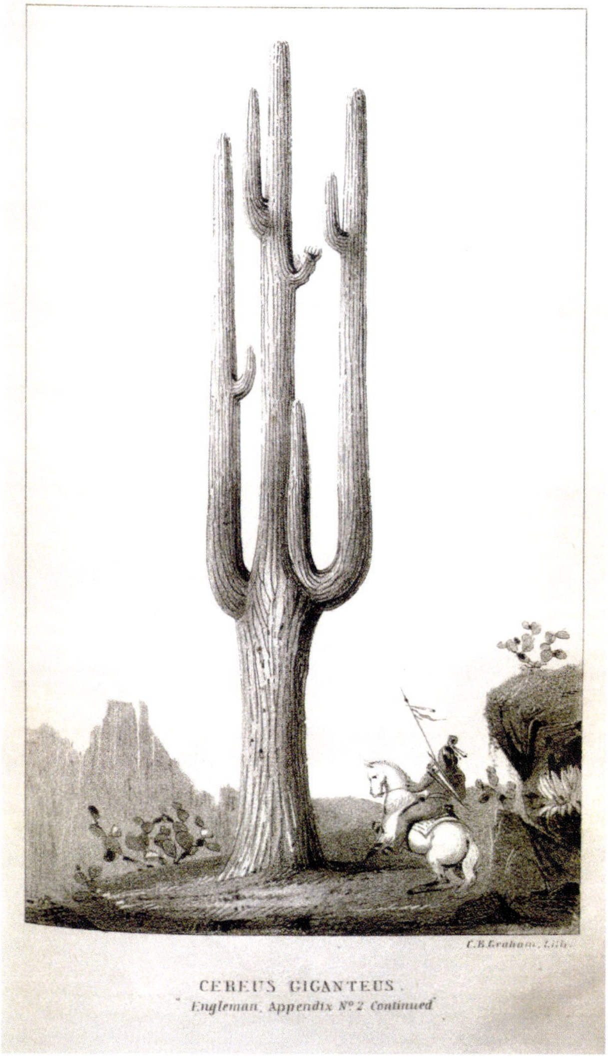

FIGURE 1.5. Lithograph. 1848. C.B. Graham, "*Cereus giganteus*," after John Mix Stanley (Emory 1848c).

Manhattan book dealer, amateur ethnologist, and artist led the new commission. The Bartlett party included John Bigelow, a surgeon who served as botanist; Augustus de Vandricourt, an artist and draftsman; George Thurber, a pharmacist by training who served as botanist, quartermaster and interpreter; and Henry Cheever Pratt, an artist and friend of Bartlett. Emory, unwilling to resign his military commission to work as a civilian in the Department of the Interior, joined the survey as chief astronomer (Werne 2007:48–49, 221–223).

The fieldwork of the Bartlett commission proceeded slowly. Bartlett was well on the way to producing an illustrated work about his travels when he wandered away from the boundary line into Mexico in search of supplies. When he became sick, he and his party debarked from Acapulco for San Diego (Werne 2007:116–177, 223). Emory assumed Bartlett's responsibilities in 1851. Bartlett, after a stay in California, which included a roundtrip from San Diego to San Francisco and back, returned to the field in 1852. But wandered away Bartlett had, and Emory found himself the commission's de facto leader. Bartlett's recall followed an acrimonious debate in Congress about his placement of the survey's initial starting point on the Rio Grande above El Paso. The Gadsden Purchase rendered the point moot. Emory became survey commissioner in 1854. He completed the survey establishing the boundary of the United States and Mexico in 1857 (James 1969; Werne 2007:193).

Emory (1857, 1859) brought the full weight of the U.S. Congress to bear upon the publication of the two-volume *Report on the United States and Mexican Boundary Survey*. Emory loaded it with lithographs, appendices of botanical and mineralogical specimens compiled by John Torrey, and a comprehensive cactus catalog, *Cactacae of the Boundary*, compiled by Engelmann (1859).

Cut out of the official record, Bartlett privately published a two-volume memoir in 1854 that took parting shots at Emory. No, Bartlett stated, Emory had not been

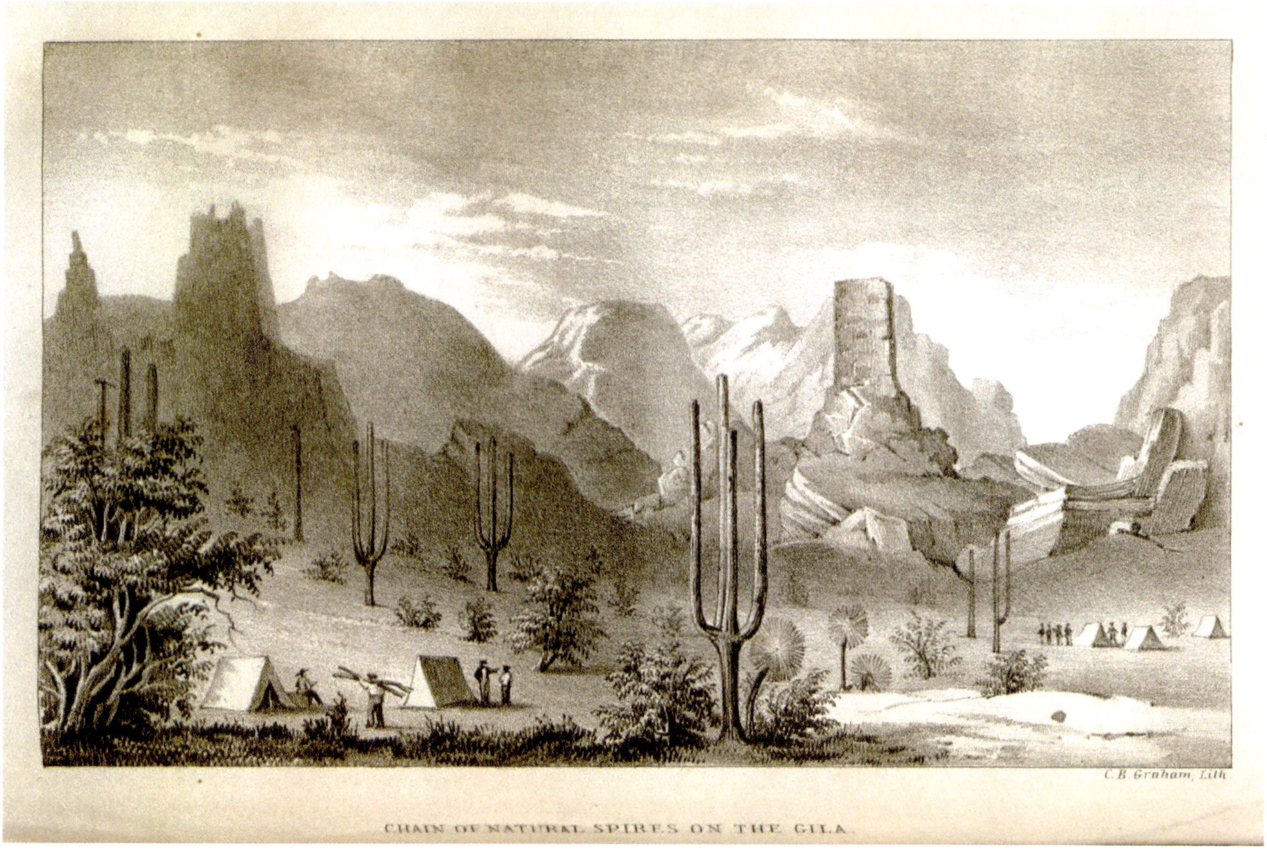

FIGURE 1.6. Lithograph. 1848. C.B. Graham. "Chain of Natural Spires on the Gila," after John Mix Stanley (Emory 1848c).

the first to describe a saguaro, as casual readers of his congressional report might believe. Ever the bibliophile, Bartlett (1854:vol. 2, pp. 191–193fn) cited the desert exploration accounts of Venegas (1759), Alcedo (Thompson 1815), Humboldt (Black 1822), Pattie (1962 [1831]), and Hardy (1829), who noted the plant and described the Indian fruit harvest.

Bartlett's scientific men, notably botanist James Thurber, had seen the saguaro flower and fruit in season. Thurber collected saguaro seeds and distributed them for the asking (Engelmann 1854). As a consequence, saguaro seedlings grew in hothouses from Cambridge to Kew.

Engelmann acknowledged Thurber's contributions in 1854, and folded them into his earlier description of the plant published in November 1852 based upon the reporting of Charles Christopher Parry, the cactus enthusiast and survey commission botanist closest to Emory. In April 1852, Emory wrote that he had sent Parry to explore the cacti in the region "passed over" in 1846–1847, with a copy of Engelmann's questions "prepared for just such an opportunity, particularly to supplying the data required for a more complete description of the *Cereus giganteus*." Emory immediately forwarded Parry's field notes to Engelmann, urging Engelmann to

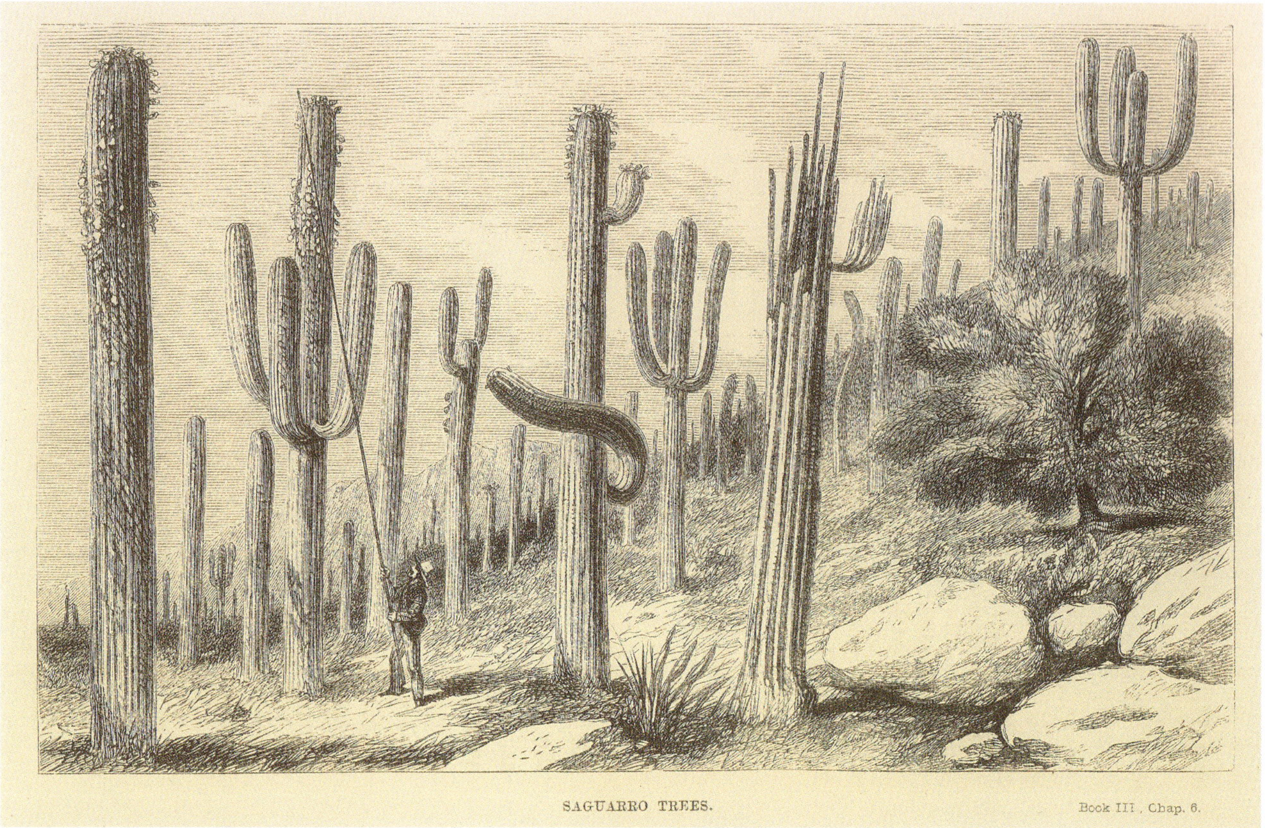

FIGURE 1.7. Lithograph. 1859. Saguaro Trees (Froebel 1859).

promptly publish an updated description of the giant cactus in "*Silliman's Journal* or any other periodical you may select." Emory (1852) continued, "I am desirous this should be done at the earliest possible moment to prevent being anticipated by others." Based on Parry's notes, Engelmann's revised description of *Cereus giganteus* appeared just seven months later in the November 1852 issue of Benjamin Silliman's *American Journal of Science and Arts*. Engelmann (1852, 1854) credited Parry, who "paid particular attention to the Cacti of that region, and made it an especial object carefully to examine the *Cereus giganteus*." Engelmann later folded the same into his *Cactacae of the Boundary* appended to Emory's *Report*.

Although the competitive saguaro field descriptions published by Engelmann and Bartlett included fruit, the German explorer Julius Froebel published the first picture of a saguaro fruit *harvest* in 1859 (Figure 1.6). Froebel traveled through the Santa Cruz River valley in the early 1850s and, though one can only speculate, knowledge of Froebel's presence in the field—like Bartlett's—may have contributed to the urgency of Emory's communication to Engelmann regarding prompt publication of a complete saguaro description. Froebel's first-person account of the Pima Indian fruit harvest pictured a figure of indeterminate ethnicity using a pole to dislodge fruit from the plants' columns. Froebel's woodcut rendered the subtle detail of dried flowers and the mechanics of dislodging the ripened fruit with a pole. Foregrounded with the harvester, a saguaro with a curved arm showed fruit and flowers. Completing the plants' life cycle from left to right, a partial, skeletal rib cage appeared to be tall enough to have provided the pole (Froebel 1859:508–509). Froebel deployed Engelmann's *Cereus giganteus* nomenclature. But Froebel, a former newspaperman, took pains to note the distinction that the Indians drew between a "pitaya" (Bartlett had used both *Cereus giganteus* and "petahaya") and a "saguarro." Froebel (1859:498) explained, "I have obtained my information respecting the difference of these two species of cactus from a good authority—the chief of the Pimas—who told me decidedly that the cactus-shafts in that part were not Pitayas, but Saguarros."

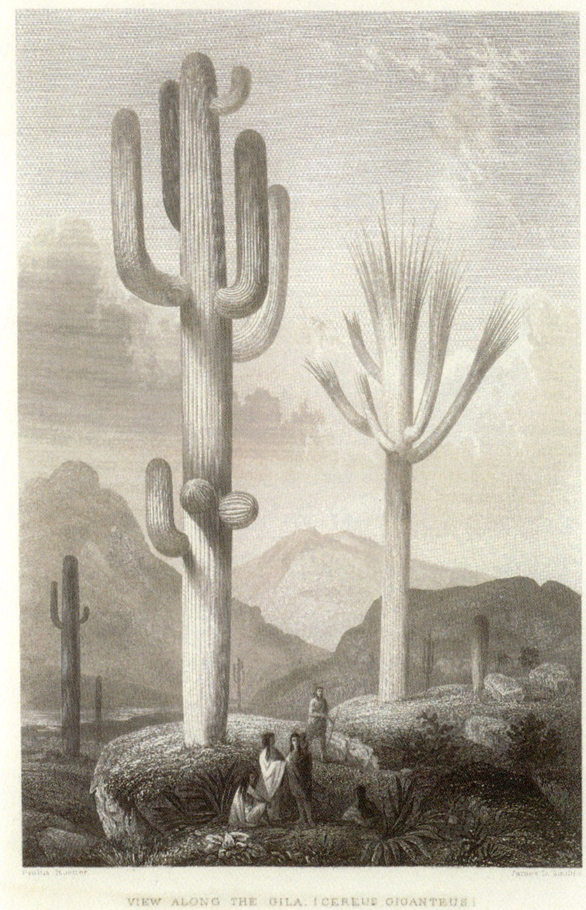

FIGURE 1.8. Engraving. 1858. Paulus Roetter based upon a field sketch by Heinrich Balduin Mollhausen. Engraving by James David Smillie. "View Along the Gila—*Cereus giganteus*" (Engelmann 1859).

The publication of Emory's and Bartlett's accounts concluded the era of government-financed discovery and publication in which saguaros grew to towering proportions that stood apart (Figure 1.7; Emory 1848c:78–79). Becoming more dramatic with each picture, the frontispiece of Engelmann's (1859:63) *Cactaceae of the Boundary*, for example, pictured two *Cereus gigantea* towering over the huddled figures of an Indian family on the banks of the Gila River (Figure 1.8). The picture is all the more puzzling inasmuch as Engelmann worked closely at St. Louis with Paulus Roetter, his hand-picked botanical illustrator for *Cactaceae of the Boundary*. Roetter painted the scene from a field sketch by Heinrich Balduin Mollhausen. Engraver James David Smillie made copies for publication (Trelease and Gray 1887:176).

As in the earlier case of the saguaro attributed to Stanley, perhaps it was simply an error of scale heightened by the shortening of the human figures posed with the plants. Or perhaps the verbal information that accompanied the field sketches likened the saguaro to a woody tree. At this late date, we can only speculate. The original artworks of Stanley, Mollhausen, and Roetter have long been unavailable to compare with the published images. The original art was considered expendable and not even the plates were saved. Asa Gray, who went looking for Engelmann's plates in the mid 1880s and could find none, concluded that they had been "mainly lost or destroyed" (Trelease and Gray 1887). Though Gray's concern was with the reproduction of what he plainly regarded as historical works, the loss of original art complicates any attempt to assign responsibility for the saguaro's inflated dramatization as a discovery. Though Stanley was perfectly capable of romanticizing western landscapes and Indian subject matter (Truettner 1991; Hassrick and Besaw 2015), neither he, Mollhausen, nor Roetter appear to have ever reviewed the lithographs or engravings of their saguaros before publication.

The rendering of saguaros as monumental artworks was hardly problematic, as few people had seen them in the field. The visualization of the saguaro presented pictures as news from a relatively inaccessible land. Bartlett (1854:vol. 1, p. 416, vol. 2, pp. 188–193) described the "petahaya" as "the monster of the cactus family." Framing his critique of Emory around anecdotal descriptions of flowering "petahayas," Bartlett said nothing about Engelmann's contribution of the plant's botanical name. For his part, Emory (1853) privately complained that Bartlett had withheld plant specimens owed the government and was "a putrid carcass...to be avoided accordingly."

Engelmann's *Cereus giganteus* stood as the saguaro's botanical name for nearly 50 years until 1908, when botanists Nathaniel Lord Britton and Joseph Nelson Rose changed it to *Carnegiea gigantea* to honor Andrew Carnegie, the founding patron of the Carnegie Institution's Desert Laboratory at Tucson, Arizona. Carnegie, then visiting Tucson, was said to have been the disappointed subject of his associates' pandering. Around the same time, the Pima-derived word *saguaro* displaced *petahaya* in everyday use. Displacing *suwarro, sahuaro, and saguarro*, saguaro became the preferred spelling (MacDougal 1905b; Washington Post 1909).

The saguaro fruit harvest became the focus of ethnobotanical study that built upon the descriptive method of Thurber and Bartlett and the narrative woodcut published by Froebel (McGee 1896; Castetter and Bell 1937; Felger and Moser 1974; Crosswhite 1980). Going forward, flowers and fruit rendered as disjoint parts contributed to the dominant perception of the saguaro's smooth, out-of-season profile, very much like the drawing that Engelmann received from Emory, who in all honesty could not remember if the plant had spines or not (Emory 1848a, 1848b).

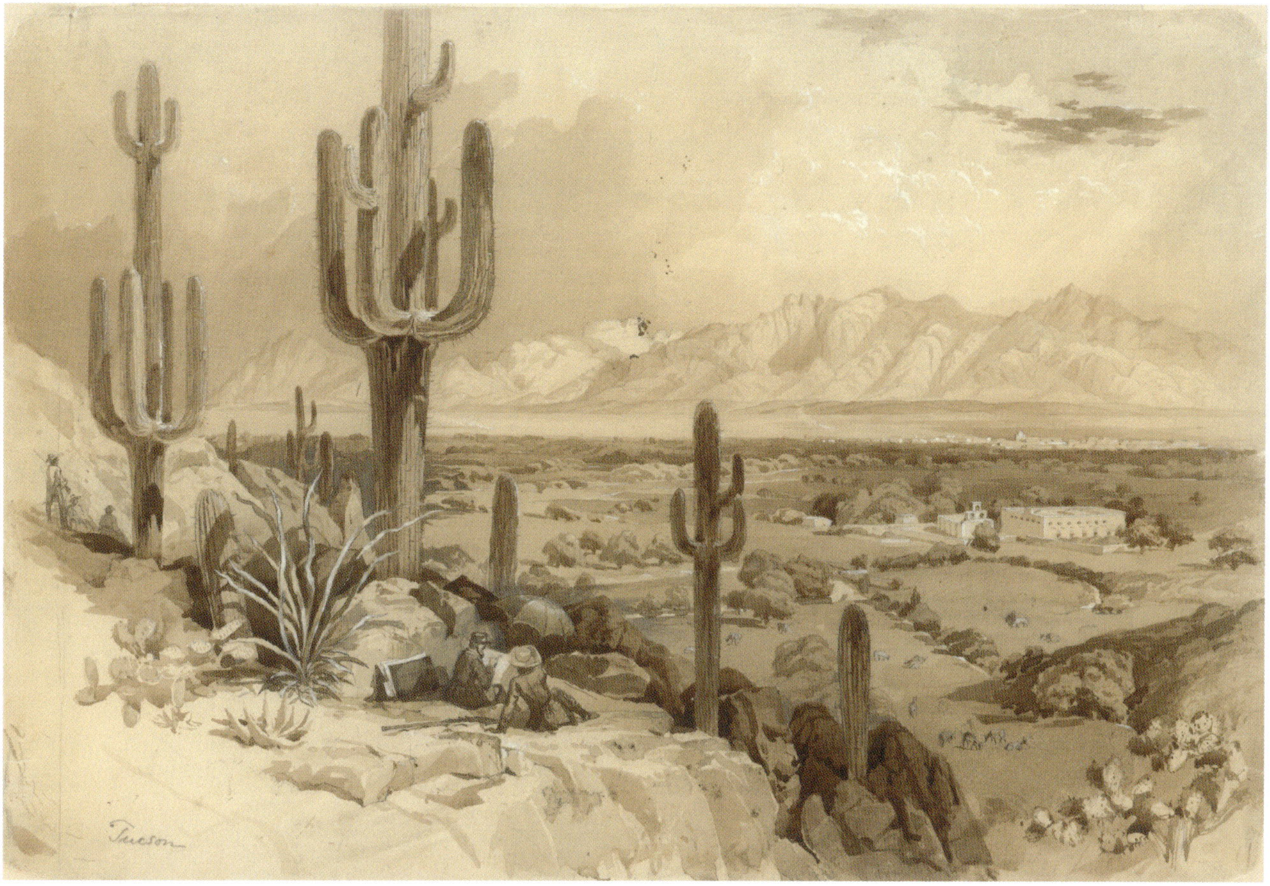

FIGURE 1.9. Pencil, sepia, and wash. 1852. John Russell Bartlett, Tucson, Arizona, from the eastern flanks of Sentinel Peak looking northeast across the Santa Cruz River Valley.

Though Emory wrote Bartlett out of the official record, Bartlett the artist had the last word in rendering an emotionally satisfying southwestern landscape dominated by the saguaro, very different from the towering plants that humbled the figures in Emory's published works (Bartlett 1854:vol. 1, p. 416). In the waning days of Bartlett's administration as Survey Boundary Commissioner, he took to the hills above Tucson to sketch and paint. Bartlett's pen-and-ink wash pictured the members of his party at rest among the saguaros on the present-day Sentinel Peak (Figure 1.9). The relaxed, social setting of people in the company of plants would become increasingly familiar to collectors, health seekers, and boosters in the years to come.

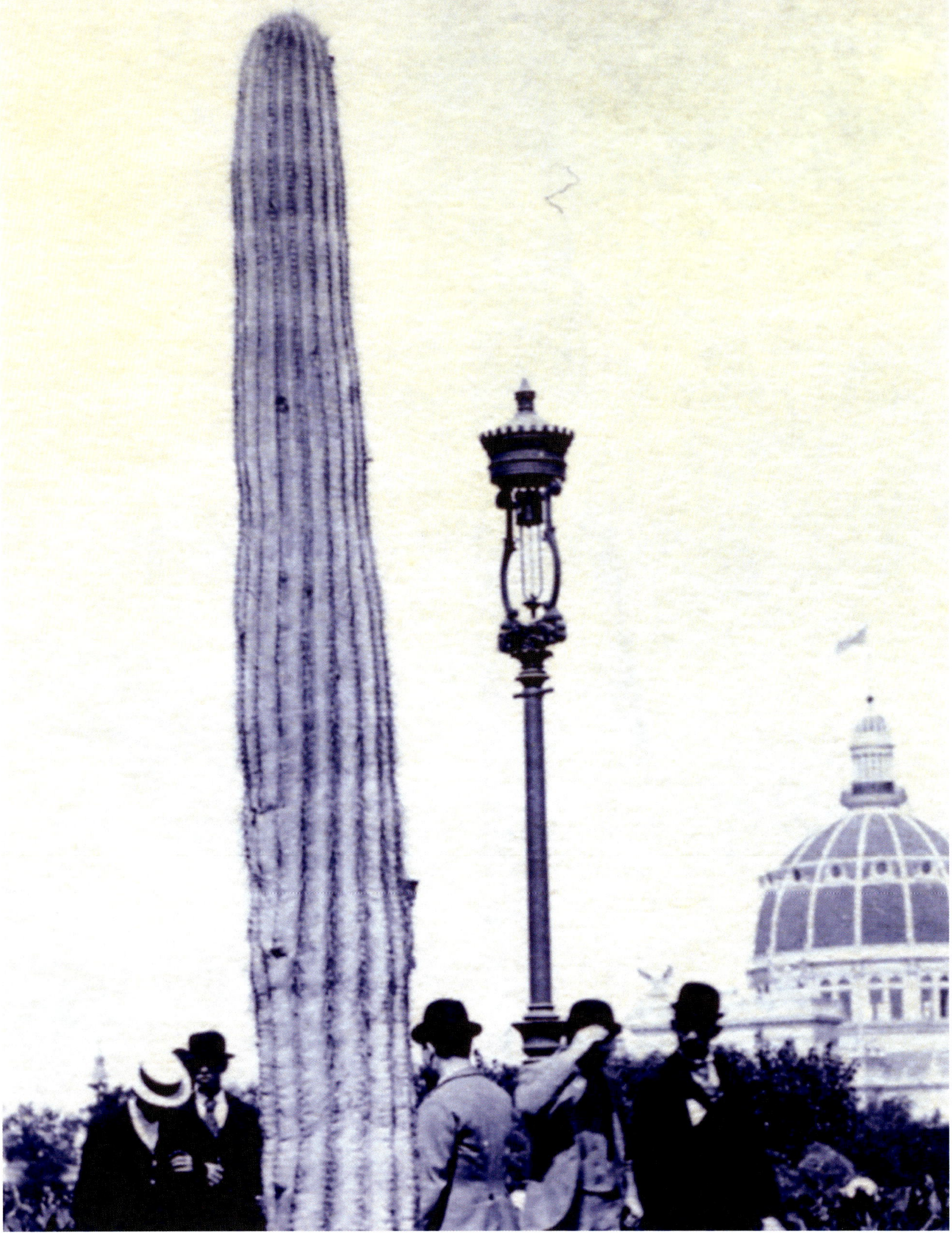

CHAPTER TWO
CACTUS DISPLAYS

The transplantation of the saguaro cactus followed the railroad. After the arrival of the Southern Pacific from the west at Tucson in 1880, the first columnar saguaros left their native habitat for California. With the completion of the railroad to the east, transplanted saguaros became perennial features at the nation's world's fairs. Hobbyists tested the limits of saguaro transplantation on a grand scale. Only gradually did cactus collectors yield to conservation concerns that curtailed the more rapacious aspects of collecting.

Built on the colonial model, the organized extraction and distribution of prime specimens from Mexico and the Arizona territory percolated through American, British, and European culture. The largest specimens fell to those with the means to maintain a greenhouse. Hobbyists with windowsills settled for more manageable succulents, including the saguaro seedlings advertised in mail order catalogs illustrated with the pictures of the mature plants first published by Emory, Bartlett, and Froebel.

The dig was on. The Southern Pacific's Charles Crocker created the earliest transplanted saguaro display of its kind, the Arizona Garden, at the Hotel Del Monte on California's Monterrey Peninsula. Designed by landscape architect Rudolph Ulrich, the Arizona Garden smoothed the desert's sharp edges with the formal conventions of European garden design practiced in Ulrich's native Germany. Ulrich emigrated to the United States in 1868 and worked in Cincinnati before moving to the San Francisco Bay area in the late 1870s. Ulrich's landscape work came to the attention of Crocker, who gave him the use of a locomotive and several boxcars for a succulent collecting trip. The party included a crew of eight men and six teams of horses. An otherwise unattainable exercise in conspicuous transplantation, the Arizona Garden debuted at the Hotel Del Monte in 1882. The garden featured columnar saguaros (without arms), barrels, prickly pears, agaves, and a decidedly non-Arizonan arrangement of palms in an ellipsis of rockwork beds and paths (Figure 2.1). The garden won popular acclaim for the hotel, passenger traffic for the railroad, and cactus garden clients for Ulrich, including Crocker's Southern

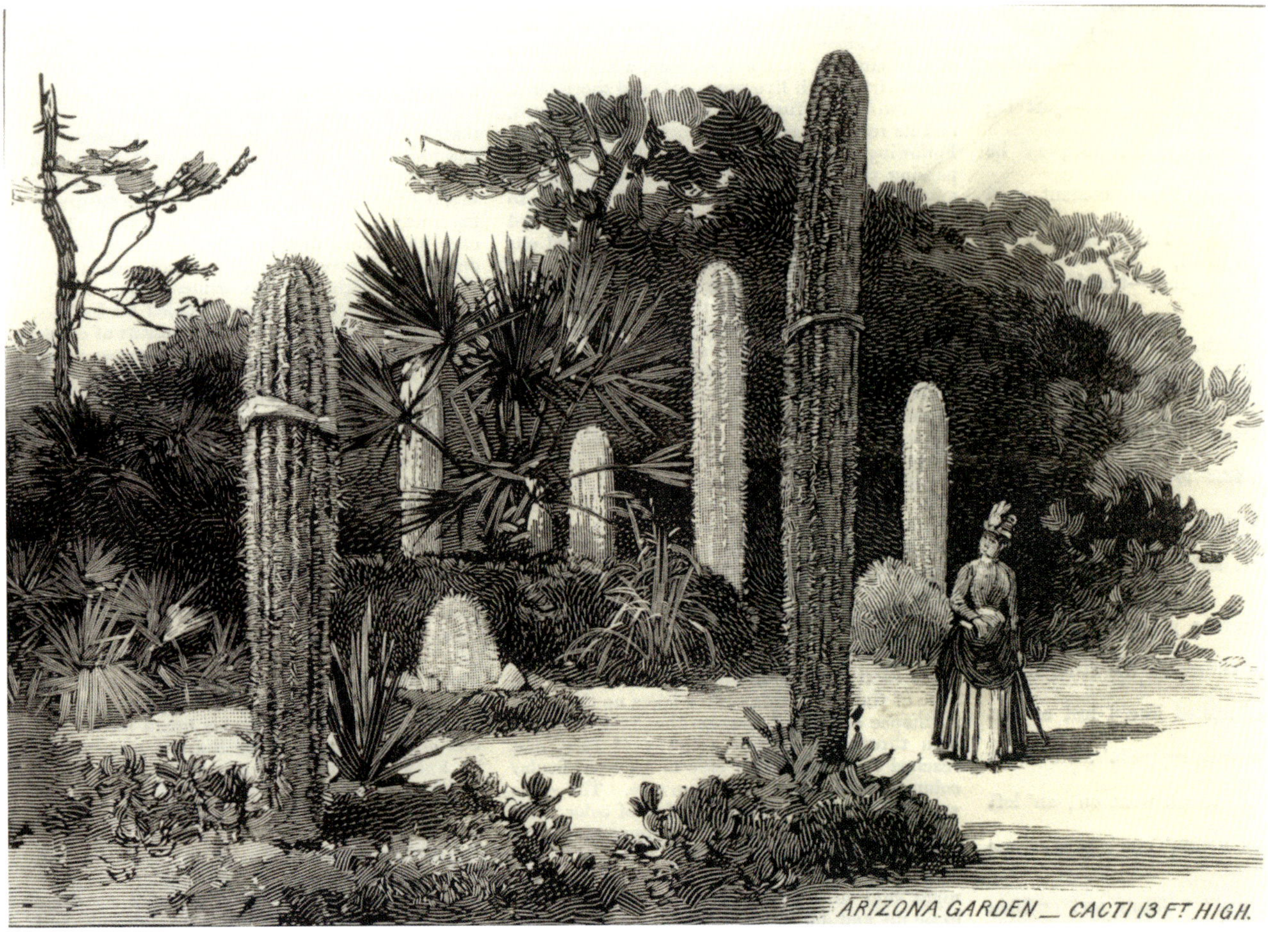

FIGURE 2.1. The earliest example of transplanted columnar saguaros for the Arizona Garden in Monterey, California (*Harper's Weekly* 1887).

Pacific partner, Leland Stanford (Olmstead 1891; Pomeroy 1957; Smith 1995: Cain and Graham 2003; Cain 2004, 2005).

The popularity of cactus collecting increased proportionately as access to the Sonoran Desert receded as a logistical problem. Cacti from Arizona and Mexico joined succulents from windowsills to greenhouses. William C.G. Ludford noted, "The cottager, who can only devote a small space to them in his window, may, and often does, grow them to greater perfection than his richer neighbours, who have greenhouses, stoves, and every convenience" (Newton 1961; Rowley 1997:196–199). Newly organized cactus clubs published newsletters and journals and held public exhibitions of members' collections (Newton 1961, 1964). The literature of collecting and display freely traded in saguaro iconography despite the unlikely prospect of growing a seedling to flowering maturity.

Botanist Lewis Castle, formerly of London's Royal Gardens at Kew, carefully explained that saguaros could only be grown from seed and not cuttings, as was the case with the jointed stems and pads of other cacti. Surprisingly though, saguaros

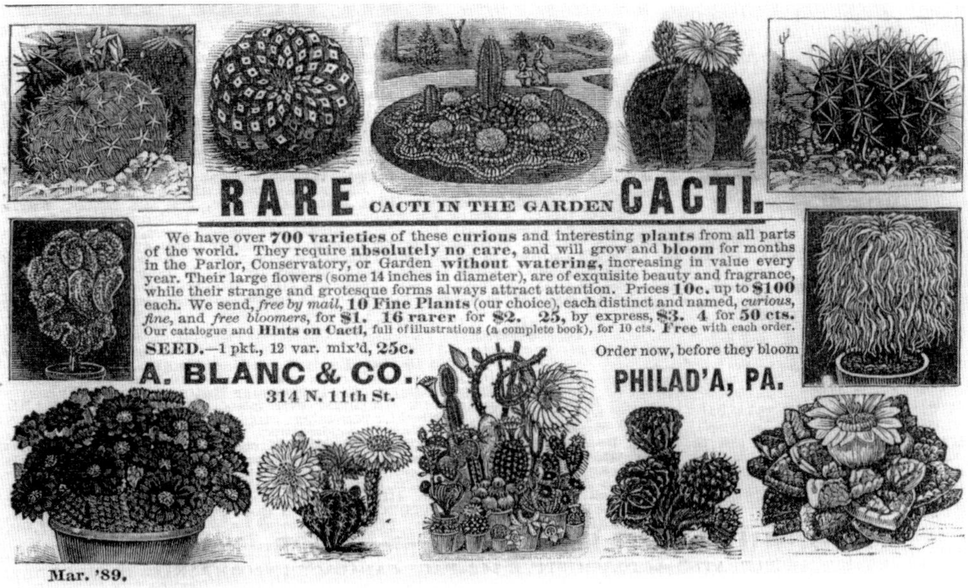

FIGURE 2.2. Advertisement for mail-order cacti offered by Philadelphia's A. Blanc and Company (1889).

were not rare, as George Thurber had distributed the saguaro seeds he collected as a member of the Bartlett survey party of the 1850s. Gardeners had propagated the seedlings ever since, including Thurber, who grew them in his Cambridge, Massachusetts, greenhouse. While most subjects in horticulture had been "thoroughly exhausted," Castle (1884:preface, 44–45) explained, "no treatise in the English language has hitherto been solely devoted to the large, peculiar, and interesting family of Cactaceous plants which long held a prominent position in popular estimation." Castle reprinted Julius Froebel's "saguaro trees" as a frontispiece, coupling the emerging hobby to the iconography of the fruit harvest (Gray 1854:302–305; Rowley 1997:200).

William Watson, an Assistant Curator at Kew, took a dim view of the mature saguaros that illustrated cactus collecting guides. In *Cactus Culture for Amateurs*, Watson (1889:76–77) cautioned, "Judging by the slowness of growth, the prospect of seeing full-size specimens of this species in English gardens is a very remote one, unless full-grown stems are imported, and this is hardly possible." Watson pictured a saguaro flower and left it at that.

Philadelphia nurseryman Albert Blanc declared his company exclusively for "cheap cacti" (A. Blanc and Company 1886). Pitching cactus collecting as a democratic hobby for all, Blanc's illustrated sales catalogs provided useful hints on cactus care, intimating that cacti thrived on the warmth of the sun, well-drained soil and benign neglect (Figure 2.2). Blanc's price sheet offered *Cereus giganteus*, "King of the cactus family." Blanc (1893) noted that "small plants of this make splendid

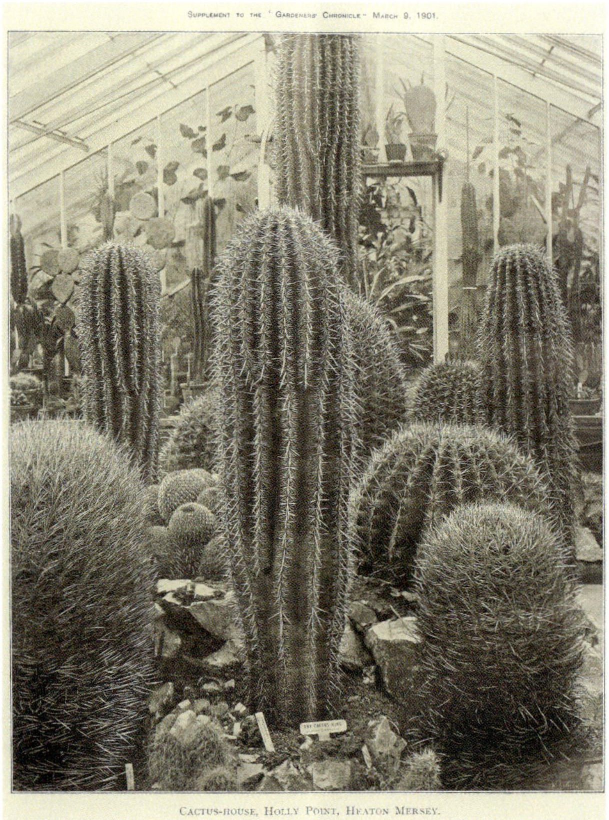

FIGURE 2.3. Saguaros front and center in the private collection of Charles Darrah (Gardeners' Chronicle 1901). In 1908, the city of Manchester, England, built a greenhouse in which to publicly display the collection.

specimens for the parlor, vestibule, or the garden; excellent for [the] center of a bed of cacti." Blanc's saguaro prices depended "much on the beauty of the subjects," beginning at 6 inches for $1.50, to six feet for $50. Blanc claimed to be "the originators of the *Cactus Craze* in the United States, and have the finest collection in existence." Blanc continued, "Cacti have much to recommend them to lovers of the curious and the beautiful...they are easily grown, so easily in fact that anyone who can only devote a small space to them in his window may grow them successfully" (A. Blanc and Company 1886). Blanc offered "Our Great One Dollar Collection of Rare Cacti," an assortment of 10 pot-size plants chosen from a list of 20 specimens (each with a Latin name) based upon availability. Blanc attributed the bump in sales of "miniature cacti" to the "magnificent exhibit" at New Orleans' World's Industrial and Cotton Centennial Exposition (A. Blanc and Company 1886; Rowley 1997:272–274).

Dr. Richard E. Kunze, a Phoenix physician, backed into the cactus mail order business. The Salt River Valley land that Kunze bought at Phoenix could not sustain the fruit orchard that he had planned, so he took up cactus instead. A German emigrant with botanical interests after George Engelmann, Kunze's cactus nursery became an international source of supply. Kunze did not so much grow the plants as gather them on spec. Transplanted to a holding bed, if they did not sell, they would be larger the following year. Kunze's clients included William Hertrich, curator of the Huntington Desert Garden at San Marino, California (Hertrich 1949; Lyons 1969, 2007:8–9; Hondagneu-Sotelo 2014; Lynch 2014). Kunze's equally remunerative clients included the United States National Museum at Washington, D.C., the Royal Botanical Gardens of England (Kew Gardens), the Royal Botanical Museum and Garden of Berlin, Germany, and the Imperial Botanical Garden of St. Petersburg, Russia. Kunze offered small, round *mamillaria*, among the least expensive plants at $0.15 per plant, to the saguaro, the most expensive, ranging from six to eight inches high for $0.50 to 15 feet high for $150.00 (Arizona Republic 1903, 1904; Kunze 1909; Munk 1912; Middleton 1914).

Cactus collecting became a competitive sport among amateur hobbyists with the means to personally gather specimens in the wild. The collecting craze swept America, England, and continental Europe. Taking stock of the collecting phenomenon in 1901, the *London Magazine*'s H.J. Shepstone (1901) stated that the largest collection in England belonged to the Swanley Nurseries of the Messrs. Cannell, noting the recent uptick in the importation of "large specimens" by "wealthy individuals interested in these strange plants last year." The [English] *Gardners' Chronicle* pictured the impressive conservatory of Charles Darrah of Holly Point, Heaton Mersey, who displayed transplanted examples of *Cereus giganteus*, the largest at eight feet, six inches. Cacti that George Engelmann had named for William Emory and George Wislizeni completed the arrangement (Figure 2.3; Gardeners' Chronicle 1901). Darrah bequeathed his collection to the city of Manchester with the condition that the city build a greenhouse in which to display it. The city-built greenhouse complex opened in 1908. The display featured 167 *Cereus* species and extensive rockwork organized into five zones. The city's new "cactus house" gave "impetus to the private cultivation of the simpler growths of the Cactus" (Journal of Horticulture and Home Farmer 1907). Yet, by the 1930s, the city of Manchester had closed in on the park, the collection was transferred to another park, and the glasshouses demolished (Hughes and Mottram 2006). What remains of Darrah's collection is under the care of the Manchester Branch of the British Cactus and Succulent Society at Wythenshawe Park just south of the city (British Cactus and Succulent Society 2011).

If additional evidence was needed of the hold that cacti had upon the imagination of the English gardener, one need look no further than F.A. Walton, a Birmingham jeweler who left his family's gold and silver business to take up collecting. In 1899, Walton traveled from Liverpool to St. Louis to tour the public garden of the late Henry Shaw and visit with members of the city's famed cactus society. From St. Louis, Walton traveled to San Bernardino, California, where he arranged an outing of several days duration to gather plants in the desert with a fellow collector and a helper with a wagon and mule team. Cautioned to not begin filling the wagon until the return trip, Walton ignored that advice and began excavating what he described as a hedgehog cactus weighing several hundred pounds. His guide objected to the inordinate heft of the specimen. Walton persevered and arranged transportation for his exotic find to his English greenhouse. Returning east from California on the Southern Pacific, Walton stopped at Tucson, where he saw newly transplanted *Cereus giganteus* on the campus of the University of Arizona. Making several excursions to the mountains surrounding Tucson, Walton succeeded in pushing over a 30-foot saguaro—to measure it. From Tucson, Walton traveled to Mexico, Texas, and New Orleans, then steamed to Liverpool by way of New York. Leaving Tucson, Walton did not mention saguaros, perhaps because they were neither rare nor unknown in the collections of his rivals. Walton seemed happy with the largest hedgehog he had ever seen, certainly the largest in England. In 1898, Walton published a journal of the National Cactus and Succulent Society, a repository of information and contact with collectors in the field. Never winning a broad subscribing readership, the journal in which Walton described his adventures folded after only two years. A journal of cactus collecting, cultivation, and display did not reemerge in Britain until the early 1930s (Walton 1899:288–296; Rowley 1966, 1995; Trennert 1988; Parker 1989).

Cactus clubs in America began as local affairs that overlapped with regional botanical and horticultural societies. New York's Torrey Botanical Club, for example, whose leadership included club president George Thurber, met at the Columbia College Herbarium. The club's focus was the botany of New York State and the as yet undeveloped environs of New York City (Barnhart 1918; Mickulas 2007:36–40).

The Baltimore Cactus Society published a journal and staged annual public exhibitions of members' collections (Figure 2.4). Formally organized in 1889 and novel enough to attract the attention of the press, the club's announcement of the arrival of two *Cereus gigantea* to a member's collection in 1890 caused a stir. The *Baltimore Sun* (1890) reported that one of the saguaros was two feet high and 10 inches thick, the other about four feet high and one foot in diameter, "the largest of the kind in the city." The *Baltimore Sun* (1889) noted that "It has only been in the past few years that popular attention has turned to the cultivation of cactaccous plants, but interest is now widespread, and few window gardens or green-houses are considered complete without a collection." In 1892, the society held what it promoted as the first "cactus exhibition" in America. Staged at Raines Hall, an event space rented from the bricklayers union, the exhibit featured 700 varieties totaling 5,000 plants. A reporter described the centerpiece of the show as "a specimen of the king cactus from Arizona. It looks like the stump of a telegraph pole painted green. It is supposed to be about 100 years old. In this time it has reached the height of seven feet and weighs 380 pounds" ([Washington] Evening Star 1893;

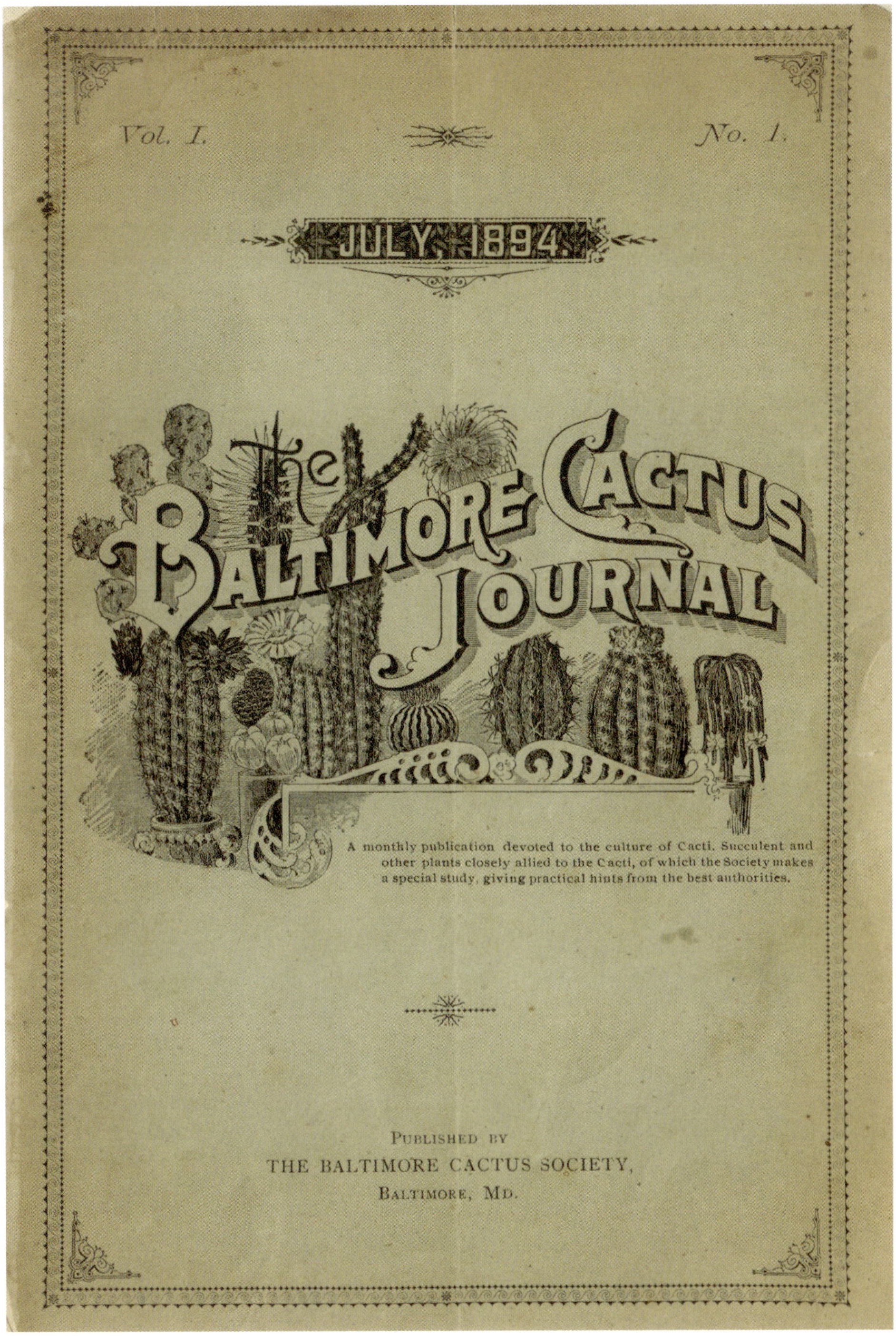

FIGURE 2.4. The journal published by the Baltimore, Maryland, Cactus Society (1894).

Ryan 1991). *Harper's Bazaar* (1894) rendered the exhibit as a social event inset with woodcuts of the "Giant Cactus" of the Gila River (Figure 2.5). John A. Becker, society president and purported owner of one of the finest cactus collections in the United States, was said to be negotiating with the Mexican Secretaria de Fomento (Secretary of Development) for its cactus display at the World's Columbian

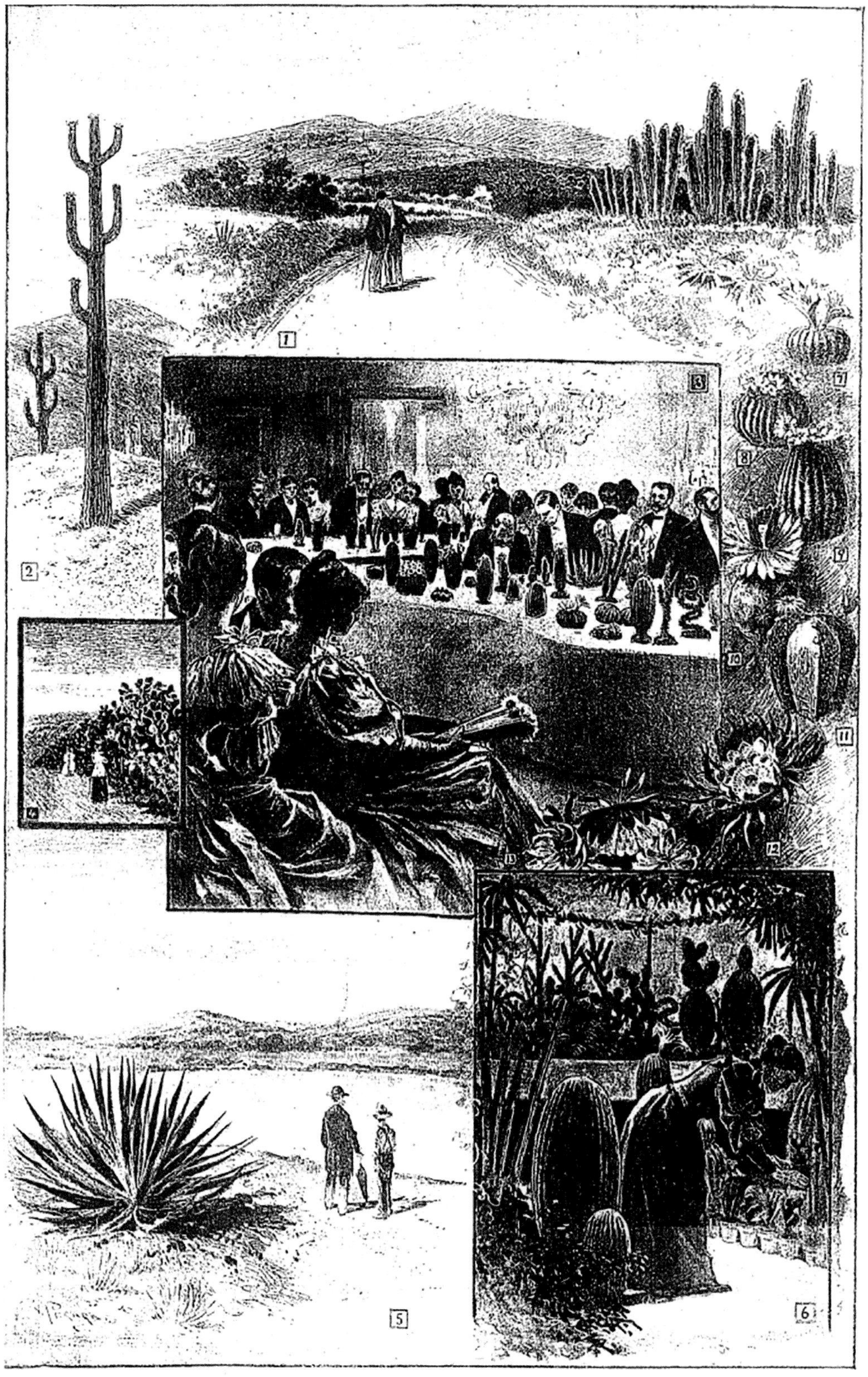

FIGURE 2.5. Woodcut. 1893. The first exhibition of the Baltimore Cactus Society in 1892 presented 5,000 specimens from members' collections, including a seven-foot saguaro. The following year, *Harper's Bazaar* (1893) rendered the exhibition as a social event picturing society members poring over their display. A collage of smaller woodcuts encircled the scene picturing saguaros, organ pipes, and agaves in the wild, with members' flowering barrels, mammillaria, and a night-blooming *Cereus*, each plant and scene identified and keyed by number.

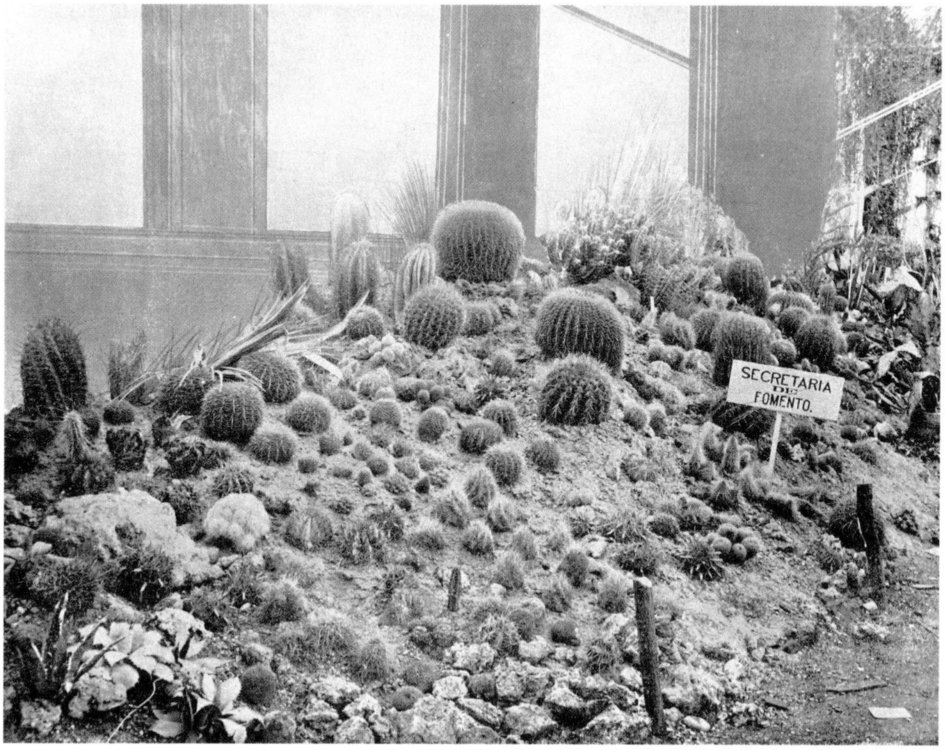

FIGURE 2.6. Cactus display mounted by Mexico's Secretary of Development, Horticulture Hall, World's Columbian Exposition, Chicago. 1893.

Exposition in Chicago (New York Tribune 1894). The Mexican display consisted of densely packed agaves, barrels, and small saguaros in a rock-strewn, linear sand pile. The garden contributed to the festive atmosphere in which business was conducted at the fair (Figure 2.6; Yeager 1977). It is now presumed that Becker was unsuccessful in his bid.

The Baltimore Cactus Society maintained a public garden display in the city's Carroll Park. Begun as a seasonal experiment, society members leant plants for the summer and retrieved them in the fall. Park Superintendent Charles L. Seybold, a cactus society member who lived in the park's Mount Clare mansion, created and watched over the display, which featured rockwork parterres radiating from a central saguaro. Seybold covered and uncovered the display to protect it from rain. The society's journal lauded the arrangement citing, "the healthfulness of Carroll Park as a summer resort for cactus" (Figure 2.7; Baltimore Cactus Journal 1894a, 1894b). Seasonal cactus landscapes included Brooklyn, New York's Prospect Park (Figure 2.8; New York Tribune 1900).

World's fairs existed to dramatize the introduction of new things, ideas, and the ultimate in display techniques. From London's Crystal Palace in 1851 to

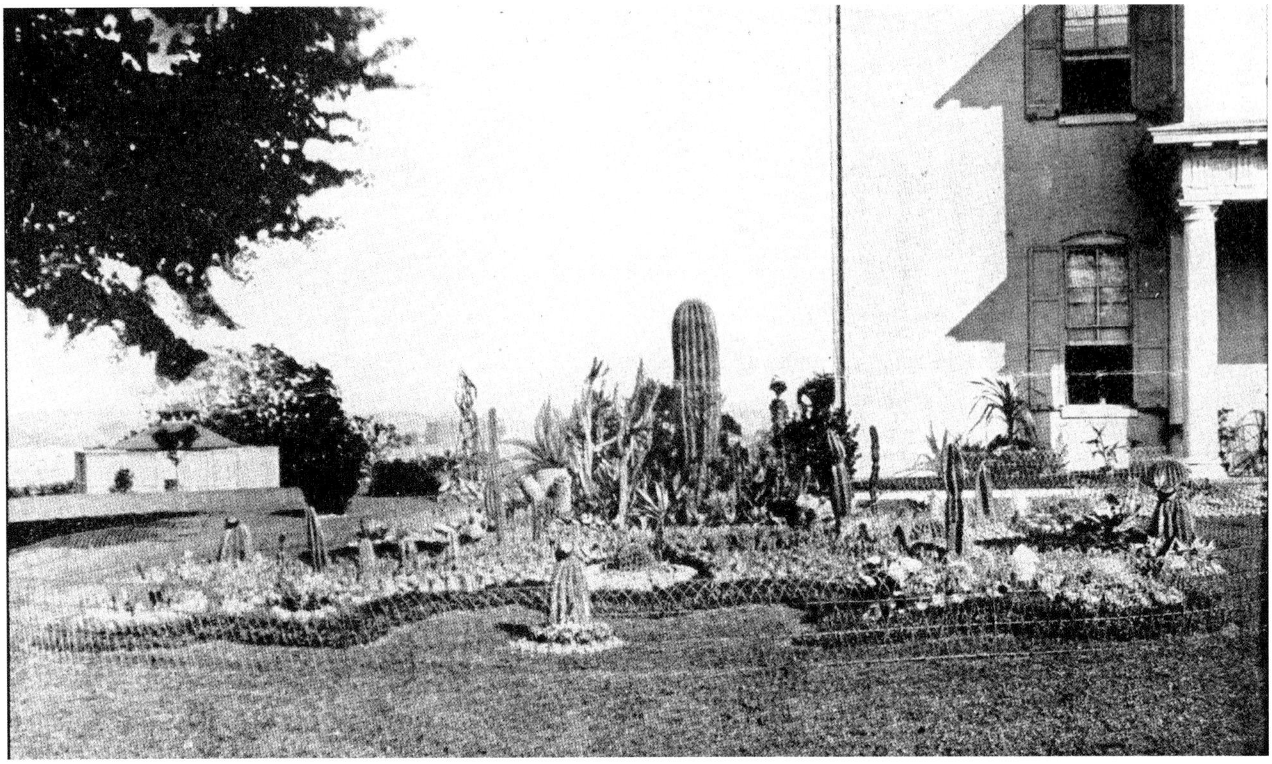

FIGURE 2.7. Baltimore Cactus Society's summer cactus bed, Carroll Park, Baltimore, Maryland. 1894.

Chicago's World's Columbian Exposition in 1893, fairs showcased horticulture and botanical displays in stand-alone, glassed-in exhibit halls. Small cactus specimens appeared at London in 1851, and at the United States' first world's fair, the Philadelphia Centennial of 1876. There a comprehensive index of the woods of America featured cross-sectioned, two-foot long wood samples showing grain and bark. The samples included *Cereus giganteus*, *Cereus thurberi*, and *Opuntia arborescens*. The U.S. Department of Agriculture collated and prepared the specimens contributed by regional specialists (Vasey 1884:383–386, 398). Dr. Edward Palmer contributed wood samples representing Arizona, southern California, and Southern Utah. Palmer traveled widely and collected in the American Southwest and Mexico, carefully documenting the specimens he sent to the Smithsonian Institution and the U.S. Department of Agriculture (McVaugh 1956). Elsewhere the Centennial's Horticultural Building displayed specimens contributed by individuals and growers identified only as "cactus" (United States Centennial Commission 1876:79).

There was never an American world's fair without a saguaro, even if it debuted at the Philadelphia Centennial as a specimen in a wood index. The saguaro's first ap-

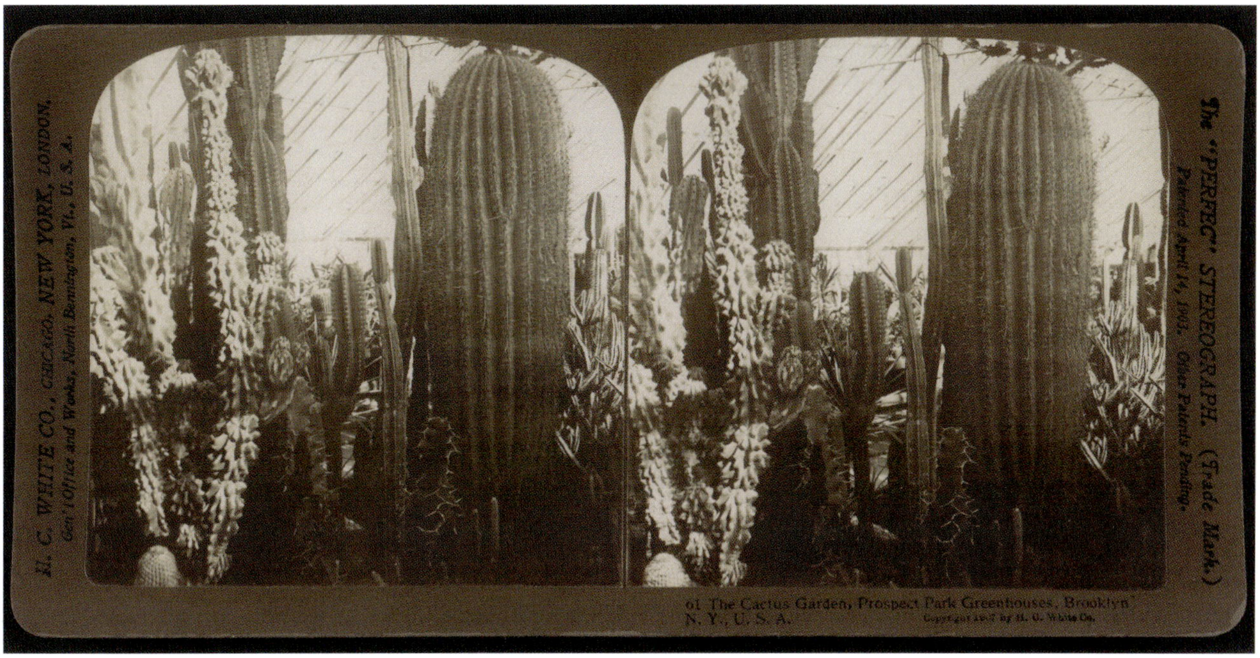

FIGURE 2.8. A stereoview picturing the Prospect Park cactus greenhouse, Brooklyn, New York. 1907.

pearance—as a columnar plant—came at the World's Industrial and Cotton Centennial Exposition at New Orleans in 1884–1885, the first fair after the Southern Pacific's penetration of the Sonoran Desert. The fair's Horticultural Hall featured an "immense cactus, twenty feet high, from Arizona" (Perkins 1885:79–80). It is not known what became of the plant at the fair's conclusion.

As ultimate displays, world's fairs advanced the professional interests of landscape design. The carefully cultivated appearance of fairgrounds complemented exhibit buildings and became public parks at the fairs' conclusion. The preparation of a fairgrounds involved moving trees and shrubs, draining swamps, and creating water features that left the impression of a natural occurrence. And then there were the saguaros.

The landscape firm of Frederick Law Olmsted designed the World's Columbian Exposition of 1893 fairgrounds. The firm hired Rudolph Ulrich of Arizona Garden fame as the fair's landscape superintendent (Maloney 2012:64–68; Olmsted 1891). The Southern Pacific contributed a complimentary railroad pass for a flatcar carrying 17 saguaros and a boxcar for barrel, prickly pear, cholla, ocotillo, and agave specimens collected within 25 miles of Tucson. The idea of outfitting the Arizona territorial building at the fair with saguaros originated with William Reid, a Tucson theater owner and World's Fair Commissioner for Pima County.

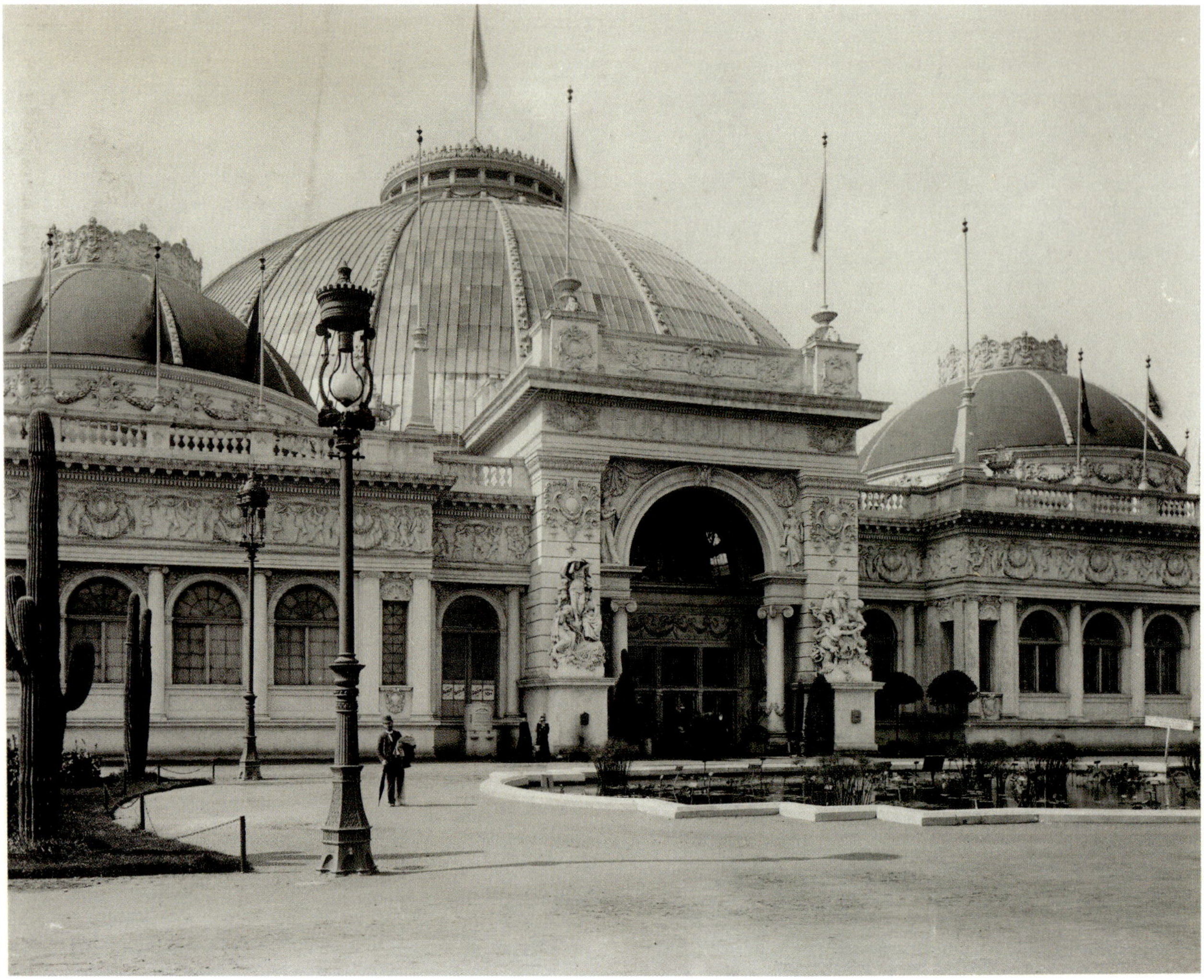

FIGURE 2.9. This photograph documenting the monumental architecture of the Horticulture Building at the World's Columbian Exposition, Chicago, captured an equally monumental saguaro at the left of the frame. 1893.

Reid took to the task as a natural showman. "Millions of Americans and foreigners have read descriptions of these extraordinary plants and have probably seen cuts or photographs of them," Reid explained, "but only those who have visited the southern territories have any clear idea of what they are really like" (Arizona Silver Belt 1893a). Moreover, saguaros would instill among fairgoers a favorable impression of southern Arizona's mild climate. Though the landscaping of the fairgrounds began more than a year in advance of the opening to allow new trees and shrubs to settle in, the cactus shipment had to wait until early spring to keep it from freezing. The saguaros were the last to arrive and be placed at the fair. Though Reid had watched over the collection of the plants in Tucson, he could not attend to their placement at Chicago. The most impressive saguaros with arms—a new development in the

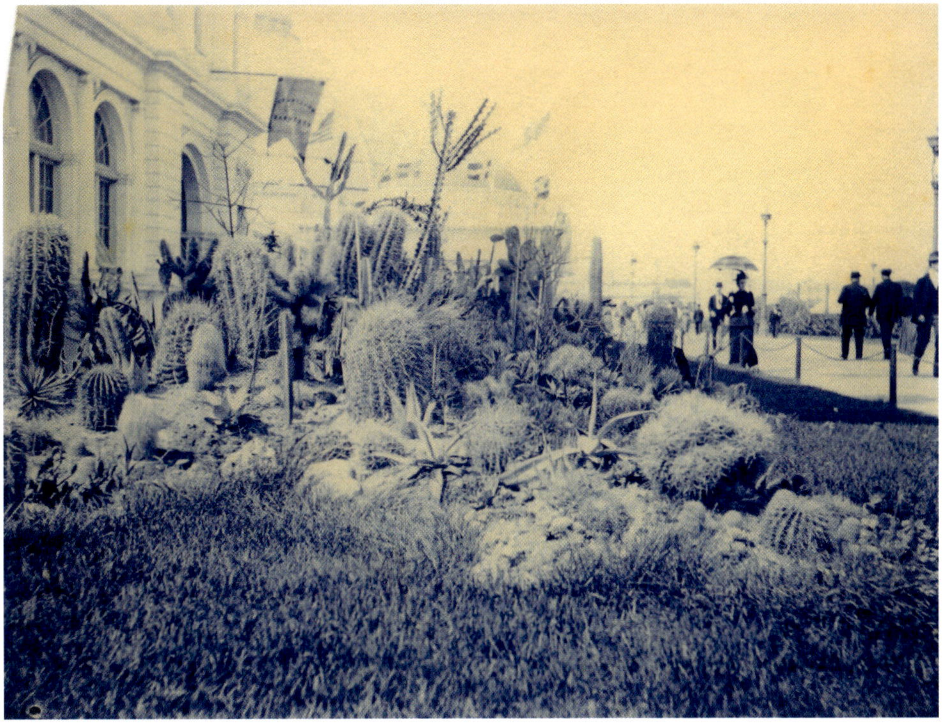

FIGURE 2.10. Cactus display tucked into a grassy corner outside the Horticulture Building, World's Columbian Exposition, Chicago. 1893.

annals of cactus transplantation—ended up outside the fair's Horticultural Hall (Maloney 2012:73). Others ended up at the Pennsylvania building (Campbell 1892; Arizona Weekly Citizen 1893a, 1893b, 1893c, 1893d, 1893e; Bailey 1893; Hilpert 1983). Saguaros loomed over the heads of visitors, extending beyond the frame of cyanotypes and stereoviews of the fairgrounds (Figures 2.9–2.12).

Six of Reid's specimens made it to the forecourt of the Arizona territorial building shared with New Mexico and Oklahoma (Figure 2.14). In part, Reid's intent had been to impress fairgoers with the clarity of the climate which had produced such plants. One territorial visitor, however, opposed the use of the cactus as a symbol, as it sent the opposite message. By installing saguaros at the fair the territory had lost an opportunity to erect "a statue of Abundance, with a horn of plenty in each hand, instead of the withered and forbidding cactus stock erected to indicate, we

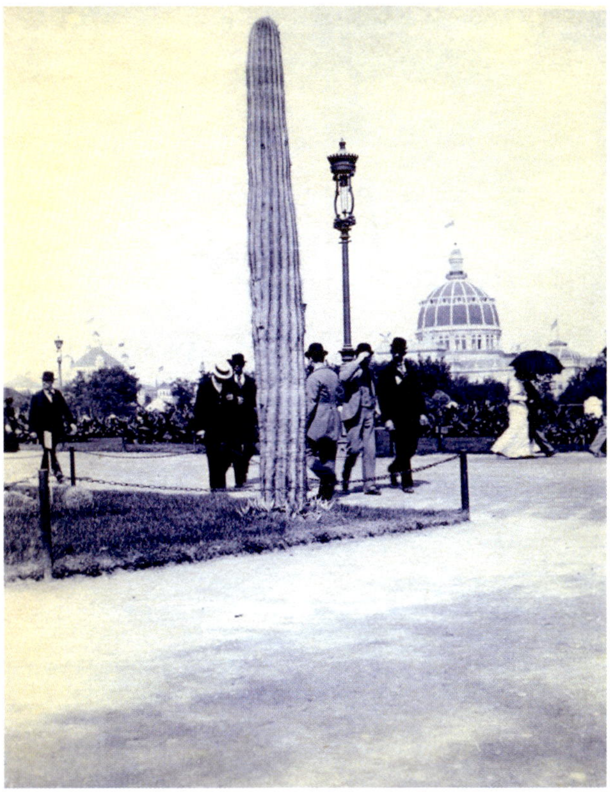

FIGURE 2.11. A columnar saguaro encircled with an arrangement of small agaves. Horticulture Building, World's Columbian Exposition, Chicago. 1893.

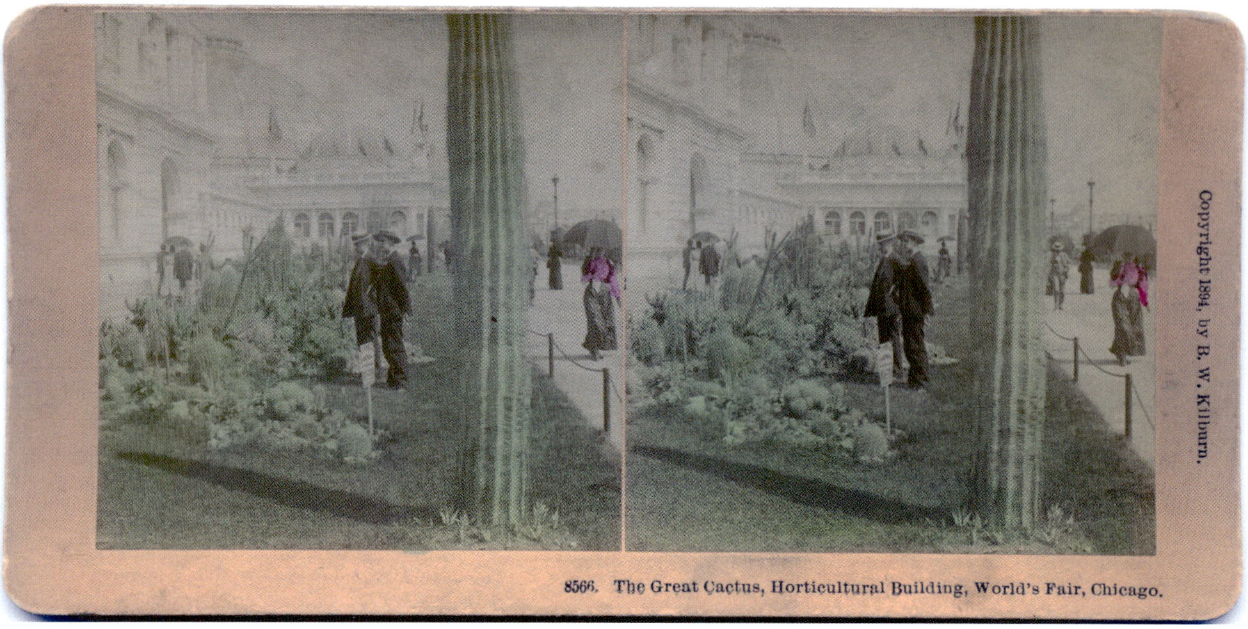

FIGURE 2.12. Foregrounding a three-dimensional stereoview with a saguaro heightened the transporting effect of the popular cards. Horticulture Building, World's Columbian Exposition, Chicago. 1893.

presume, the poverty of Arizona's soil" (Arizona Silver Belt 1893b). The territory would be more properly promoted as an agricultural paradise under irrigation, not a desert.

The tension of cactus-or-not coursed beneath the highly competitive claims of primacy advanced by the boosters of Tucson and Phoenix. The latter envisioned a latticework of irrigation that bore fruit and alfalfa. An early set-aside of park land in Phoenix, the Papago Saguaro National Monument, quickly devolved from the wilderness it might have been. Postcards of the park's most popular geological formation, the Hole in the Rock, show the saguaro forest yielding to the irrigated features of a public park (Figures 2.14 and 2.15). Tucson boosters, fortified by the community of Desert Laboratory and university botanists, envisioned a ring of saguaro parks that withstood the pressure of further development. But in Tucson's case the parks were hardly assured (Burtner 2011:23, 155–157).

Fairgoers who encountered saguaros might be excused for believing that they grew throughout the Southwest. The Southern Pacific Railroad selectively decorated passenger stations with prime specimens. In the year of the World's Columbian Exposition, for example, press accounts noted the carloads of Arizona cacti destined for the Southern Pacific's Los Angeles Arcade Depot passenger station (Los Angeles Times 1893). A photograph of the station's cactus garden pictures

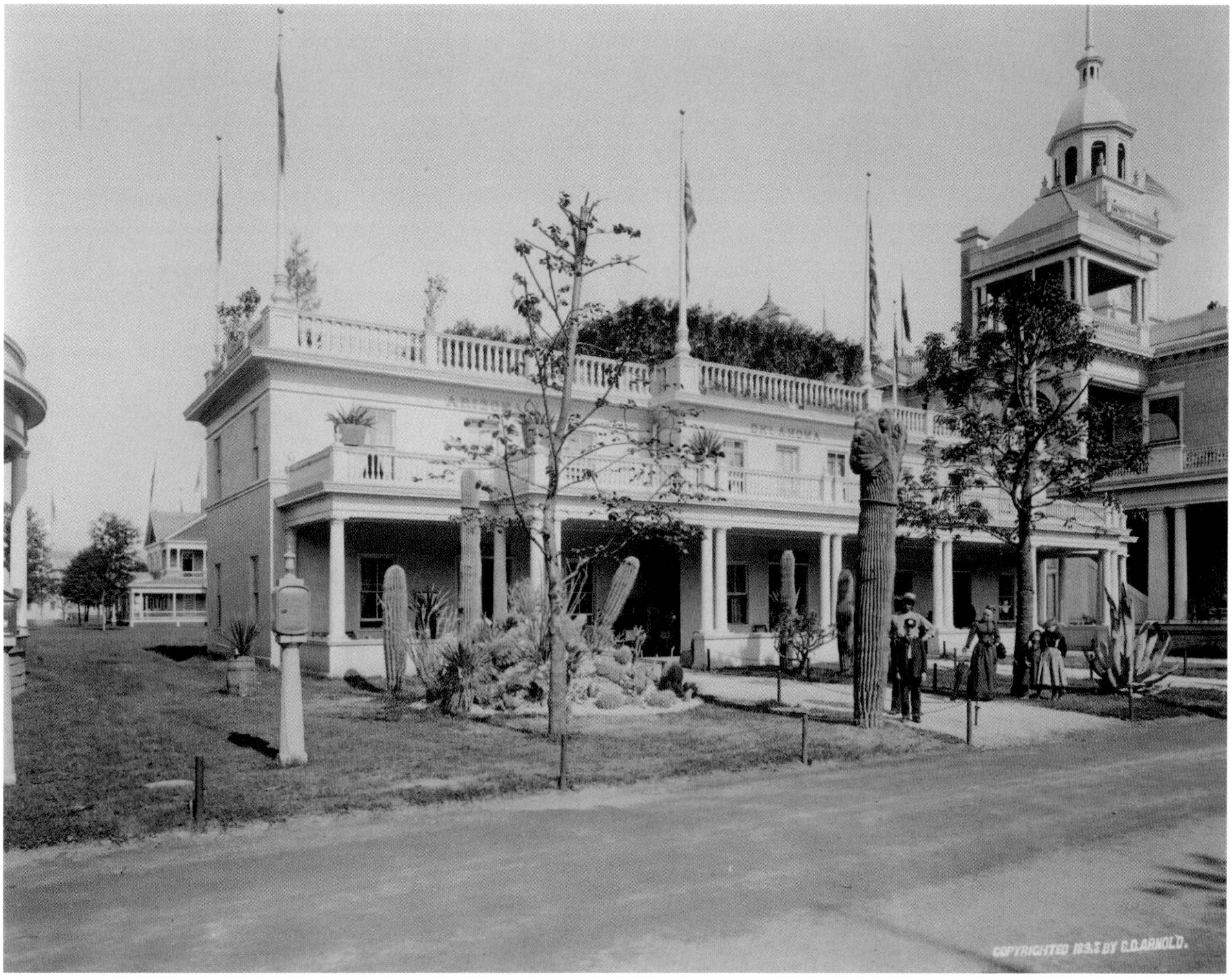

FIGURE 2.13. A fan-shaped cristate saguaro stabilized with a brace identified the Arizona portion of the Arizona-New Mexico-Oklahoma Territorial Building, World's Columbian Exposition. 1893.

no less than 12 columnar saguaros (Figure 2.16). To the east, the railroad installed a saguaro as the focal point of its San Antonio, Texas, station, hundreds of miles before—or after—passengers became acquainted with the plants in Arizona (Figure 2.17).

The Southern Pacific continued to test the limits of transplantation well into the early 1900s. In 1915, for example, the railroad provided saguaros for the garden of the California state building at San Francisco's Pacific Panama International Exposition (Figure 2.18; San Francisco Examiner 1915; Raymond 1915:182–183). By this time it was widely understood that the saguaro could sustain itself through a fair season. Rather than convey ideas about the poverty of Arizona's soil, cactus displays promoted California's climate, in which it was said that anything could

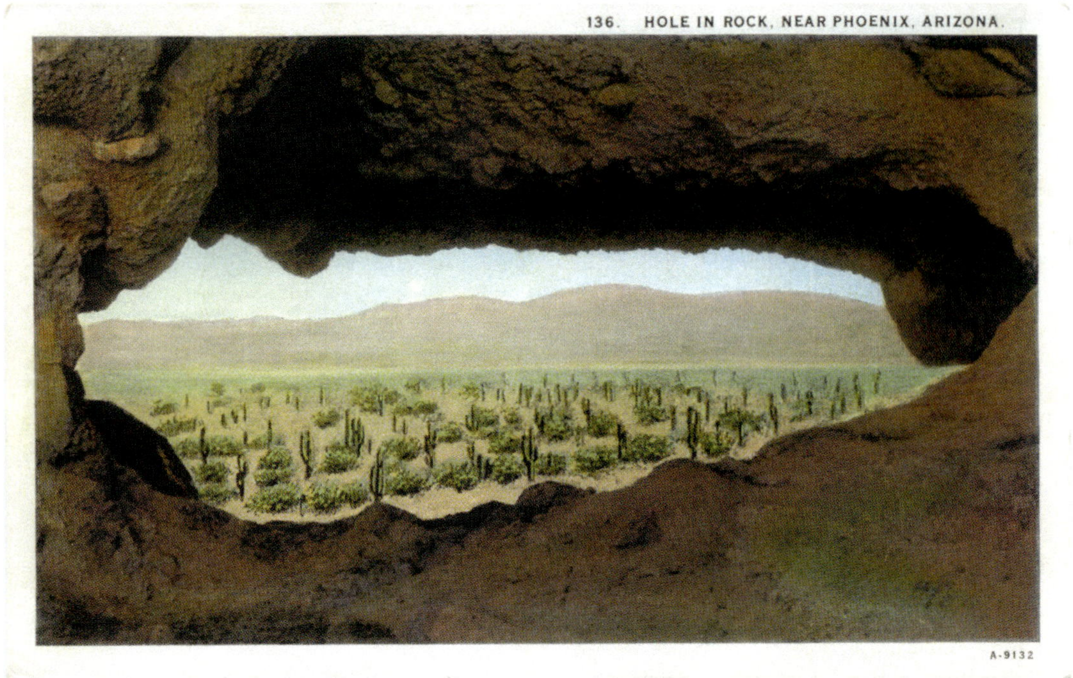

FIGURE 2.14. Postcard. 1922. "Hole in Rock, Phoenix, Arizona." The window of this popular geological formation framed the changes taking place in the park over time.

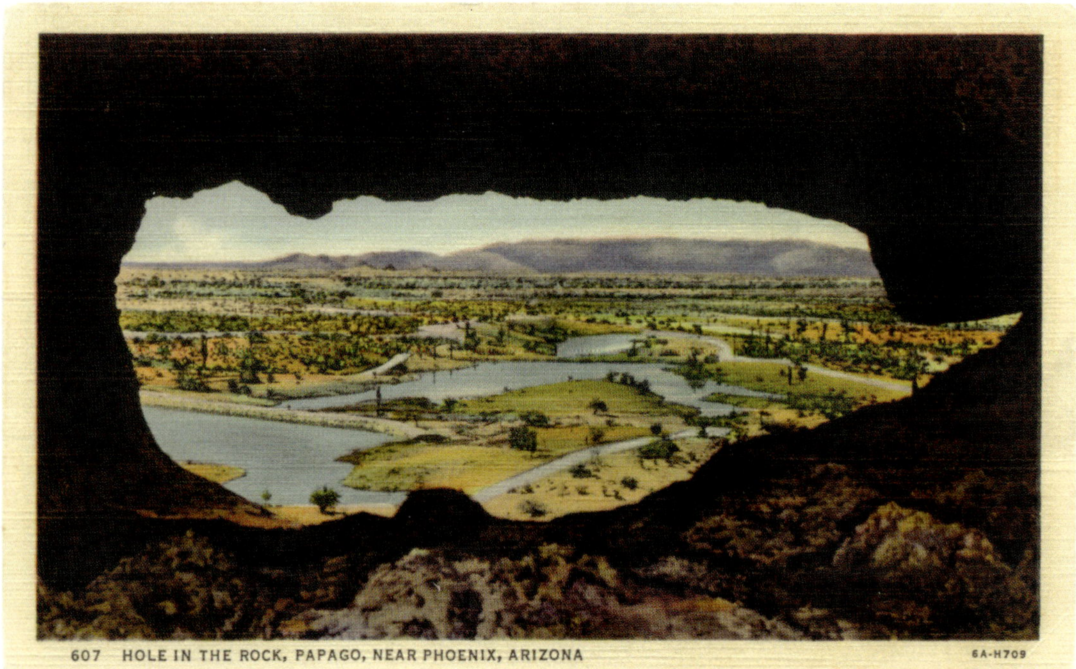

FIGURE 2.15. Postcard. 1936. "Hole in the Rock, Papago Park, Phoenix, Arizona [with water]."

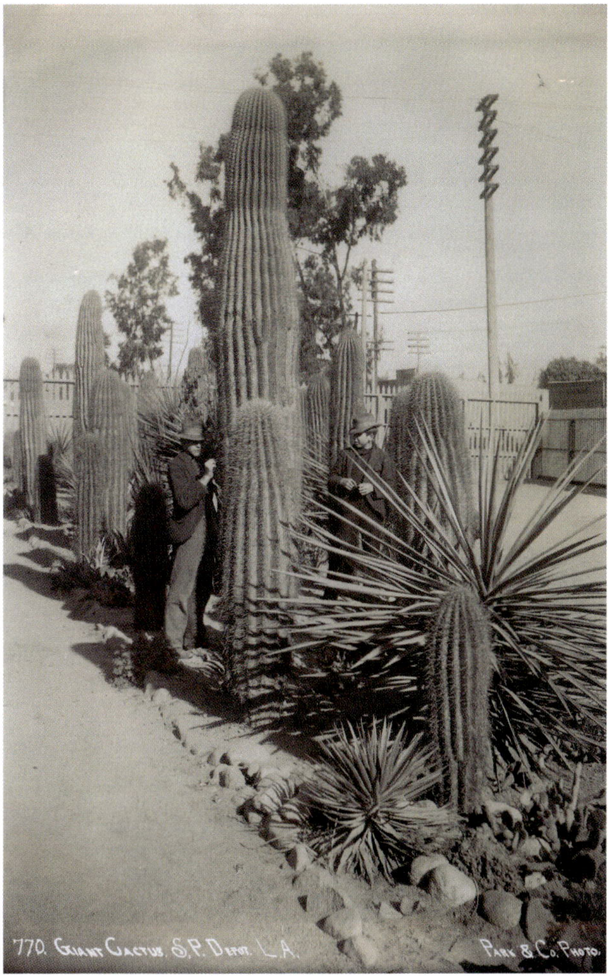

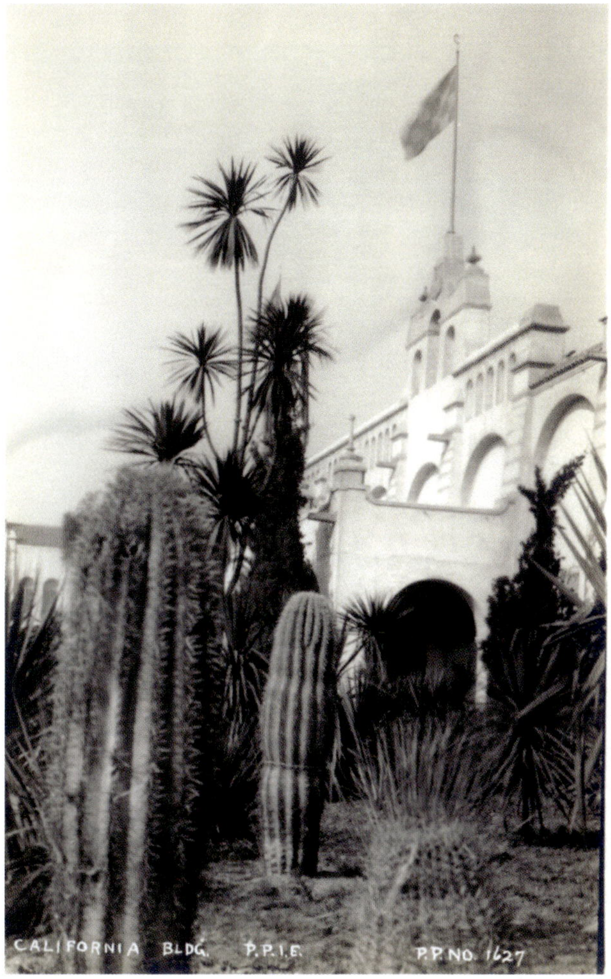

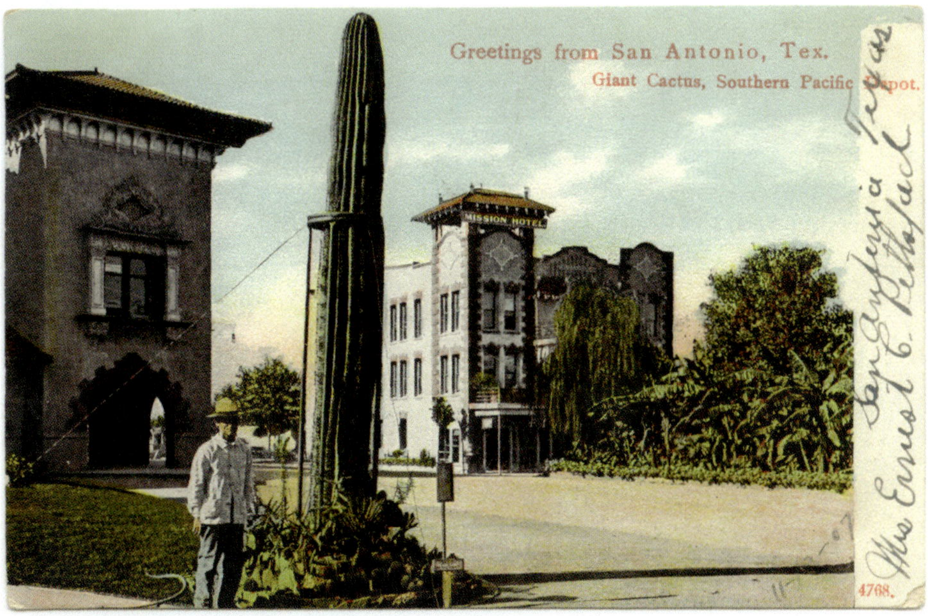

FIGURE 2.16 (ABOVE LEFT). Cactus garden. 1893. Southern Pacific Railroad Arcade Depot, Los Angeles, California.

FIGURE 2.17 (LEFT). Giant cactus with brace. 1907. Southern Pacific Railroad Depot, San Antonio, Texas. The sender wrote, "The cactus was brought here from New Mexico [territory] and is as you see a giant."

FIGURE 2.18 (ABOVE RIGHT). Realphoto postcard. 1915. Saguaro transplants fronting the California Building, Panama Pacific International Exposition, San Francisco.

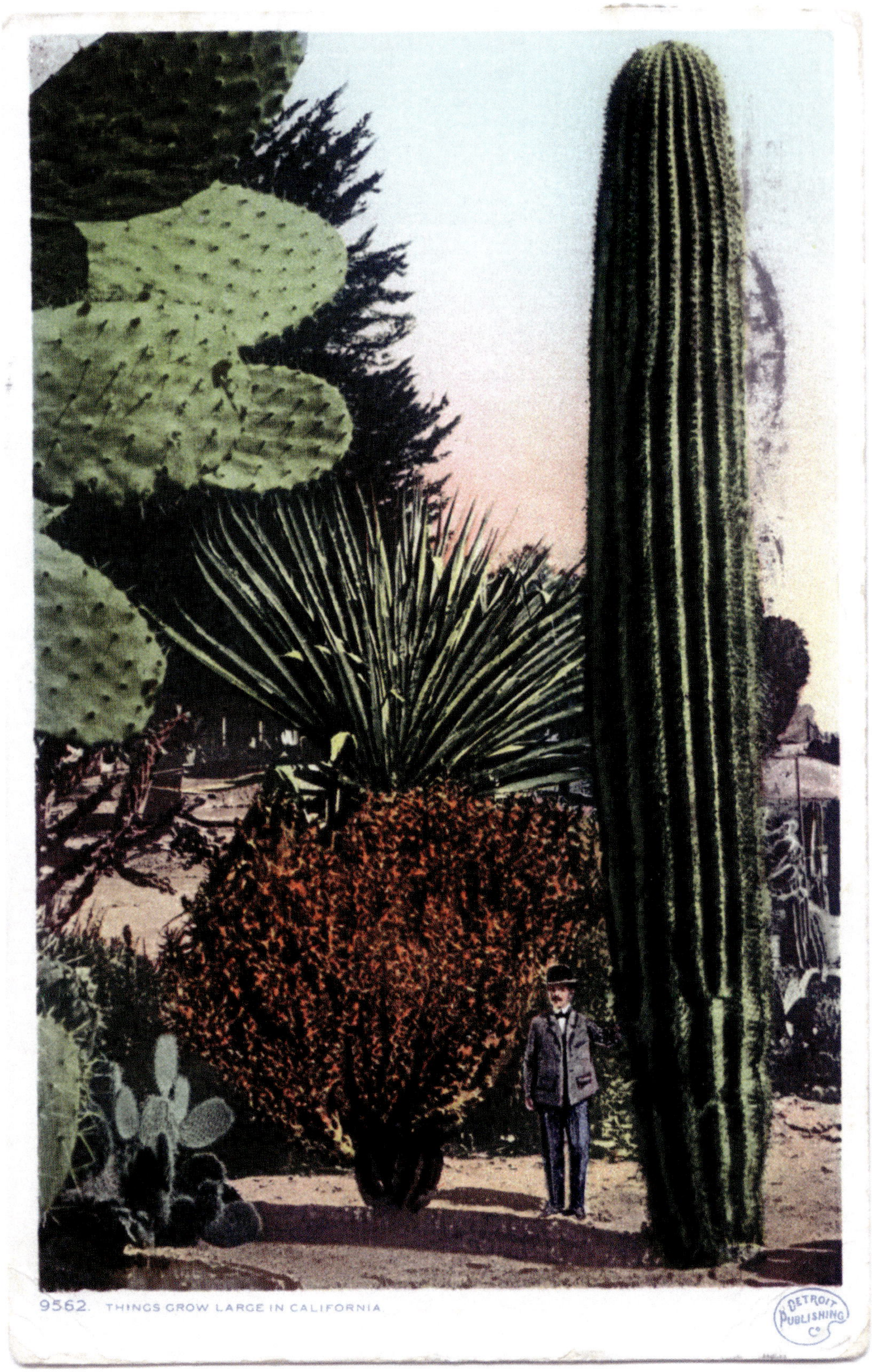

FIGURE 2.19
Tall tale postcard. 1913. Tall tale postcards playfully exaggerated the proportions of locally produced potatoes, tomatoes, corn, fruit, fish, and game—and a saguaro promoting the fecundity of California's oddly productive climate.

grow. A tall-tale postcard playfully confirmed the sentiment: "Things grow large in California" (Figure 2.19)

The transplantation of saguaros anticipated a highly imaginative commercial climate in which anything could be *sold* by putting a cactus on it. But the railroad's transplanted saguaros did not last. At the Hotel Del Monte in Monterrey, the home of the largest and most well-documented cactus garden in the west, postcards pictured the circular braces that propped up the saguaros, and later, the plants' complete absence (Figures 2.20 and 2.21). The box-like brace propping up the Southern Pacific's San Antonio, Texas, saguaro portended the plant's doom and eventual removal.

Merely because the plants did not last did not mean that their persistent imagery failed to take hold as a symbol, particularly in and of the Southwest. Rather the opposite. As late as 2018, a *Texas Monthly* reader complained that the saguaro had worked its way into the Lone Star state's pop culture lexicon, even appearing on a map (Courtney 2018). Critics seeking to define a Southwest without it wrote off the saguaro's pervasive imagery as "hackneyed" (Byrkit 1992:342).

At the Hotel Del Monte's Arizona Garden on California's Monterey Peninsula, the Southern Pacific Railroad fell back upon hybrid rose beds and a hedgerow maze topped with topiary figures. Out on the craggy Pacific coast, a windswept cypress on a high cliff overlooking the ocean became the peninsula's most memorably photogenic tree (Farmer 2013:xvii–xix; Los Angeles Times 2013).

From the Pacific Ocean in April 1934, the officers of the U.S.S. Arizona requested cactus specimens for their ship's newly remodeled lounges. The Tucson Chamber of Commerce engaged University of Arizona ecologist Andrew Alexander Nichol to assemble a collection for the ship. Nichol, a young assistant professor then compiling a monograph on Arizona flora, went by the initials A.A. (Mitich 1996). Nichol's seagoing display featured "symmetrical clumps" of purple Santa Rita prickly pears, an organ pipe, and two four-foot saguaros. Nichol did not know how the plants might fare. He volunteered only that some had "sufficient adaptability" to live (Arizona Republic 1934a; Mitich 1996).

Later that month, University of Arizona president Homer LeRoy Shantz announced that Nichol and the university would lead the design and installation of "two immense cactus gardens" at the Chicago Century of Progress adjacent to the Arizona State building (Arizona Republic 1934b; Casa Grande Dispatch 1934). Scaled back in the intervening weeks from a 4,000 square foot proposal that might

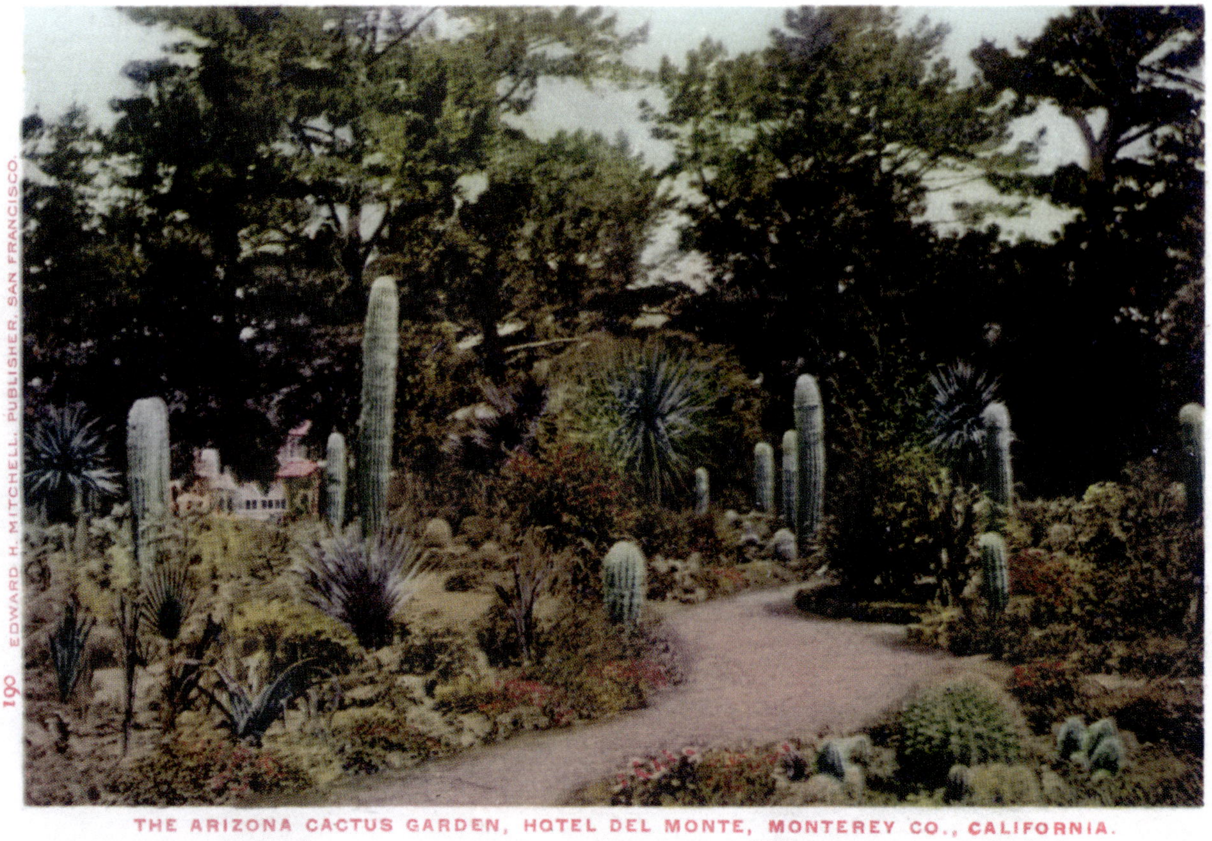

FIGURE 2.20. Postcard. 1901–1907. Arizona Cactus Garden at the Hotel Del Monte, Monterey, California. The producer of this hand-tinted postcard colored in the braces propping up the garden's transplanted saguaros.

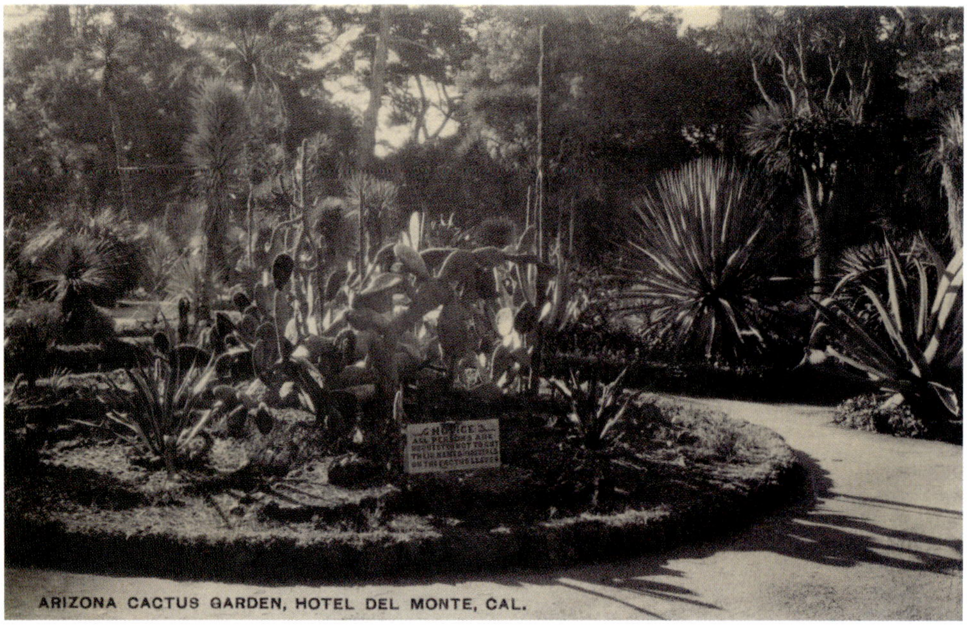

FIGURE 2.21. Postcard. 1898–1906. Arizona Cactus Garden, Hotel Del Monte, Monterey, California. After its saguaros died, the Arizona Garden lingered on with a collection of palms and succulents. A sign posted next to a prickly pear asked visitors "not to cut their names or initials on the cactus leaves."

rival even the university's own campus cactus garden, to a more easily accomplished display, the exhibit that Nichol eventually installed at Chicago featured symmetrical clumps of hedgehogs, barrels, chollas, ocotillos, prickly pears, and large armed saguaros. Inside the Arizona exhibit proper, a fenced-off floor area continued the display of the state's native flora under a copper ceiling with a close arrangement of ocotillos and saguaros courtesy of the Southern Pacific Railroad's partner, the Rock Island Railroad. With allowances for limited botanical knowledge, or perhaps because of it, the eye-catching display confirmed the saguaro's iconographic range without objection or complaint.

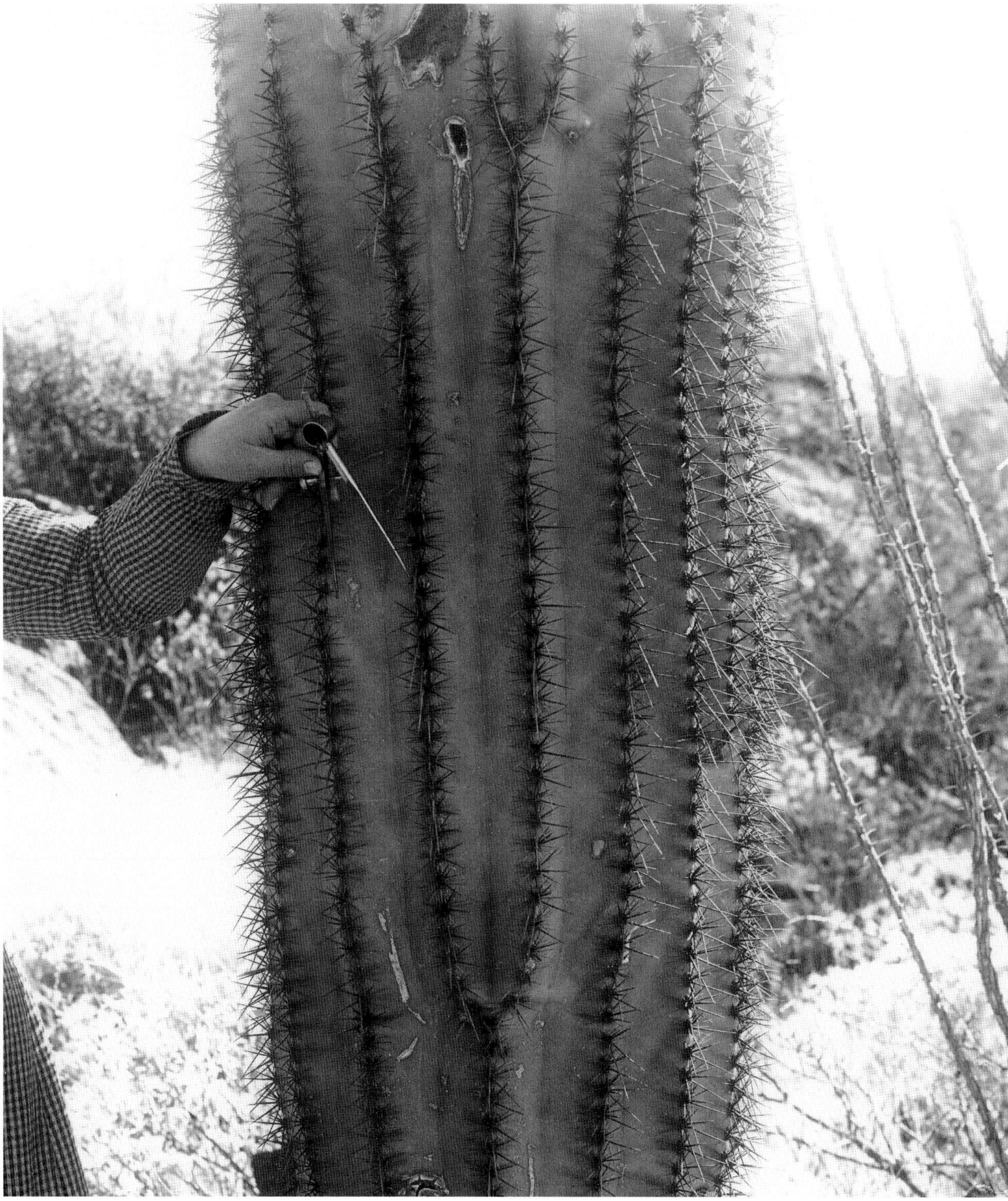

CHAPTER THREE
THE ARBOREAL DESERT

At the turn of the twentieth century, saguaros found new audiences in favorable climates and conservatories as living botanical trophies—attractions that closed the loop-trail of scientific investigation with a powerful public display. Enthusiasm for new knowledge gained from collections of all kinds extended from world's fairs to the endowment of museums, zoos, and botanical gardens. Saguaros in and around the World's Columbian Exposition at Chicago in 1893 prefigured their curation in the new monumental glass and steel conservatories erected in Pittsburgh, St. Louis and New York (Lemmon 1962:11, 46, 138, 153, 168; Koppelkamm 1981; Grove 2005; Dunkak 2007; Mickulas 2007:109–115, 137–143).

At the conclusion of the 1893 Chicago fair, over 6,000 cactus plants traveled to Pittsburgh as the leading collection of the Phipps Conservatory opened to the public in December of that year (New York Tribune 1893). Endowed by steel baron Henry Phipps, partner of Andrew Carnegie, the conservatory soon augmented its cactus display with purchases from Philadelphia's Albert Blanc, then disposing of his collection, said to be one of the most complete in the country. A separate cactus house at Phipps designed by New York's Lord and Burnham opened in 1902. The installation featured raised beds with natural rockwork and informal paths of "undulated" design for cactus and kindred plants (Pittsburgh Post-Gazette 1902). A newspaper reporter noted that Conservatory Director J. Guy McCandless "determined that Pittsburgh shall not be behind other cities in the matter of cacti any more than in that of any other conservatory plants" (Pittsburgh Post-Gazette 1902). The photographic record of the conservatory's new building pictures columnar saguaros, suggesting that the armed saguaro specimen seen on the fairgrounds at Chicago had since died or did not move to Pittsburgh with the other specimens (Figure 3.1). Phipps officials rejected exhibiting the plants outdoors during the summer in the Monongahela Valley, "on account of the smoky atmosphere, as a clear, dry air is necessary to cacti" (Pittsburgh Post-Gazette 1902).

St. Louis philanthropist Henry Shaw acquired the herbarium of his late friend and mentor George Engelmann. The Missouri Botanical garden that Shaw en-

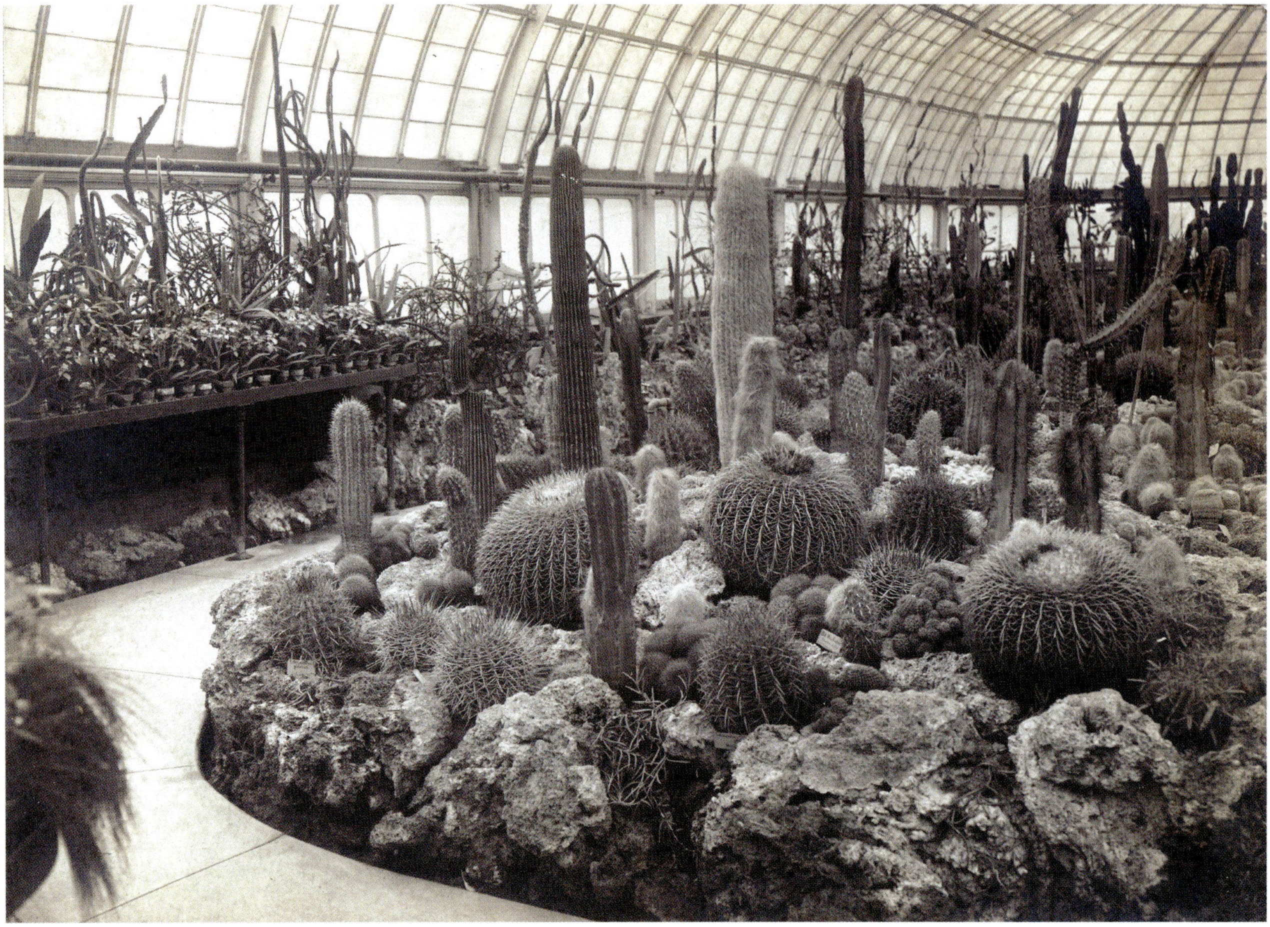

FIGURE 3.1. Phipps Conservatory Cactus House, Pittsburgh, Pennsylvania. 1903.

dowed continued after his death in 1898 as a leader in botanical research and Victorian display values. Popularly known as "Shaw's Garden," period photographs show potted cacti displayed outdoors in metal-edge parterres, labeled and arranged by type. Two columnar saguaros flank the garden's gated portico inspired by the formal conventions of English garden design that Shaw experienced at Chatsworth House and at London's Kew Gardens, whose groundskeeper he soon hired. Shaw's outdoor saguaro displays occupied the garden's laborers who moved the plants in and out of the safety of their winter greenhouses for the summer (Figures 3.2 and 3.3). A Shaw photographer documented the receipt of an 11-foot saguaro in 1914. The open crate shows the saguaro cushioned with creosote branches (Figure 3.4). Officials placed the saguaro indoors in the center of the garden's Yucca dome (Dimmock 1890; New York Evangelist 1873; White 1896; St. Louis Post-Dispatch 1914; Styles 2012).

As in the cases of Phipps and Shaw, a conservatory dependent upon public support could not fail to dramatize the largest and most curious plant of its

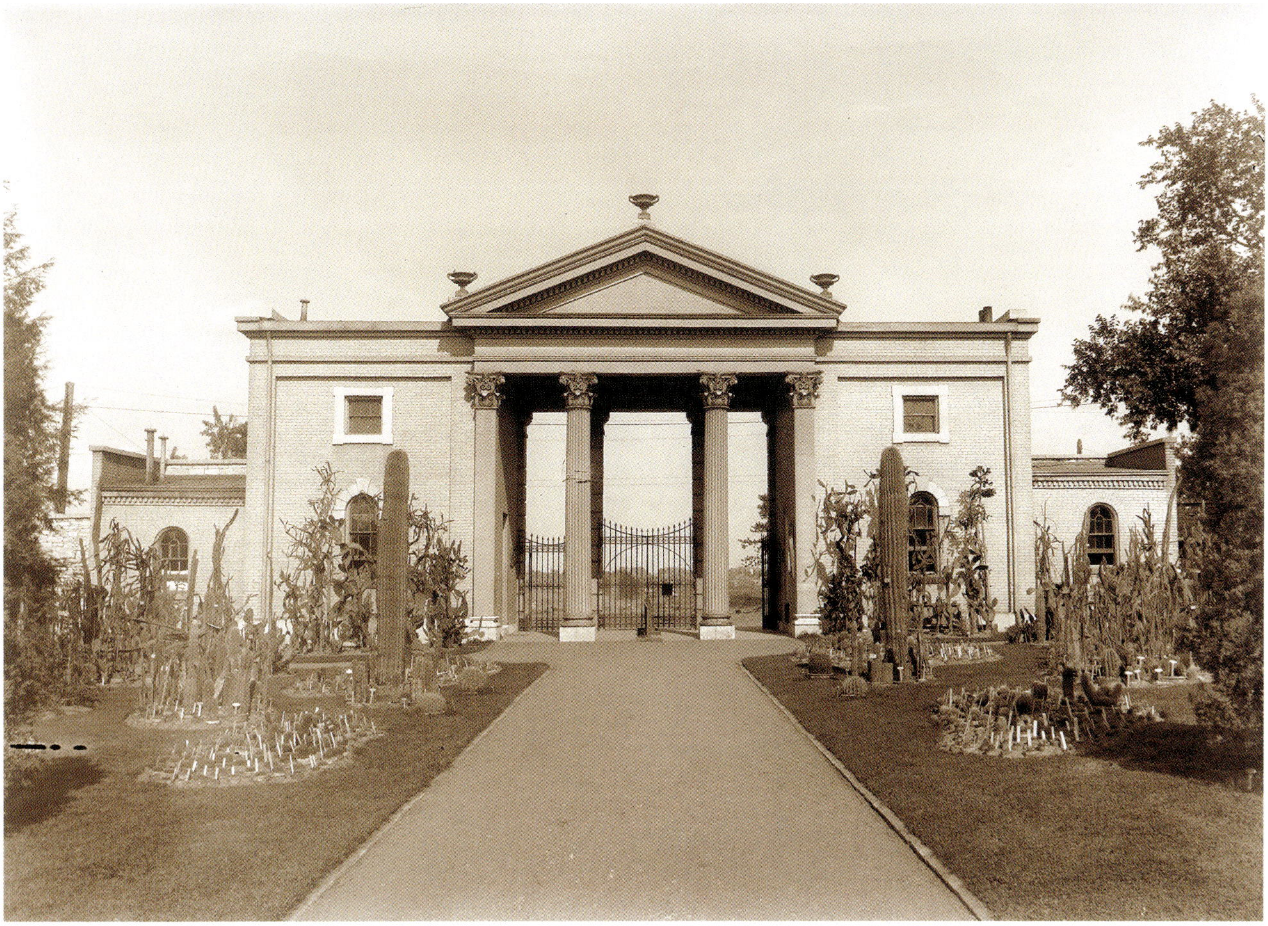

FIGURE 3.2. Main gate flanked by columnar saguaros, Shaw's Garden, St. Louis, Missouri. 1897.

kind. Yet like the earliest transplanted saguaros that failed, their appearance as curated specimens begged the question of what was actually being displayed. A saguaro could cheat death and bloom for years in the worst climactic conditions, in or out of the ground. The same was not true for other cactus specimens (Rose 1911–1916).

In 1899, the New York Botanical Garden, whose leadership came from the city's Torrey Botanical Club led by George Thurber and Nathaniel Lord Britton, embarked on the construction of a new series of connected steel and glass hothouses. The largest, a domed structure, displayed palm trees. Four glassed-in environments extending from the sides held arid plants, two for cacti, and two for yuccas and agaves. Garden director Britton filled three of them with small specimens exhibited at the recent Paris Exposition purchased from cactus collector M.C. Simon of Saint-Ouen. That left the fourth for the exclusive display of the cacti of the desert Southwest (American Gardening 1900; Harper's Weekly 1902; New York Times 1921).

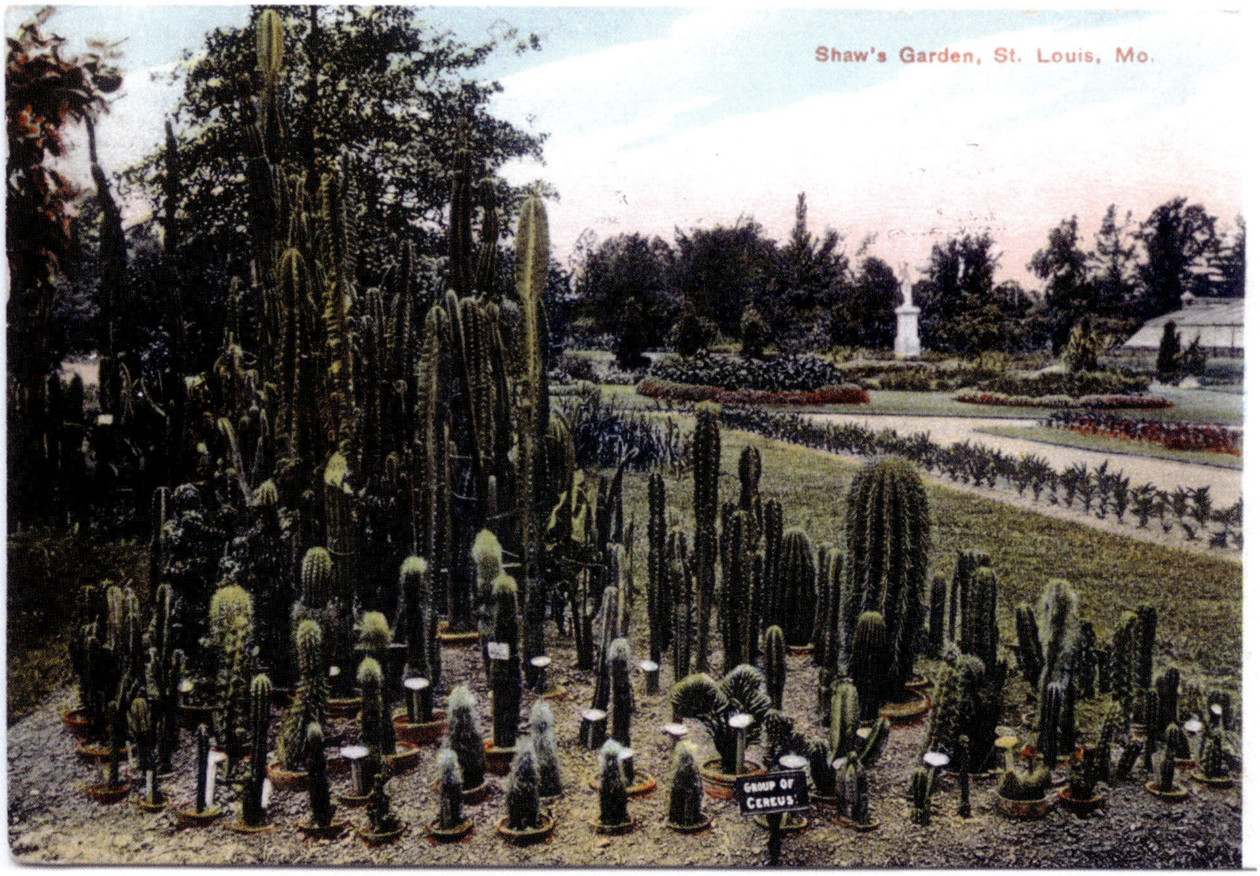

FIGURE 3.3. Hand-tinted postcard picturing potted *Cereus* cactus specimens at Shaw's Garden, St. Louis, Missouri. Printed in Germany, published by the St. Louis News Company. Postmarked 1908.

Britton had ambitions for the garden as a leader among competitive cultural attractions in New York City (Britton 1896). Britton dispatched botanist and First Assistant Daniel Trembly MacDougal to the Southwest to collect cactus specimens specifically for the new building. Bringing MacDougal under his wing as construction began, Britton (1899) confided, "of course we will have a leaning toward rare things."

MacDougal traveled with crates of camera equipment with which to document the plants *in situ* before and during their removal. On MacDougal's collecting trip to Tucson and Sonora, Mexico, in February 1902, he famously excavated a 25-year old, 13 ½-foot saguaro specimen for the new cactus house (American Gardening 1902; Harper's Weekly 1902; MacDougal 1902; Boom 1996). In this particular instance, an unidentified man in a pith helmet poses with a saguaro undergoing excavation exposing the topmost portion of its taproot. The extraction required six men to steady the plant, dig out its supporting taproot, and lower the specimen into a custom-made crate cushioned with creosote branches (Figure 3.5; American Gardening 1902; MacDougal 1902).

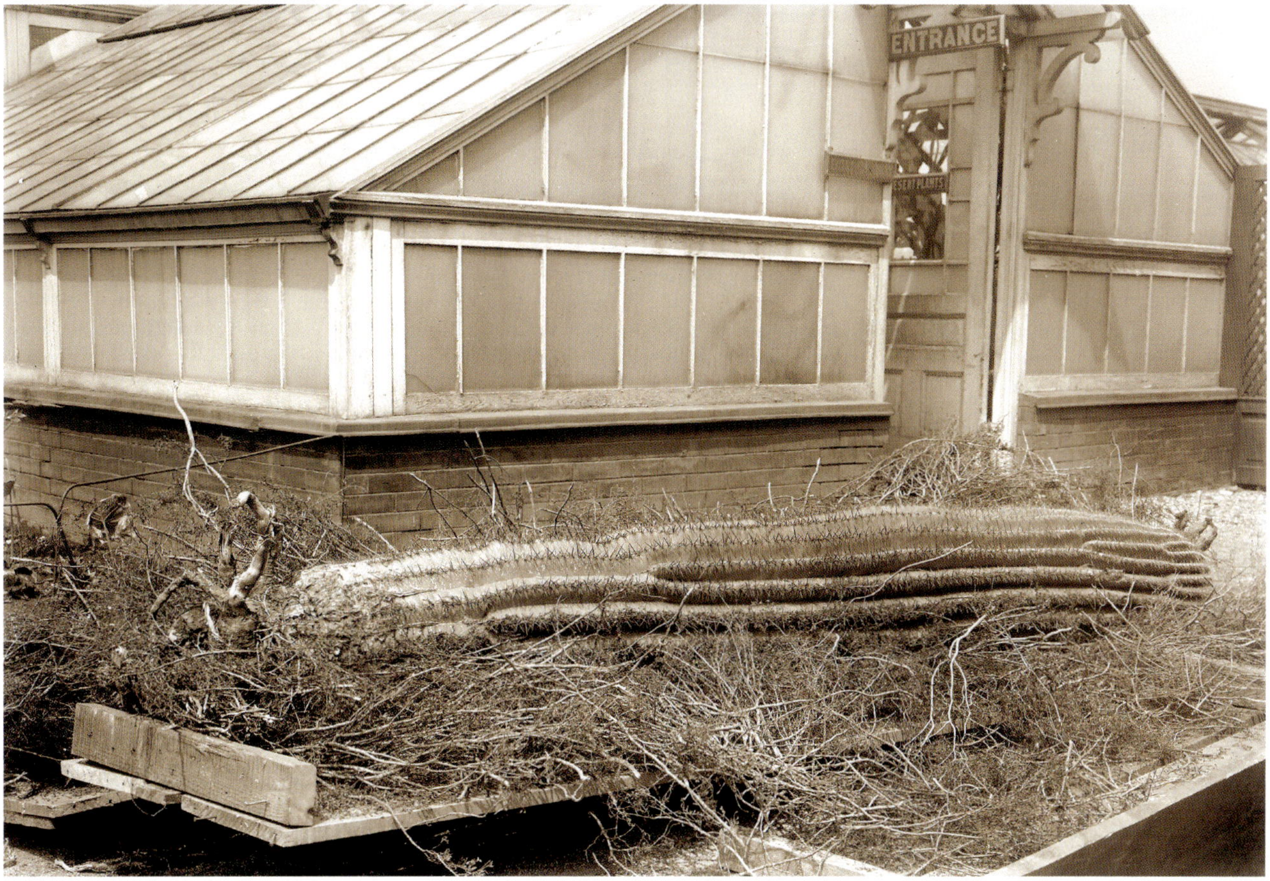

FIGURE 3.4. A newly arrived saguaro from Tucson with an abbreviated tap root in its wooden crate padded with creosote brush, Shaw's Garden, St. Louis, Missouri. 1914.

It is hard to overstate the impact of the display of ostensibly living saguaros upon the garden-going public. First, there was the tracery of the conservatory's light and glass; and then the overwhelming effect of the plants. "Sometimes the mountain comes to Mahomet" [*sic*], one New York reporter enthused. "We of the East cannot all of us go to the arid regions of Arizona to see the giant Cacti which flourish there, but we can go to the Cactus house of the New York Botanical Garden." The "strange plants . . . are a center of much curious attention from visitors to the Garden" (American Gardening 1902).

Dead or alive, in or out of the ground, MacDougal's saguaros put up a good front and bloomed regardless. University of Nebraska students and their professor to whom MacDougal sent a saguaro arm observed the unusual flowering phenomenon first hand. The flowers on the tip of the sawed-off arm blossomed in Nebraska for two consecutive years (Pool 1916). MacDougal did not attempt to transplant mature saguaros with arms as seen at the Chicago world's fair. The plant's arms appeared around the age of 30 years old when its water-absorbing height and girth would have complicated removal.

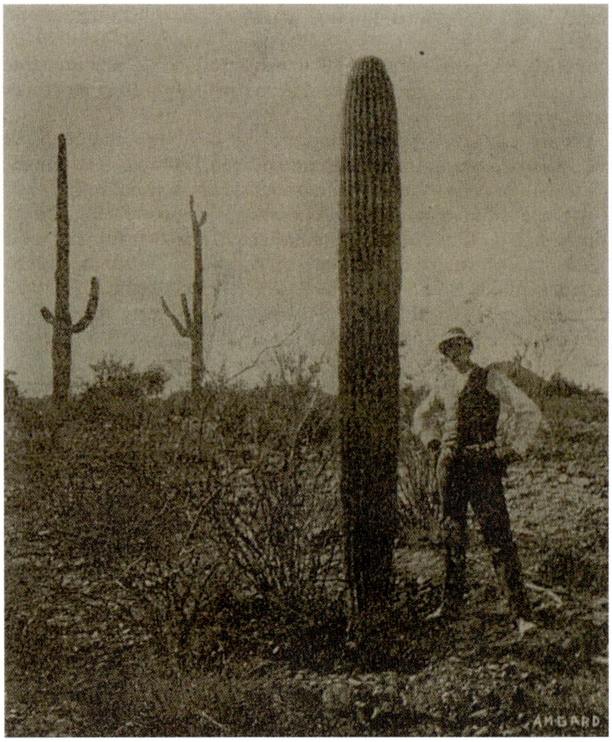

FIGURE 3.5. An unidentified man lends scale to a saguaro specimen collected by Daniel T. MacDougal near Tucson, Arizona, for the New York Botanical Garden. 1902. Visible here, the excavation has exposed the topmost portion of the saguaro's carrot-shaped taproot.

The hole in the desert, however, did not go unnoticed. *Youth's Companion* (1902) reported that the Arizona territorial legislature had named the saguaro as the territory's flower and sarcastically quipped, "that it will not be many years before it will be necessary to make a new choice," for alfalfa under irrigation now covered the saguaro's native habitat. Believing irrigation to *be* a habitat, others consigned the saguaro to the rocks, the rills and the sun-bleached hills unavailable to irrigation (Youth's Companion 1902). One critic took issue. H. Tullsen (1902) decried the recent complaint in the *New York Sun* about the irrigation of the "arid wastes of Arizona" and its consequent effect on the *Cereus giganteus*. He took the *Sun* to task, claiming that the plant's habitat was not in the valley, but the rocky hillsides "where water cannot be got" (Tullsen 1902).

MacDougal took a similar defensive position. Addressing criticism that his collecting denuded the desert, MacDougal maintained that most of the saguaro population grew on hillsides inaccessible to predation, both man-made and natural. Yet Sonora and Tucson never experienced the kind of irrigation that MacDougal foretold in his defense. The saguaro he collected came from neither an inaccessible nor a rocky hillside, as the picture of his find clearly shows. MacDougal stated that the Tucson saguaro came from a "comparative small area of the region capable of irrigation." He might have added that like his specimens loaded at Torres, the plants came from tableland within easy reach of the railroad (American Gardening 1902).

Saguaros played a symbolic role as living specimens in the endowment of botanical and climate studies. The board members of the New York Botanical Garden supported MacDougal's expeditions to collect display specimens for their newest conservatory. Garden board member and philanthropist Andrew Carnegie went on to endow the Desert Laboratory in Tucson. The laboratory's research program gathered data about how plants adapt to aridity. After a two-month search for a suitable laboratory location during February and March 1903, the newly organized Carnegie Institution settled upon Tumamoc Hill, a mountain site adjacent to

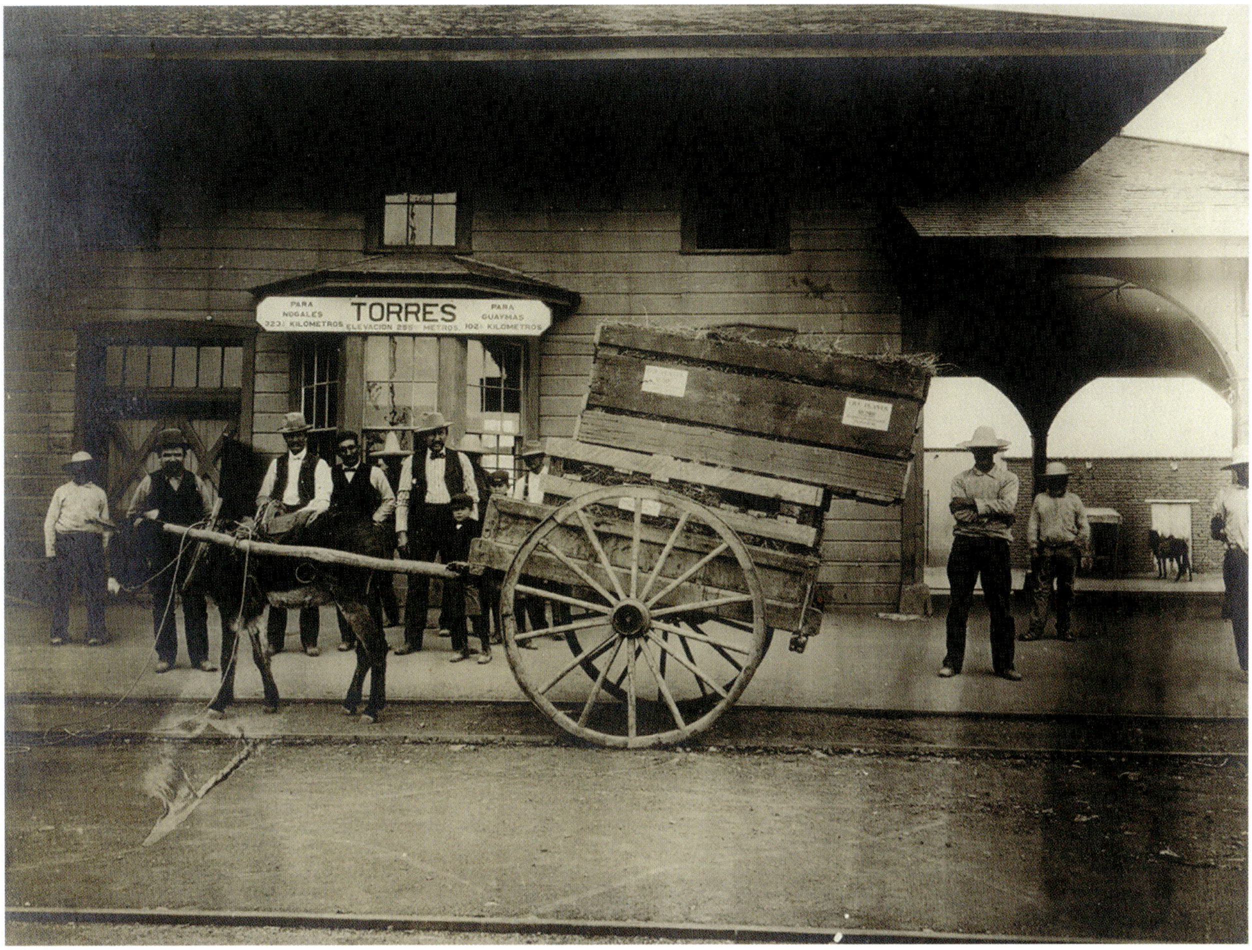

FIGURE 3.6. A mule-drawn load of MacDougal's cacti collected for the New York Botanical Garden and crated in Torres, Sonora. 1902.

Tucson. The search provided a second opportunity in as many years for MacDougal to collect cacti throughout the borderlands of the Sonoran Desert. The principal investigators included MacDougal and botanist Frederick Vernon Coville, an authority on Death Valley. Coville administered the laboratory from the Carnegie Institution's Washington, D.C., headquarters in concert with MacDougal, who in the first years of the laboratory fulfilled his obligations as director from his office at the New York Botanical Garden (Wilder 1967; McGinnies 1981:56–58; Craig 2005:1–9).

Conducting their site search, MacDougal and Coville weighed access to transportation, communication, and abiding botanical interest, ruling out areas expected to undergo irrigation including Phoenix's Salt River Valley (Rogers 1899; Coville and MacDougal 1903). Collecting cacti, MacDougal did not repeat his previous year's success (Figure 3.6). The specimens he crated at Torres, Sonora, "were almost

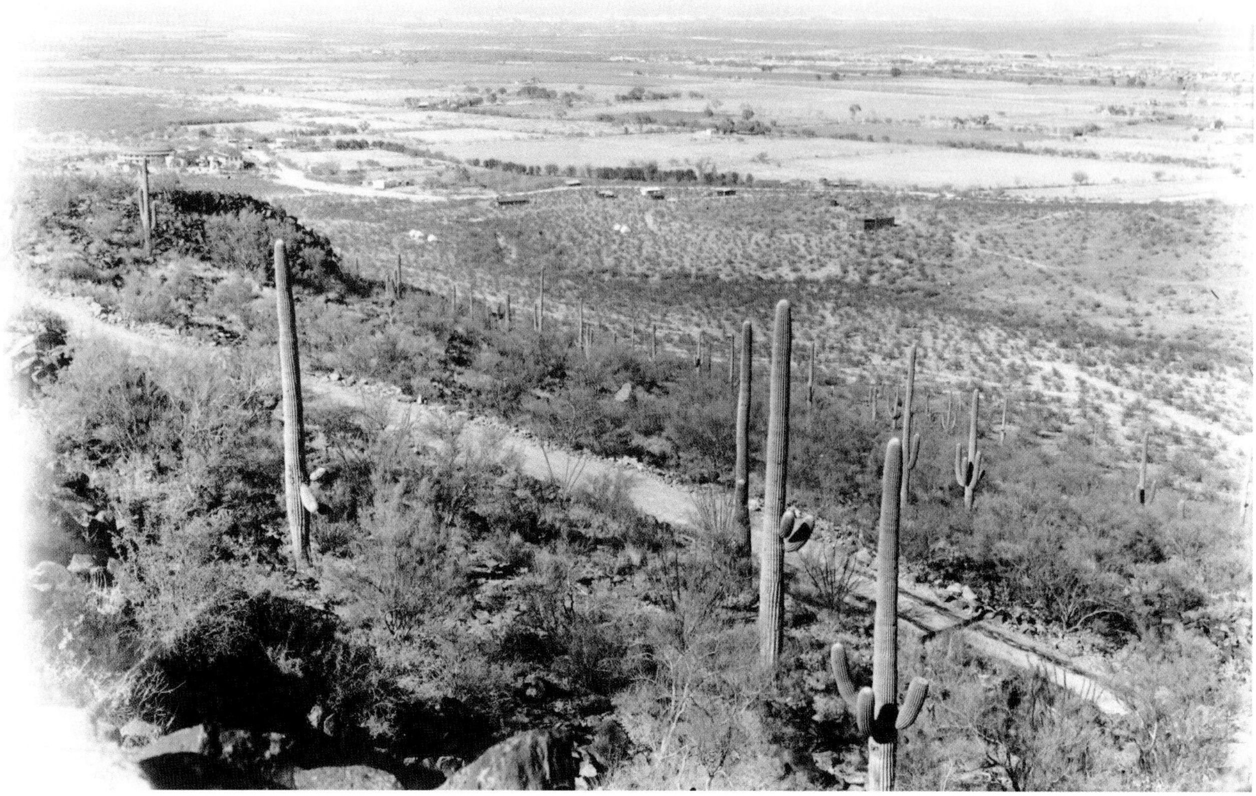

FIGURE 3.7. The Santa Catalina mountains and the city of Tucson looking east from the Desert Laboratory. 1914.

all ruined" upon receipt in New York, including a small *Cereus*. Even his shipment from Tucson arrived with an "injured" specimen. He thought that he could salvage some of them to pose for photographs in the Carnegie report announcing the laboratory. Instead, the announcement fell back upon the landscape photographs that MacDougal made, pointing up the hit-or-miss travel of cacti and other botanical specimens even at this late date (MacDougal 1903a).

Though it is hard to imagine where the Carnegie Institution might have more profitably located a laboratory for desert study, the contribution of the Tucson Chamber of Commerce proved decisive. The Chamber offered a semi-remote, 40-acre, tract of "stone land" on Tumamoc Hill above the Santa Cruz River, not far from that painted by John Russell Bartlett in 1852 (Figure 3.7; Wilcox and Larson 1979). The Chamber provided the site; the blasting of a road to a buildable mesa; light, power and telephone lines; and a rudimentary water line and a pump. The land had recently become available when sold for delinquent taxes (Coville 1903; Jaynes 1903; Sykes 1944:268–284).

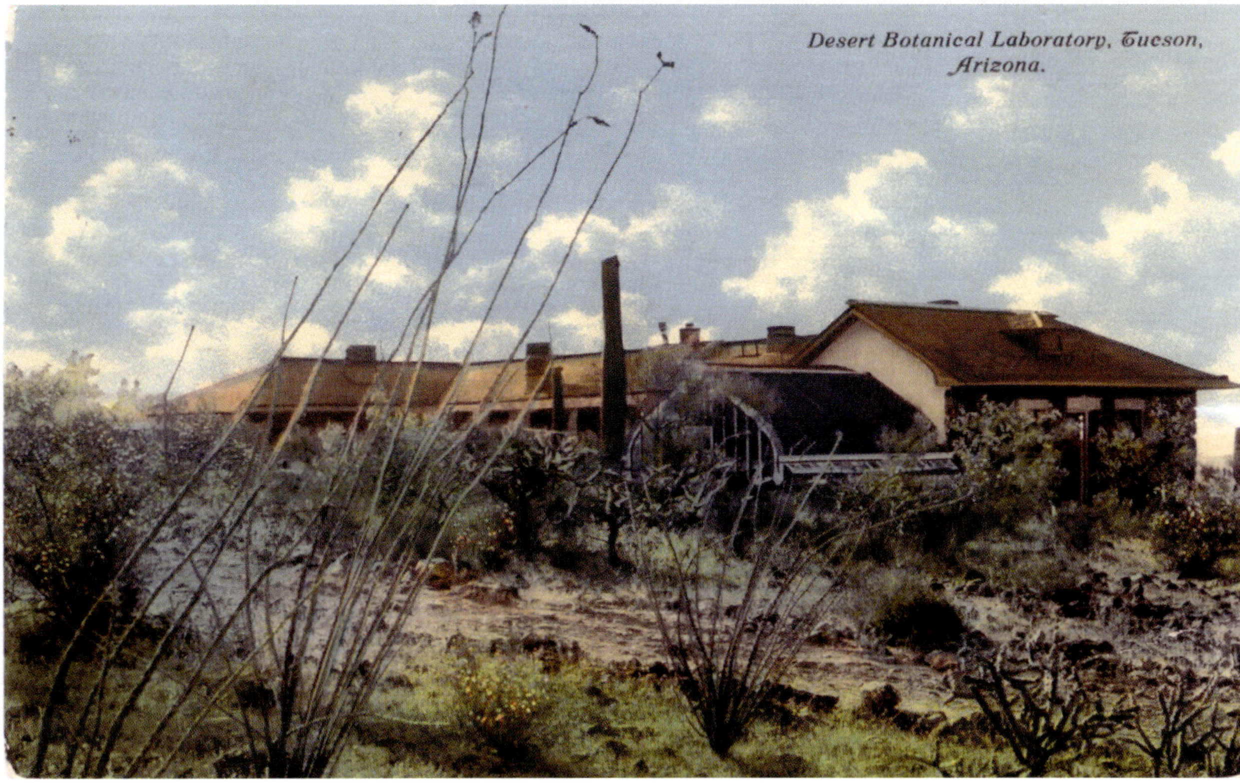

FIGURE 3.8. Postcard. Desert Botanical Laboratory, Tucson, Arizona, ca. 1912.

The Carnegie Institution set up shop under the immediate direction of William Austin Cannon, the Desert Laboratory's first resident investigator. Cannon busied himself locating the boundaries of the cactus-filled tract on Tumamoc Hill. Cannon oversaw the construction of a sturdy stone building with a wooden roof that still houses the active laboratory today (Figure 3.8). The finished laboratory featured a library, work and office spaces, and a darkroom that turned out a stream of professionally produced photo publicity. During construction, Cannon documented the desert vegetation with photographs sent to newspapers and journals. The photos could also be made into slides for use on the illustrated lecture circuit (MacDougal 1903b). A photograph taken by Cannon documenting the property in February 1903 shows a saguaro fruit picking pole found on the site, presumably left for the season. Cannon repurposed the crook-topped pole as a corner boundary marker and took its picture (Figure 3.9).

By November 1903, Cannon (1903) had explored "the summit of our mountain," alerting MacDougal to the existence of "an old Indian village ruin…Pottery and instruments for grinding (all broken of course) are rather plentiful. Much of it is in the Laboratory domain. In the course of events I shall photograph this for the Laboratory." It is not known if he returned to document the scene with

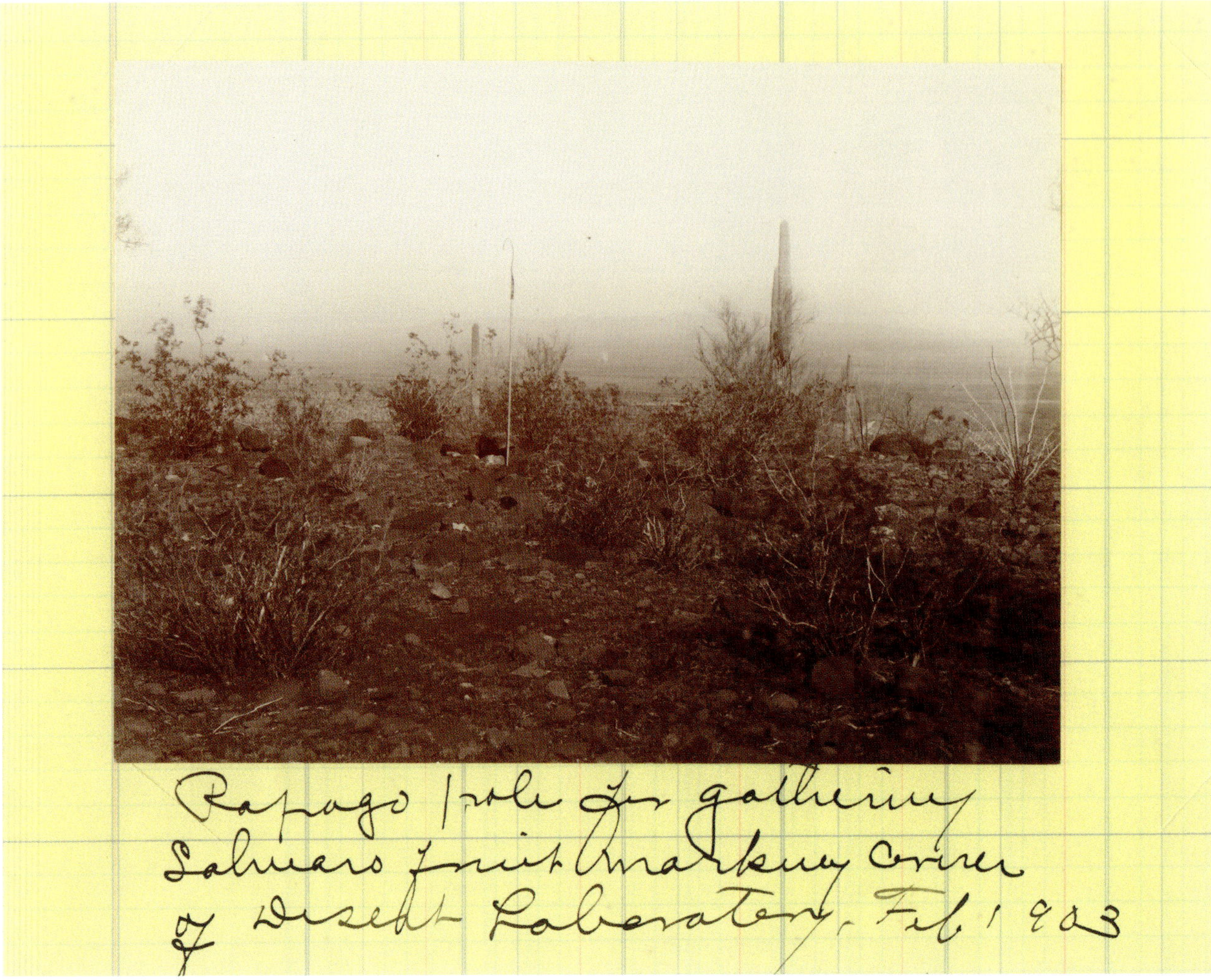

FIGURE 3.9. Annotated scrapbook page and photograph attributed to W.A. Cannon. 1903. Cannon pictured the saguaro fruit harvest pole he discovered while exploring the property and repurposed it as a corner boundary marker.

his camera or not. The subject failed to appear in his subsequent correspondence. Ironically, that very month MacDougal published the illustrated announcement of the Carnegie Institution's Desert Laboratory program at Tucson. In an effort to showcase the site's historical, cultural and botanical assets, MacDougal excerpted ethnologist William McGee's descriptive account of Tohono O'odham Indian life at the *base* of Tumamoc Hill (Carnegie 1903). It was left to a later generation of ethnologists to investigate the significance of Tumamoc's visible trail systems, rockwork walls, petroglyphs, and sherds (Ferg 1979; Hartmann and Hartmann 1979; Masse 1979; Wilcox and Larson 1979; University of Arizona 2008; Downum et al. 2009; Wilder and O'Meara 2015).

Siting the building among the saguaros was of great importance to MacDougal. At the outset of construction he cautioned Cannon, "The preservation of the plants saves the building from the look of barreness [*sic*] seen around so many places. Every

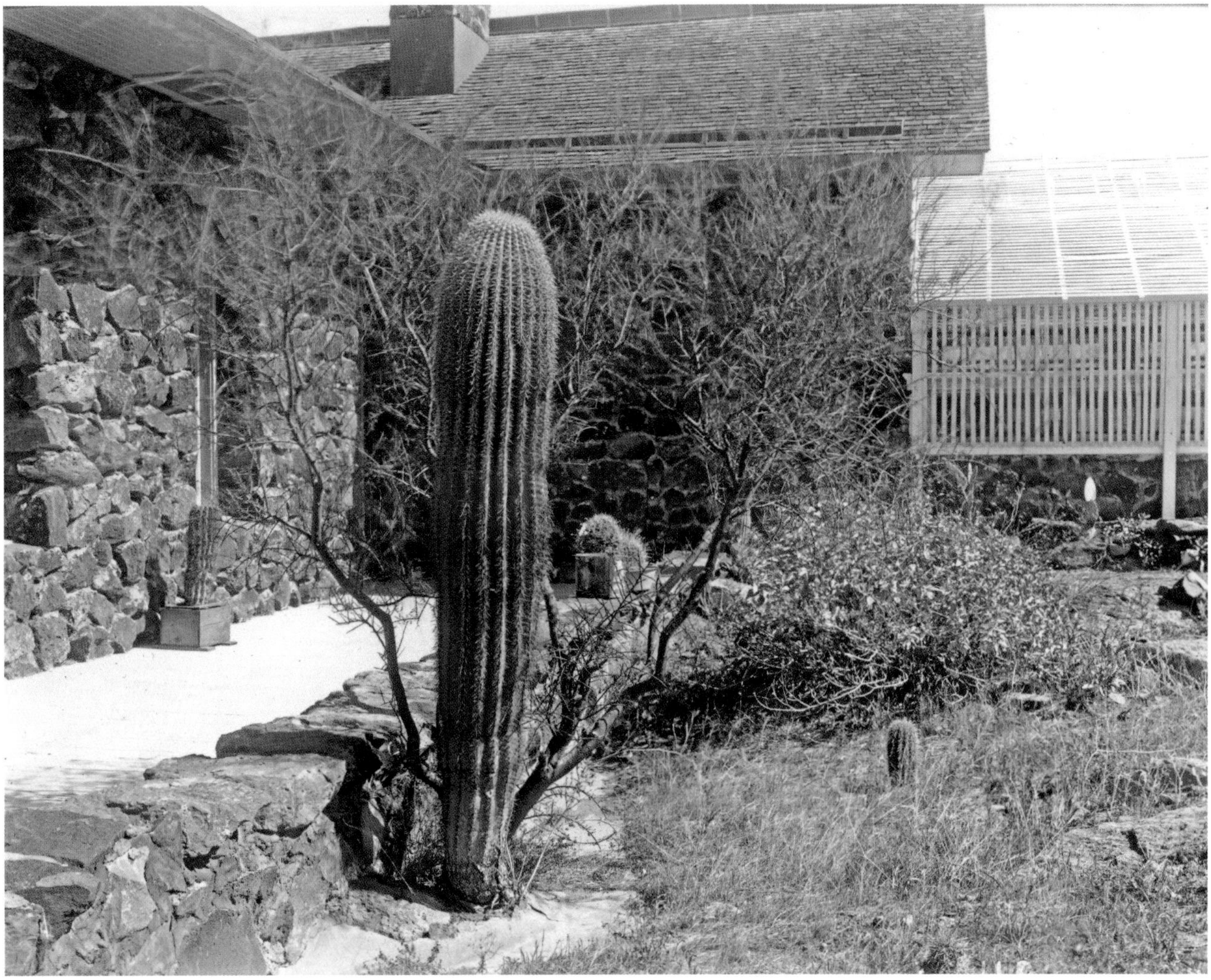

FIGURE 3.10. MacDougal cautioned Cannon to retain as much close-in vegetation as possible during the building's construction.

plant on the mountain ought to be regarded as if had been placed there for some special purpose and should not be destroyed without good cause" (Figure 3.10; MacDougal 1903c). Botanists working in the new building needed not travel to pursue research on light, soil, water, or the process of transpiration that accounted for the saguaro's accordion-like expansion and contraction (Figure 3.11). Botanists publicized their research with photographs processed at the laboratory and pictured in the Carnegie Institution's annual reports. The Institution also invested in displays and exhibits highlighting its research. MacDougal, for example, created an exhibit for the 1915 Panama Pacific International Exposition at San Francisco. He had recently moved from Tucson to take up residence at the Institution's Coastal Laboratory at Carmel, California. MacDougal explained that the exhibit would be graphic, novel, and unique, advertising Tucson as an educational and scientific center (Arizona Daily Star 1914b).

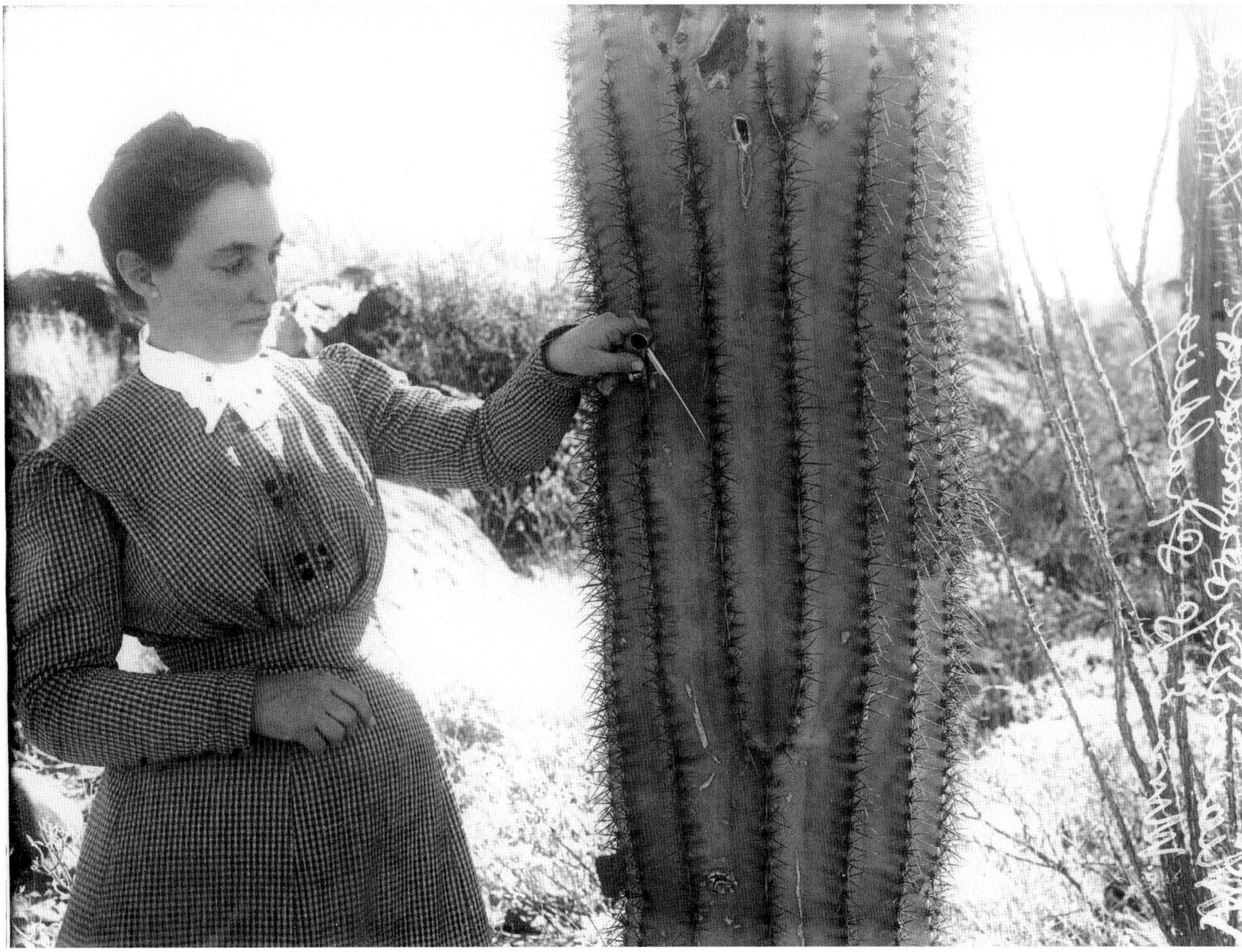

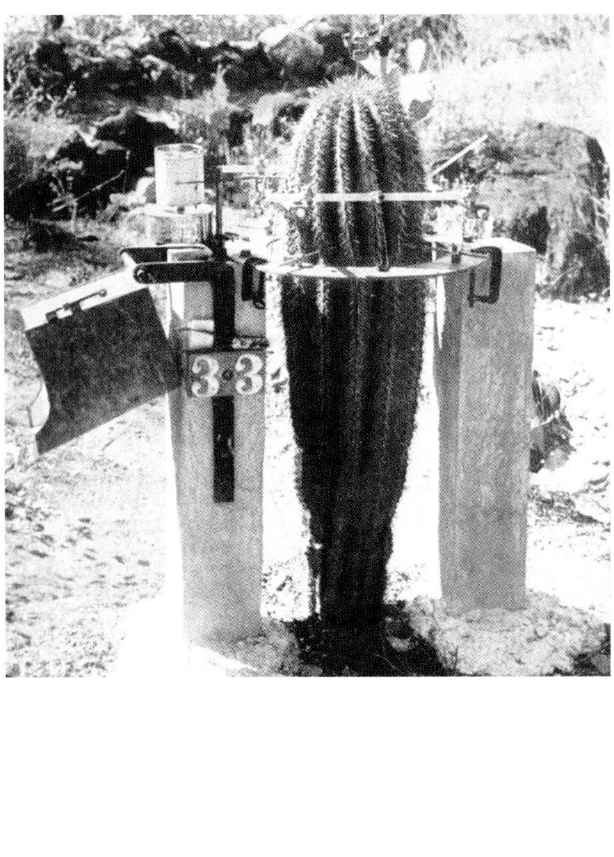

ABOVE. FIGURE 3.11. Botanist Effie Southworth Spalding measuring the folds of a saguaro, Desert Laboratory. 1909. Spalding's engaging work portrait has since become a touchstone in the history of science. But in its first use as an illustration in her "The Water-Balance of Succulent Plants" coauthored with Daniel T. MacDougal, 1910, it showed only her hand.

LEFT. FIGURE 3.12. Dendrograph recording growth and changes in volume of young saguaro, Desert Laboratory. 1925.

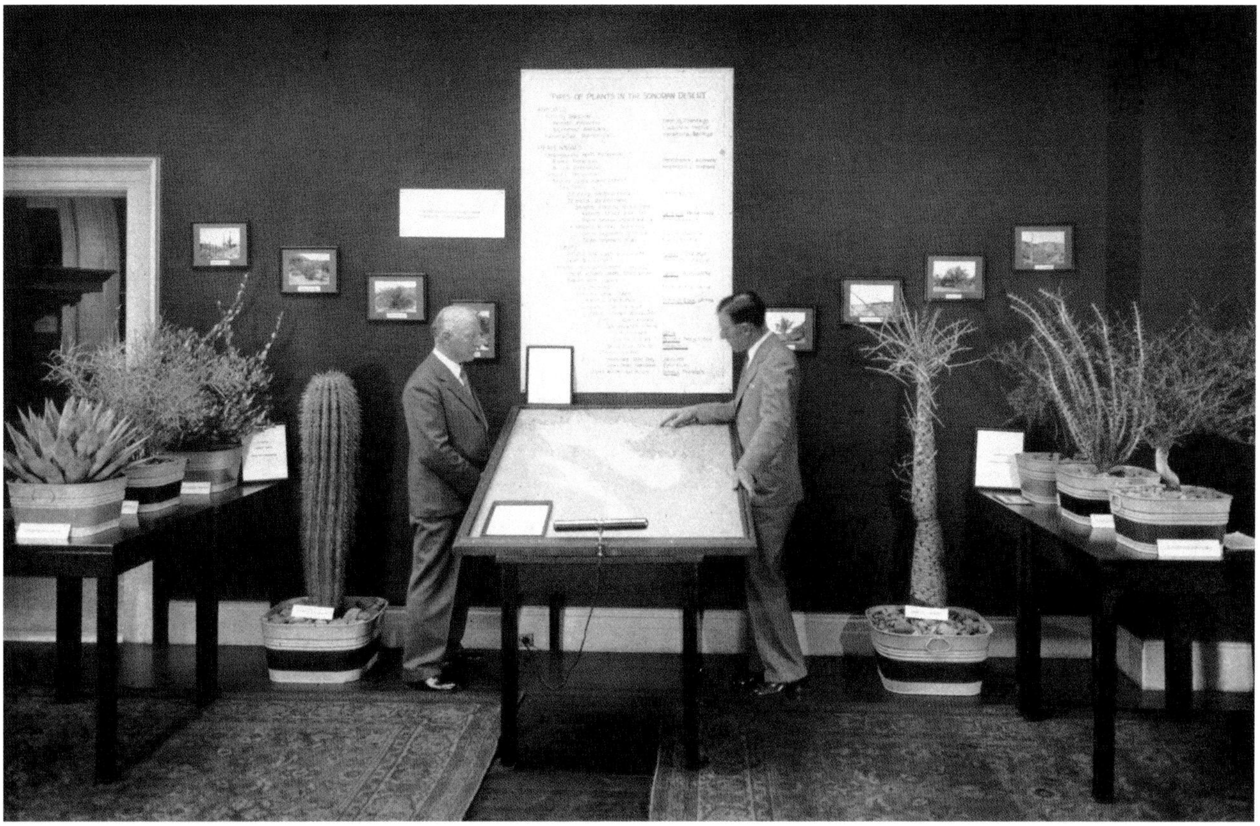

FIGURE 3.13. Laboratory for Plant Physiology (Desert Laboratory) Director Forrest Shreeve points to a map of the Sonoran Desert in conversation with an unidentified man at the Washington, D.C., installation of the laboratory's traveling exhibit, "Plants of the Sand." 1936.

The name Desert Laboratory changed to Laboratory for Plant Physiology in 1923, reflecting the expansive nature of its programs and MacDougal's new interests at Carmel. But the focus of the laboratory's most-pictured work remained the saguaro, the subject of experiments in water measurement with a dendrograph bracketing the plant in photographs (Figure 3.12; MacDougal 1925).

Near the end of its administrative existence the Desert Laboratory mounted a traveling exhibit of native plants for display at the Carnegie Institution's headquarters in Washington, D.C. Overseen by laboratory director Forrest Shreeve, the exhibit, entitled "Types of Plants in the Sonoran Desert," featured a map flanked by a columnar saguaro, a boojum tree, agaves, and other specimens in galvanized containers (Figure 3.13; Arizona Daily Star 1936; Carnegie Institution 1937; Shreeve 1937). In 1922, the Desert Laboratory's Godfrey Sykes named the curious, upside-down, carrot-shaped plant a "boojum"—an allusion to Lewis Carroll's *The Hunting of the Snark* (Humphries 1974:31). The protagonist of Carroll's (1876:24, 30, 33) poem describes a "a dreamy delirious fight" in which the sight of a "boojum"—should it appear—would cause his life to "softly and suddenly vanish away" (Carroll 1876:24, 30, 33).

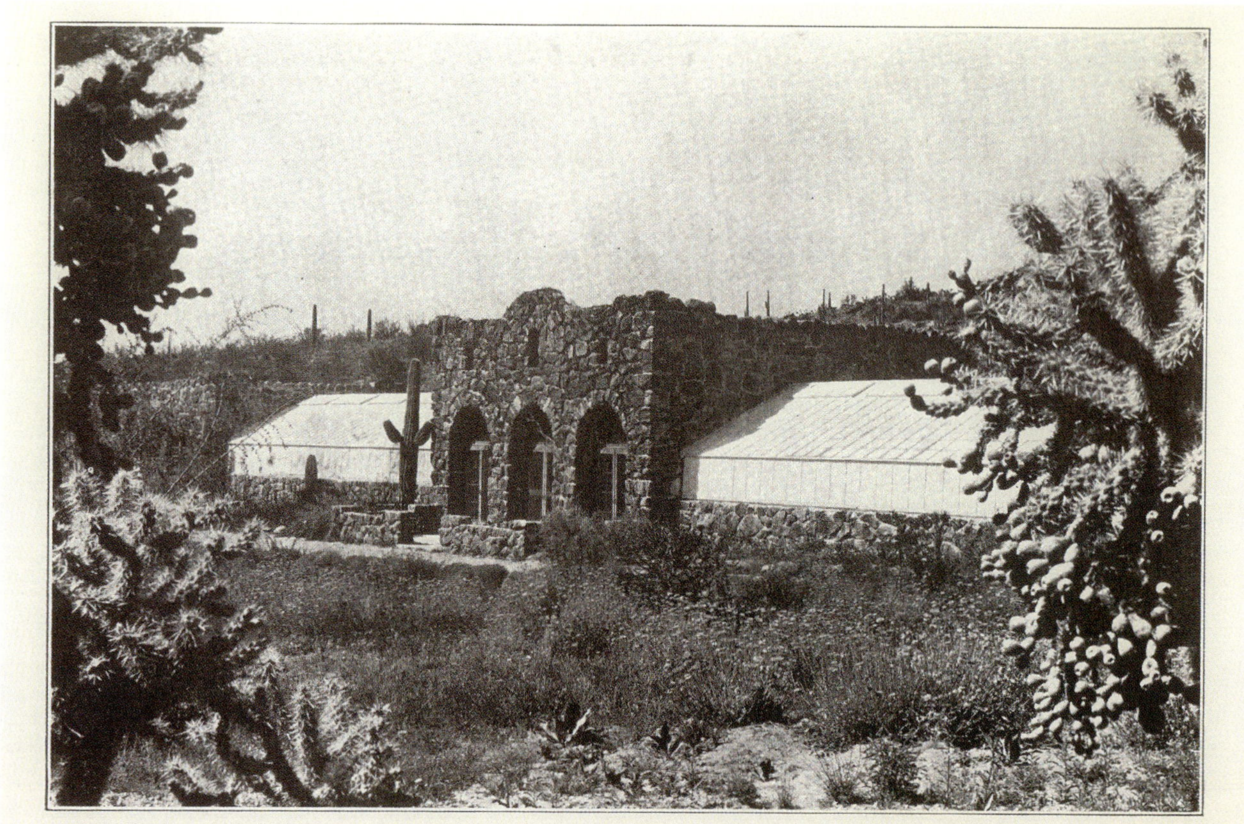

FIGURE 3.14. Saguaro and Boyce Thompson Arboretum stone greenhouse, Superior, Arizona. 1930.

In the display culture of gardens, fairs, zoos, and museums that valued rare things there would be no vanishing away. The sentiment that the desert had value anticipated the ecological sensibility of writers John C. Van Dyke, Mary Austin, William Carr, Joseph Wood Krutch, and more recently, Edward Abbey, Lawrence Clark Powell, Gary Paul Nabhan, and Tom Miller. Once the spot for explorations documented with specimens, artwork, and photographs, Tumamoc Hill came full circle as a jumping-off point for new knowledge that piqued hearts and curious minds.

Perhaps it was not surprising when individuals—learning what they could from literature, photographs, and displays—experienced the Sonoran Desert firsthand, they championed it as an "arboreal desert." Such was the experience of William Temple Hornaday, who coined the phrase on holiday in Tucson in 1907. Hornaday, the director of the New York Zoological Garden (the Bronx Zoo) came to Tucson at the invitation of MacDougal, to hunt bighorn sheep in Mexico's Pinacate region. A zoo director, Hornaday had a modern understanding of publicity. Only recently had Hornaday exhibited a Congolese pygmy, Ota Benga, with an orangutan at the zoo. The presentation of a human being and an orangutan in a cage with a label

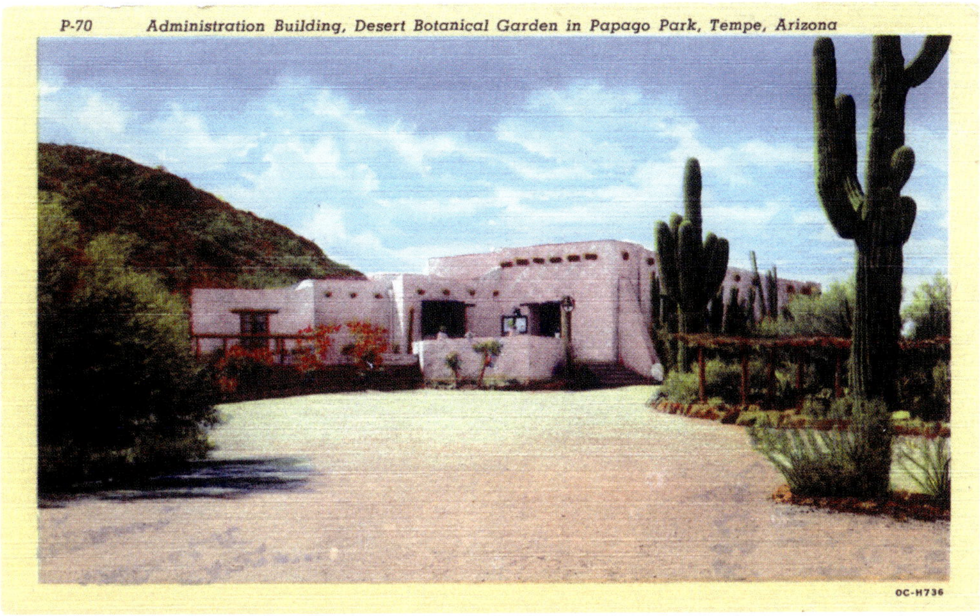

FIGURE 3.15. Postcard of the Spanish Colonial Revival style Administration Building, Desert Botanical Garden in Papago Park, Tempe, Arizona. 1950.

like any other—first in a cage, then on the grounds—shattered zoo attendance records. Facing mounting criticism in the press, Hornaday ended the sad show in September 1906 just short of three weeks into its run (Bradford 1992:176–190; Dehler 2013:96–98).

Arriving at Tucson in November 1907, Hornaday enthused about the unanticipated verdant quality of his desert surroundings. Hornaday talked up the area as an "arboreal desert" and "botanical garden." He later devoted chapters to each in a memoir of his hunt entitled *Camp-Fires on Desert and Lava* (Hornaday 1908).

The Desert Laboratory's Godfrey Sykes would have published a personal account of the hunt himself, but demurred to the more prominent Hornaday's literary ambitions. As a youth, Sykes (1944:55, 163, 272), inspired by the "flamboyant literature current at the time which purported to be descriptive of the Great West," had worked his way to Arizona from his native England. Sykes recalled that at the turn of the century American scientific bodies became "'Arizona conscious,'" sending "platoons" of investigators like his companions MacDougal, Coville, Cannon, and others, "to look the country over and make examinations and reports." Sykes judged Hornaday no differently, regarding Hornaday's description of the "'Arboreal Desert'" with detached amusement. As Sykes (1944:272) put it, "The giant cacti, the wide-branching ocatillas [*sic*], the untouchable chollas, and other large-scale xerophytic vegetables so impressed him that he declared that no other term would adequately describe this kind of scenery."

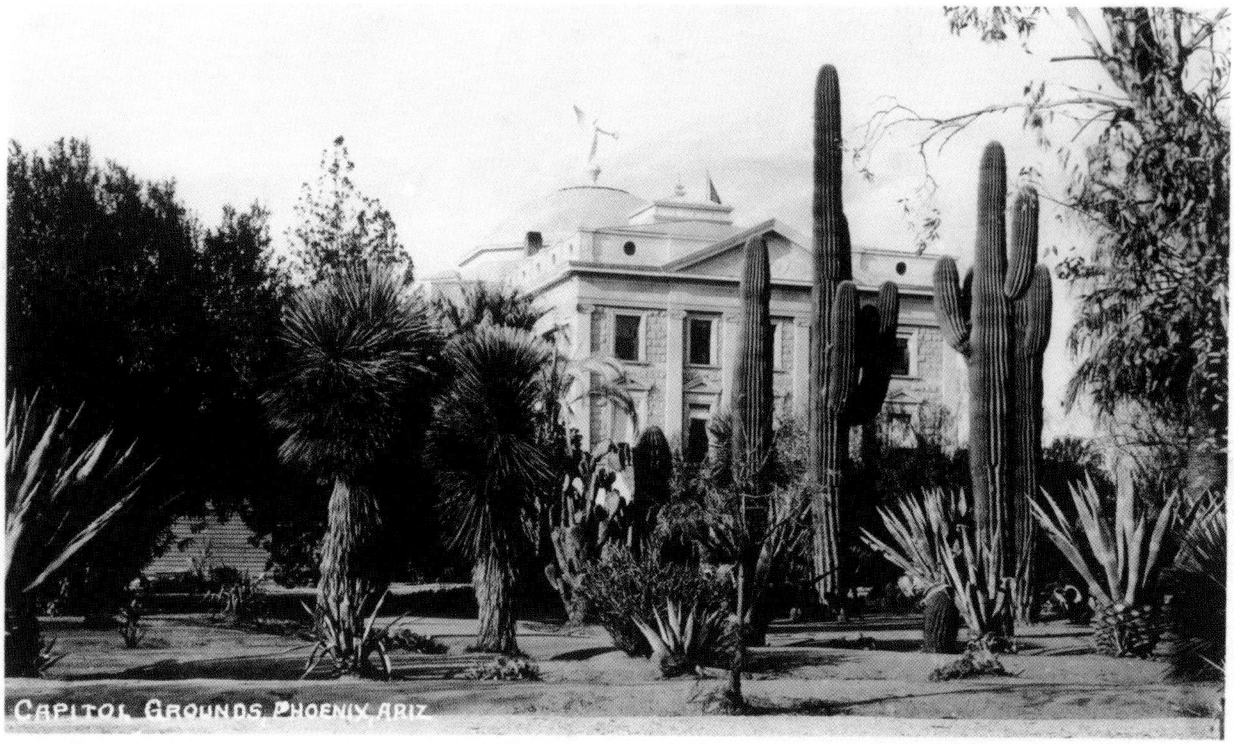

FIGURE 3.16. Realphoto postcard. Cactus Garden, Capitol Grounds, Phoenix, Arizona, ca. 1925.

In his memoir, Hornaday celebrated the apparent fact that the saguaro had no commercial value. Likened to "a sort of green-vegetable stand-pipe," Hornaday wrote, "the saguaro seems to serve only two important functions—to entertain and cheer the desert traveler, and to furnish high places for the nests of woodpeckers." There was nothing about the desert that Hornaday did not endorse. MacDougal introduced Hornaday to cactus candy made from the barrel cactus in Tucson and Phoenix. Hornaday (1908:214, 218–219) "ransacked the city for candy-stores," finding two pounds to take home, endorsing it as "the most delicious product of the South-west, not even excepting the preserved figs of California." With the successful publication of *Camp-Fires* and in subsequent conversations with friends, including President Theodore Roosevelt, Hornaday's trip to the arboreal desert never ended (Hornaday 1908:28, 35–42; McGinnies 1981:56–58; Broyles 1987; Dehler 2013:101–103).

The arboreal desert struck a chord with writers and publicists who used it to distinguish the Sonoran Desert from others. The popularity of cactus gardens now included, of all places, Arizona. Private gardens became public educational institutions, led by the dedication in 1929 of the Boyce Thompson Southwestern

Arboretum with a symbolic armed saguaro fronting the entrance of an experimental hothouse (Figure 3.14) and the founding of the Arizona Cactus and Native Flora Society at Phoenix in 1934 and its development of the Desert Botanical Garden at Papago Park in 1936–1939 (Figure 3.15; Arizona Daily Star 1929a; Boyce Thompson Southwestern Arboretum 1930; Arizona Republic 1937; Blanc 2000; Lee 2019:86–90).

Early on, the changing appearance of the Territorial Capitol grounds in Phoenix displayed the contradictions of native versus non-native plants. Before the construction of the capitol began in the late 1900s, landscape designer George Hough Smith opted for non-native Russian mulberries, Carolina poplars, elms, mountain ash, pepper, weeping willow, eucalyptus, orange, lemon, olive, and varieties of dates and palms (Arizona Republic 1890; Carman 2001:7–8). A reporter noted that the land "was most of it covered with mesquite and other desert growths, though part of it was covered with alfalfa" (Arizona Republic 1909). An Englishman who had lived in the territory for many years, Smith introduced the "depressed lawn" to his Phoenix clients, that is, a lawn sustained by flooding. The sunken lawn became a feature of Smith's Capitol plan in 1890 (Arizona Republic 1890). By 1899, however, a corner of the capitol tract featured some thirty varieties of cacti, "the most interesting corner of the capitol grounds," an exception to deciduous trees, plants and grasses (Arizona Republic 1899). The discovery surprised and pleased the reporter who noted, "As unsightly as these plants appear on the desert they become interesting and beautiful when transplanted and skillfully arranged by the experienced gardener." The reporter attributed the effect to Smith.

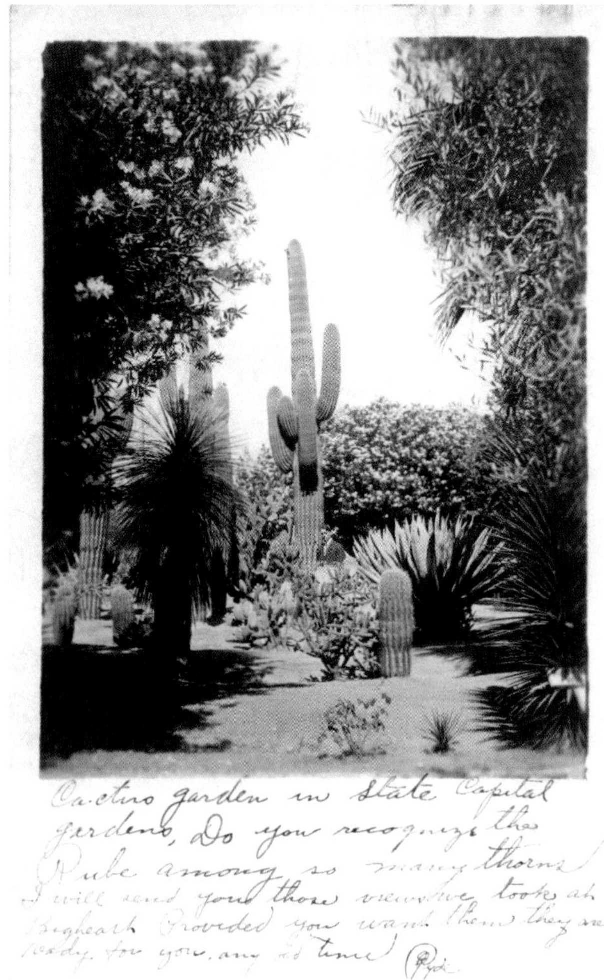

FIGURE 3.17. Realphoto postcard, Capitol Grounds, Phoenix, Arizona. ca 1912.

The first-time Capitol visitor might wonder where he had landed. The Honorable Frank E. Guernsey, the former congressman from Maine and member of the House Committee on Territories who had been instrumental in helping to enact statehood for Arizona, visited for the first time in 1919. Guernsey found the capital's

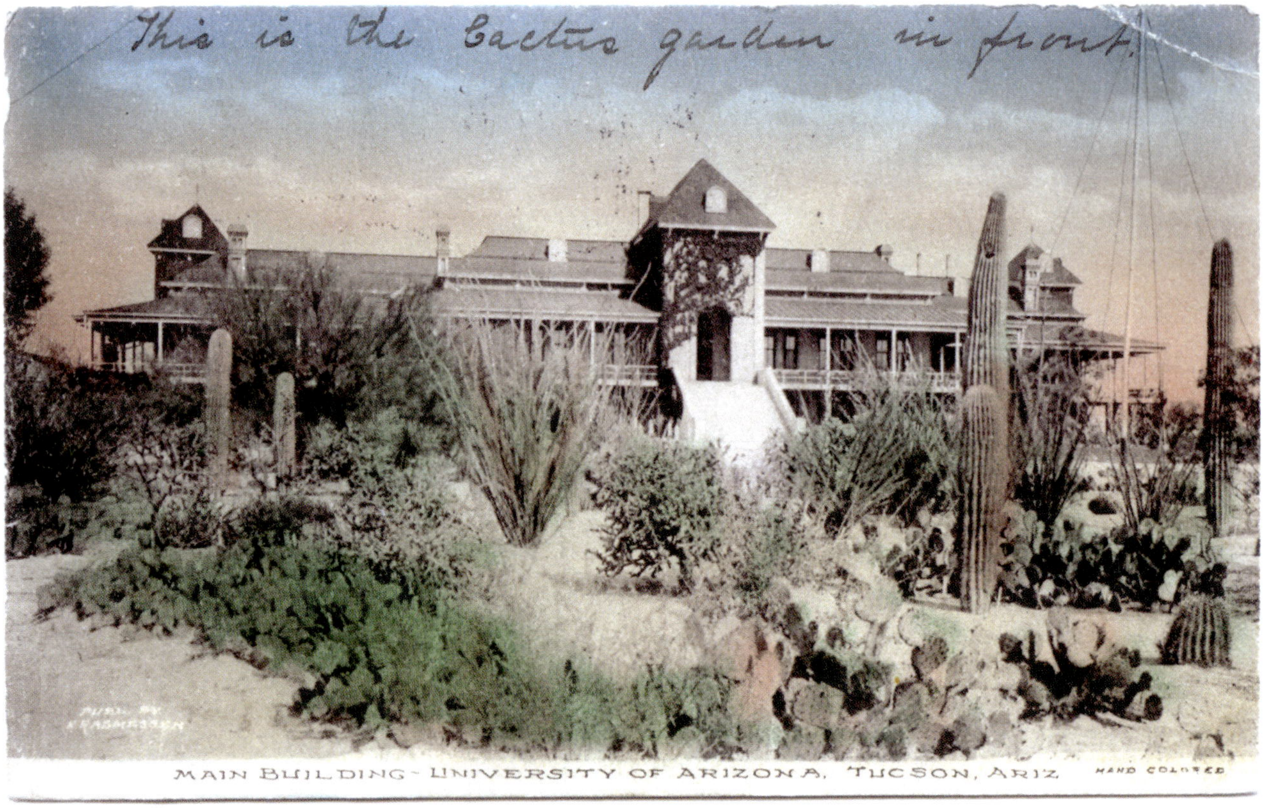

FIGURE 3.18. Hand-colored postcard of the Main Building, University of Arizona, Tucson, published by Rudolph Rasmessen. Handwritten note reads "This is the cactus garden in front." Postmarked 1909.

landscaping at odds with the desert conditions that he had expected to find. He did not anticipate green fields, much less such a "solidly built" city. The only cacti Guernsey saw were around the Capitol (Figures 3.16 and 3.17; Arizona Republic 1919).

At the University of Arizona, the campus cactus garden arranged by Professor James W. Toumey, botanist at the Arizona Experimental Station, became known as "a desert in an oasis" (Thornber 1906). The university surrounded the garden's heart-shaped outline with deciduous trees and Smith-type depressed lawns. One of the first accounts of Toumey's cactus garden, published in 1906, described the saguaro as "easily the most striking plant . . . perhaps the best known to the visitor, so far as it is possible for one to become acquainted with plants through literature" (Figure 3.18; Thornber 1906). After Toumey's departure for Yale in 1899 the university's College of Agriculture oversaw the display. The college had recently begun experimenting with the trench cultivation of cactus pads for use as cattle fodder. Desert rabbits and rodents frustrated the experiment by destroying the most edible plants (Agricultural Experiment Station 1908). Beyond the ethnographic documentation of the annual saguaro fruit harvest and the use of the plant's dried ribs in fence and roof construction, further study of the saguaro did not involve its investigation as a crop.

Increasingly the documentation of saguaros went from east to west. In 1904 Daniel MacDougal wrote from the New York Botanical Garden to Professor Robert L. Forbes, the Dean of the University of Arizona's Agricultural College. "The *Cereus giganteus* which I procured near Tucson two years ago are coming into bloom for the third time," MacDougal wrote, "and the plants are showing every evidence of being thoroughly suited in our glass house" (MacDougal 1904). In reply, Forbes thanked MacDougal for the photos picturing the saguaros displayed in the botanical garden's new conservatory. "It is interesting to me to observe that your giant cacti are coming into bloom at the present time," Forbes wrote. "Yesterday I observed them just beginning to bloom here. It looks as though these fellows were creatures of habit and went by the clock regardless of surrounding conditions" (Figure 3.19; Forbes 1904).

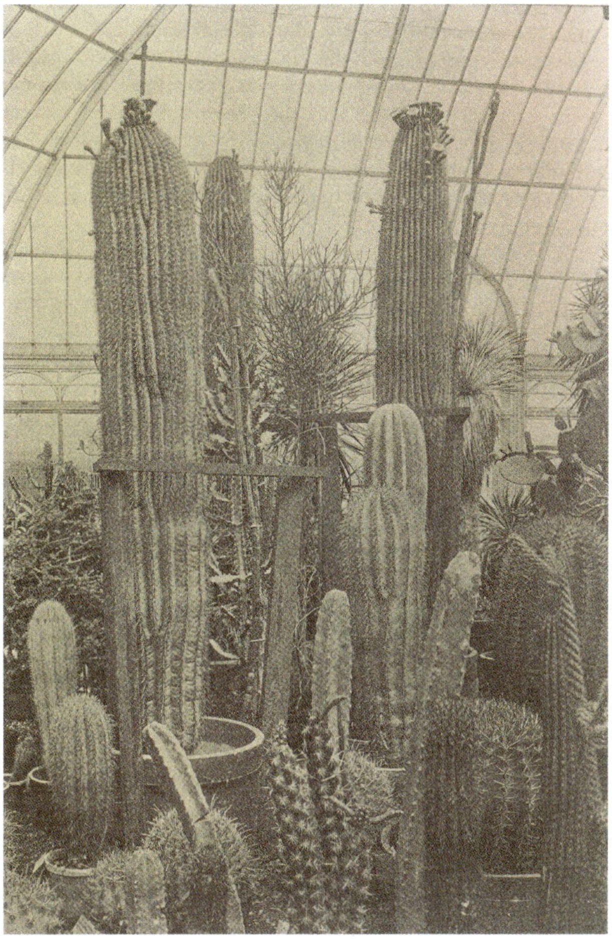

FIGURE 3.19 Photogravure. 1905. Columnar saguaros collected by MacDougal from 1902 to 1903, replanted in pots and carefully braced, bloom for the camera at the New York Botanical Garden.

The New York saguaros continued blooming. Writing about the saguaro's desert habitat in 1908, Nathaniel Lord Britton and J.N. Rose noted, "Travelers through these regions are always impressed by its very unusual form, and many thousands of people have become familiar with it since three plants were brought to the New York Botanical Garden by Dr. MacDougal in the spring of 1902, where they have since been successfully maintained, flowering every year in late spring and early summer." It now appeared that in giving the plant its botanical name, the late George Engelmann had erred in drawing the *Cereus* genus too broadly. It required a genus of its own. In 1908 Britton and Rose publicly announced by publication that they had renamed it, *Carnegiea gigantea*, in honor of garden and Desert Laboratory patron Andrew Carnegie (Britton and Rose 1908; Washington Post 1909). While Britton won Carnegie's approval to use his name (Britton 1908a and 1908b), Carnegie's experience with the actual plant was limited to the few he had seen at the New York Botanical Garden. In 1990, the Carnegie Institution *Newsletter* noted that Mr. Carnegie visited the saguaro-studded environs of the Desert

Laboratory in 1909 and only then learned, in MacDougal's words, "the grandeur of the honor intended in the dedication of the new species" (Bowers 1990; Craig 2005). According to the folklore surrounding Carnegie's brief laboratory visit in 1909, he dismissed the honor (Crosswhite 1980:6; Yetman 2007:51–58). Neither photographs nor contemporary press accounts documenting Carnegie's visit to Tumamoc Hill can be found today.

The name stuck, confirmed in Britton and Rose's (1919–1920) monumental work, *The Cactaceae*, published in four volumes with color plates beginning in 1919. The noted English botanical illustrator Mary Emily Eaton rendered the artwork for the plates, which featured a flowering saguaro that Eaton rendered from life at the New York Botanical Garden during her residence there in 1912. An unpublished note penciled on the top margin of the art states that MacDougal collected the specimen in 1903. This would have been during MacDougal's second trip to the Southwest while scouting locations for the Desert Laboratory, a memorable experience broken by damaged specimens upon his return. Perhaps this one had survived and flowered for Eaton to represent the best and most beautiful of its kind (Figure 3.20; Britton and Rose 1919–1920:vol. 1, p. 3, 1920, vol. 2, pp. 164–167; Angell 2003).

Providing visual information about context and setting, the earliest saguaro images from the Sonoran Desert focused upon the most dramatic, the tallest, the largest and, like Eaton's, easily passed through the eye of botanical significance to the most striking and eye-catching. Never subject to the economics of horticulture with the status of a crop, the saguaro took its place in the economy of travel and leisure with the status of a symbol.

The arrival of photography as a tool for gathering information about the sustainability of botanical life in the desert yielded to the promotion of living in the desert. This had been anticipated by Britton and MacDougal and company, who framed laboratory research as carefully as the desert and its unique plants. The geography of rare things became the fodder of countless illustrated newspaper reports, magazine features and picture postcards, all of which recommended themselves to the promoters of passenger railroads and leisure travel. In these images the arboreal desert was not something one rode through; rather a destination that enveloped and sustained one's curious nature.

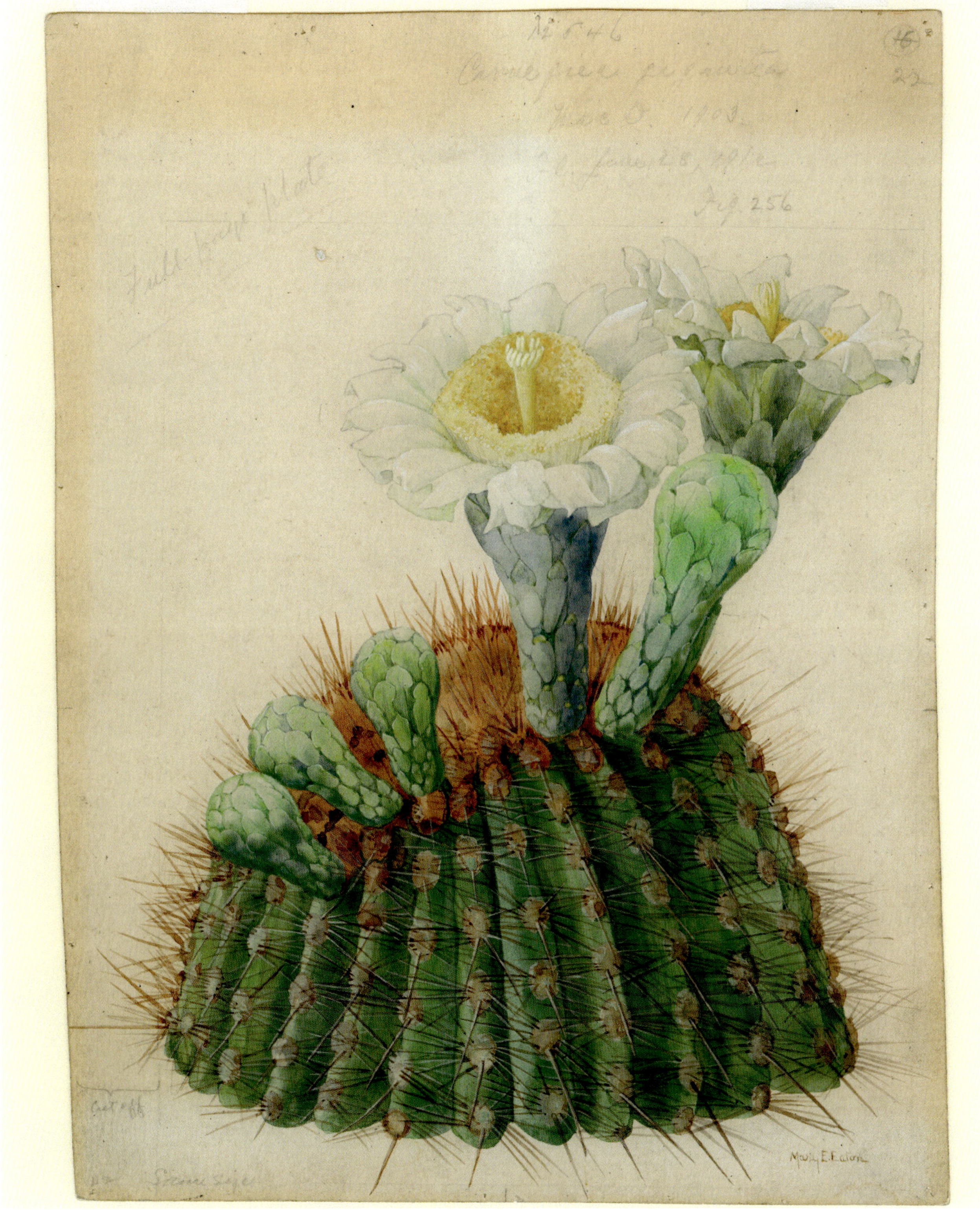

FIGURE 3.20. Mary Emily Eaton, pen and ink watercolor. 1912. Eaton painted this flowering saguaro in the collection of the New York Botanical Garden. Eaton's watercolor became the specimen plate for *Carnegiea gigantea* in *The Cactaceae: Descriptions and Illustrations of Plants of the Cactus Family* (Britton and Rose 1919–1920). A light pencil annotation, top right of center, attributes the specimen to MacDougal, who collected it for the garden in 1903.

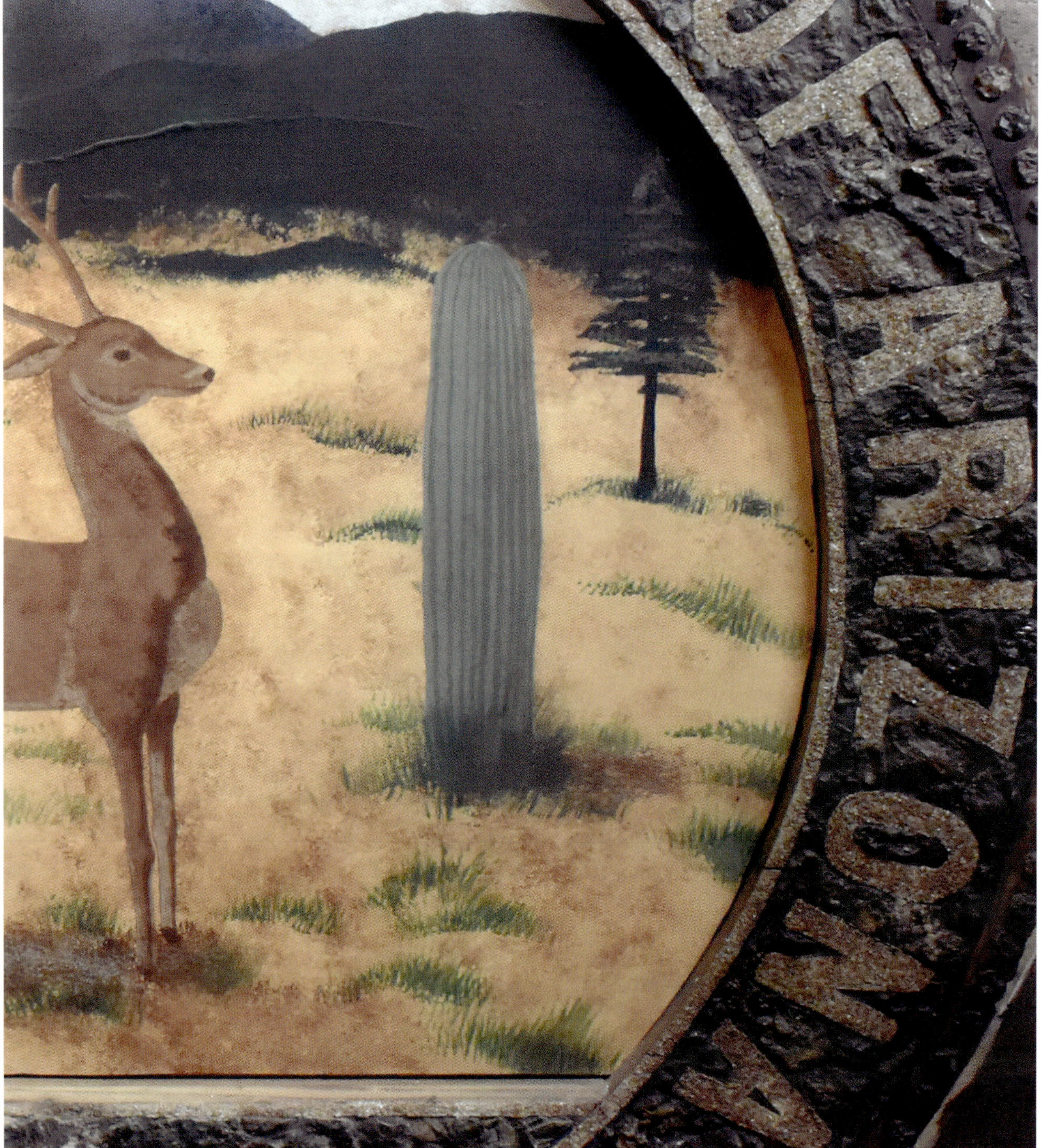
OF ARIZONA

CHAPTER FOUR
ON THE MARGINS

The iconography of botanical discovery did not automatically convey to the promotion of travel and tourism in the Southwest. Railroad passengers at Tucson might find postcards picturing an unusually formed saguaro with drooping arms, a result of severe freeze events over the 100-year-plus life of the plant (Figures 4.1–4.4). "Here is a strange plant that grows on the desert," wrote one correspondent. "There is [*sic*] a lot of them to be seen from the train." Not all passengers warmed to the sight. An eastern reporter taking a transcontinental trip to California in 1883, for example, dismissed cacti as "the reptile of the vegetable world" (Blake 1883:83; Trennert 1988; Hyde 1990:209, n63).

That same year, Tucson merchant Sam Hughes created a small poster for the newly organized Arizona Pioneers Historical Society. Hughes' poster pictured a saguaro *and* reptiles—a Gila monster and a horned lizard. In the distance he placed the ridgelines of two hills, visible from his Main Street home. Placing the saguaro at the center of the composition around which everything revolved anticipated the social saguaro of barbeques, bathing suits, and swimming pools that came to be favored by the city's mid-twentieth century boosters (Figure 4.5). This landscape owed a debt to the curation of the desert with locally transplanted saguaros and the importation of non-native plants and trees. For at the same time that Hughes created his poster, he and his neighbors—notably, the other Hughes—L.C. Hughes—were well on the way in experiments with sunken lawns sewn with Bermuda grass and trees placed at the railroad station, the gateway to the city featured in promotional pamphlets, brochures and postcards (Figure 4.6; Arizona Daily Star 1908; Lockwood 1932:330).

Railroad passengers at Los Angeles and San Antonio saw transplanted saguaros at their respective stations (Duncan 1929). But not Tucson's. A postcard of the Southern Pacific Tucson depot about 1911 pictures saguaros some distance away in a park. It is not clear if they are recent transplants or, incongruent cutouts on a sunken lawn, tipped-in for effect (Figure 4.7; Meikle 2015:65).

With its trees, lawns, and cactus-at-a-remove, the landscaping of the Tucson station reflected the arboreal tension underlying the pictures that promoted Arizona

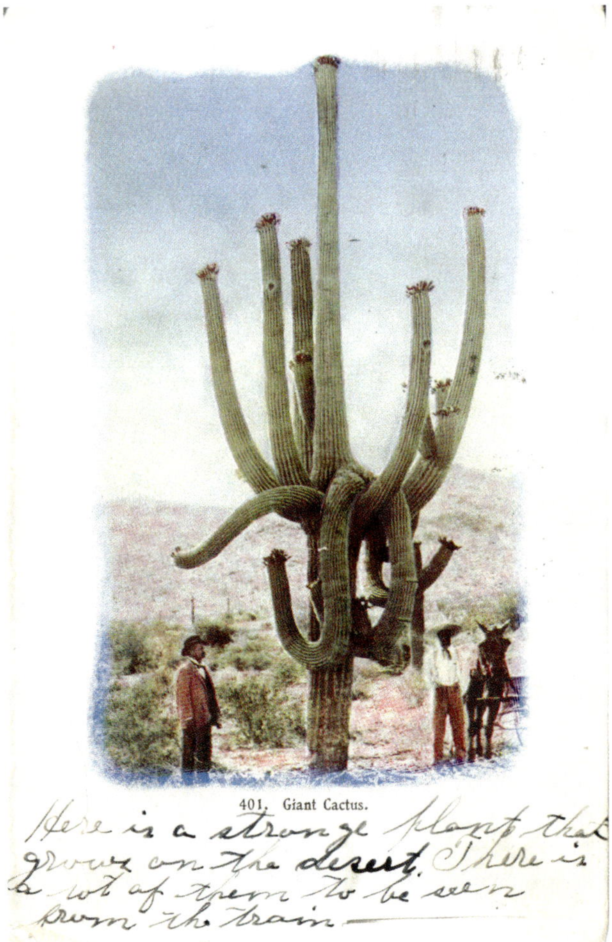

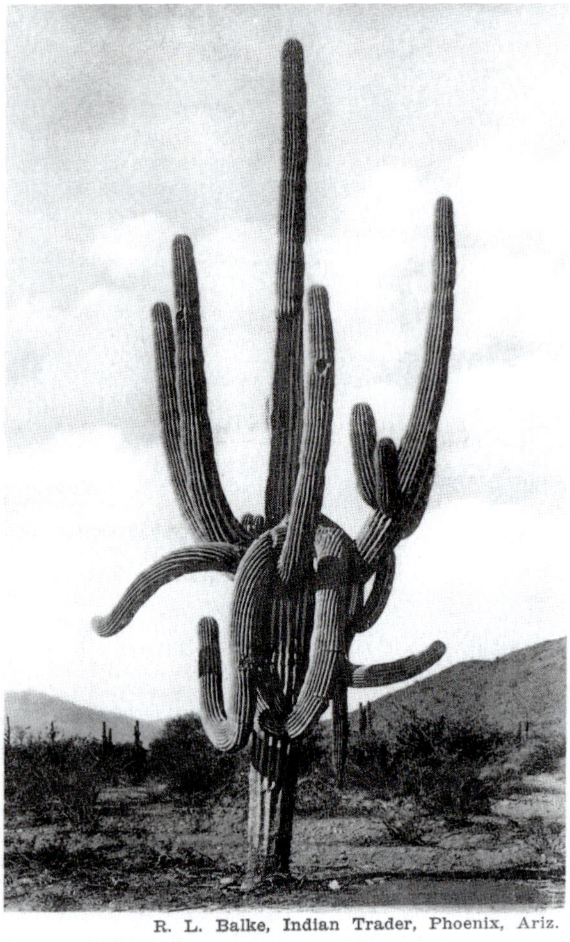

FIGURE 4.1. Postcard. Giant Cactus. Published before 1907, postmarked Phoenix 1909. The sender of this card wrote, "Here is a strange plant that grows on the desert. There is [*sic*] a lot of them to be seen from the train." Many businesses, cities, and regions claimed this most unusually formed saguaro. Repeatedly photographed during the first decade of the twentieth century, the actual location of the plant is unknown.

FIGURE 4.2. Postcard. Giant Cactus Salt River Valley Phoenix, Arizona, ca. 1908. Printed by Albertype Co., Brooklyn, New York. Published by R.L. Balke, Indian Trader, Phoenix, Arizona.

to the outside world. The territory was open for development. But what did development look like and talk about? For many the answer came in pictures of extractive industries such as mining and logging, the peripheral vision of an open range, and agriculture under irrigation. From hilltop to valley, development had nothing to do with turning the saguaro into a fixture of regional identity, yet there it was.

The tendency to understate—cactus is here, but—permeated the era's land sales propaganda. A review of immigration literature published between 1863 and 1900 reveals few mentions of cacti (Mowry 1864:160–161; Hodge 1877:242–248; Hinton 1878:342–343). Authors frequently failed to acknowledge it at all (Mowry 1863; McCormick 1865; Blake 1870; Farish 1889a, 1889b). The guides for immigrants published by the Arizona territorial legislature shared the ambivalence of railroad publications in which cacti figured as low-lying casualties of irrigation (Black 1890:58; Southern Pacific Company Passenger Department 1900:29, 31).

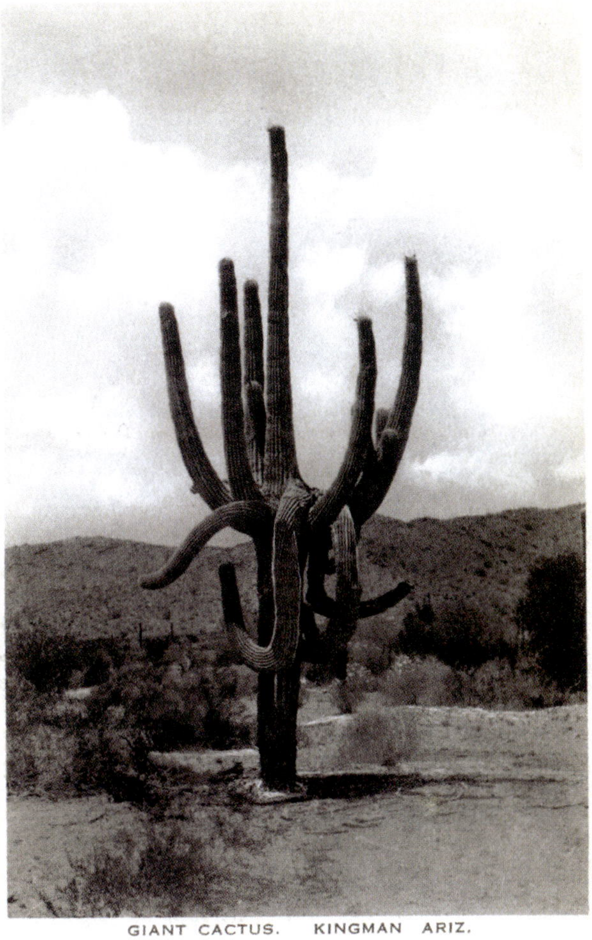

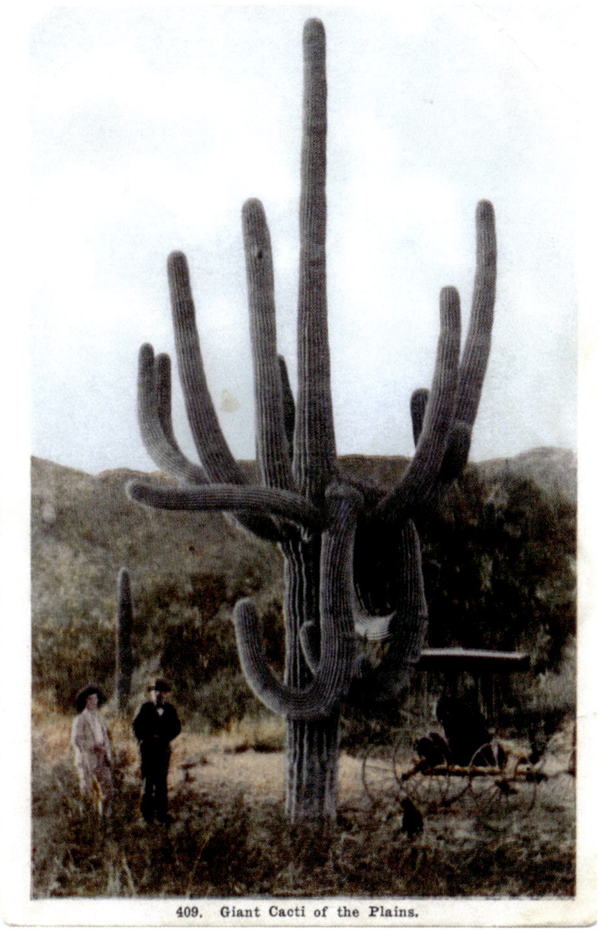

FIGURE 4.3. Postcard. Giant Cactus Kingman, Arizona, ca. 1907. Printed by Albertype Co., Brooklyn, N.Y. Published by Hotel Beale, Kingman, Arizona.

FIGURE 4.4. Postcard. Giant Cacti of the Plains, ca. 1910. Printed and published by Thayer Publishing Co., Denver, Colorado.

Nevertheless, by the mid-1910s a baseline of accessible photographs joined the lithographs and engravings of botanical discovery in picturing the Sonoran Desert. First-person encounters with saguaros by photographers led to second-person familiarity with the plant through the publication of popular stereoviews and penny postcards (Sandweiss 2006; Rowe 1997:106–108).

Tucson photographer Henry A. Buehman exemplified the itinerant desert traveler who maintained an active interest in landscape photography after settling into studio work. Buehman arrived in Tucson in 1874, six years before the railroad, qualifying him as a bona fide member of Sam Hughes' society of pioneers. Buehman augmented his income working as a dentist while employed at the Rigorio brothers' Photographic Parlor. Buying out the brothers in 1876, Buehman changed the business' name to the Buehman Studio and advertised himself as a "landscape and general photographer." He became well known for the quality and finish of his studio portrait work, especially babies. He built a two-story downtown block,

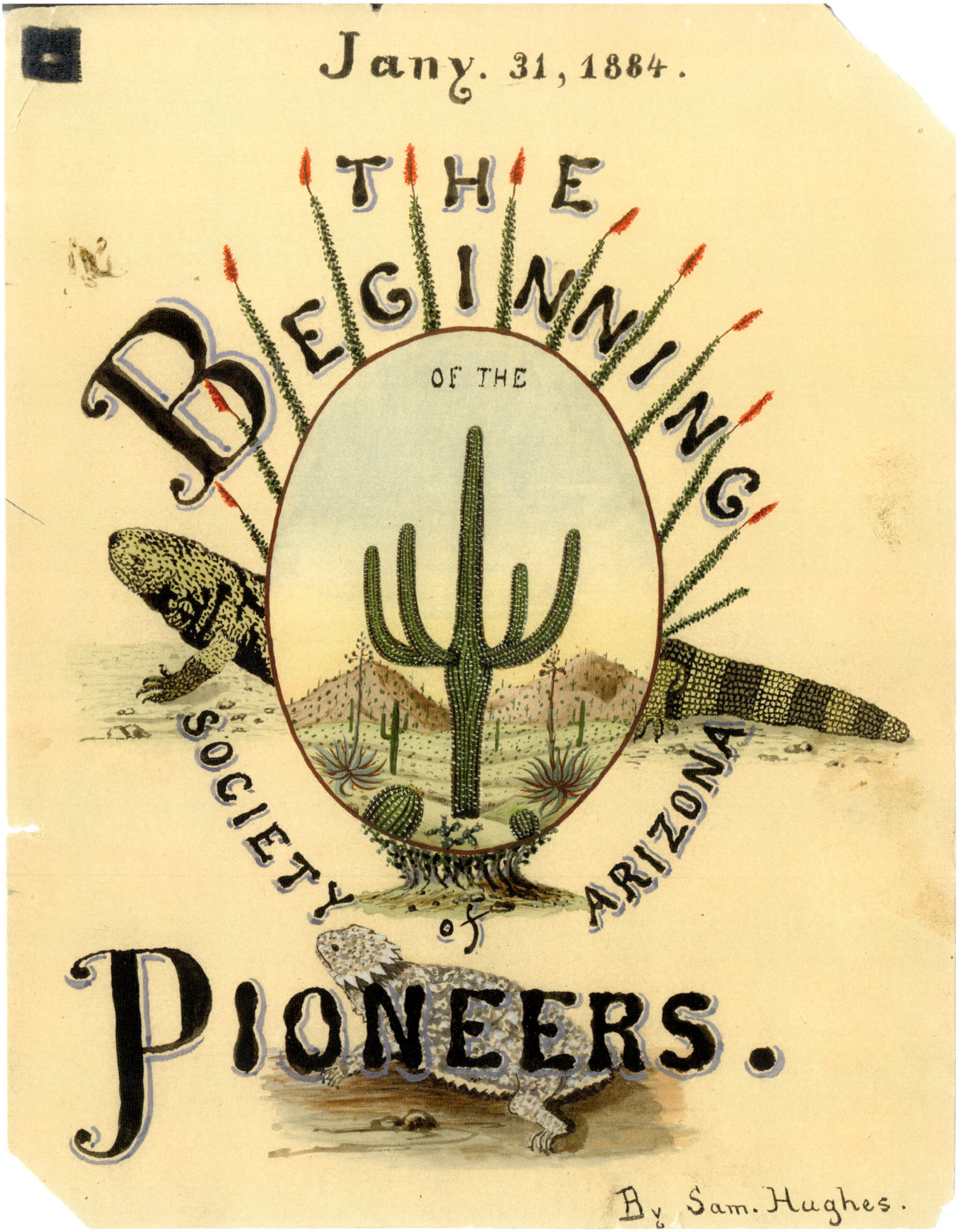

FIGURE 4.5. Pen and ink watercolor. "Jan'y 31, 1884, The Beginning of the Society of Arizona Pioneers, by Sam Hughes."

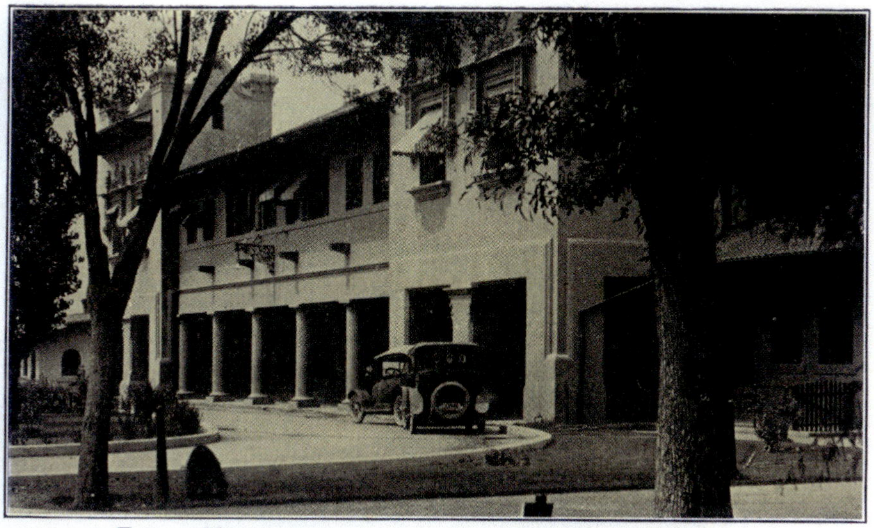

Trees and lawns greet you on arrival—Southern Pacific Station

FIGURE 4.6. Booklet photogravure. "Trees and lawns greet you on arrival—Southern Pacific Station." "Man-Building in the Sunshine Climate." Tucson Sunshine Climate Club. 1922–1924.

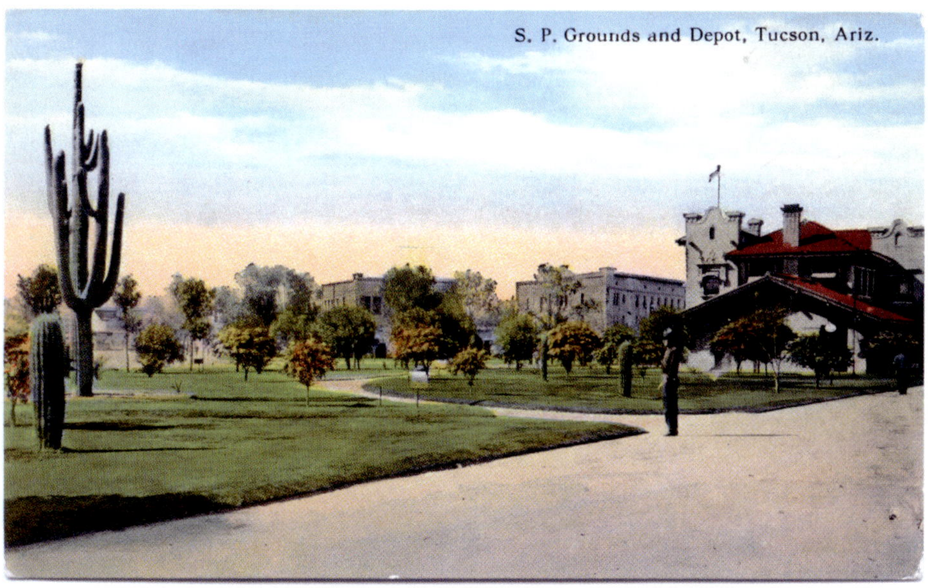

FIGURE 4.7. Postcard. Southern Pacific Grounds and Depot, Tucson, Arizona. 1915. Printed by Curt Teich Co.

an investment that housed his studio on the second floor, with a slanted glass wall to capture northern sunlight. An exterior sign advertised his line of stereograph "Arizona Views" that documented native flora, including the cacti staged in his studio (Figure 4.8). Among his civic accomplishments Buehman held office as Pima County assessor and often took his camera along on property inspections to photograph the "beauty spots" he encountered (Figure 4.9). He served two terms as major from 1895–1899. Buehman's downtown studio block also housed the offices of the *Arizona Daily Citizen*. Buehman implicitly understood the importance of location in establishing a studio not only for the repeat customers he kept supplied

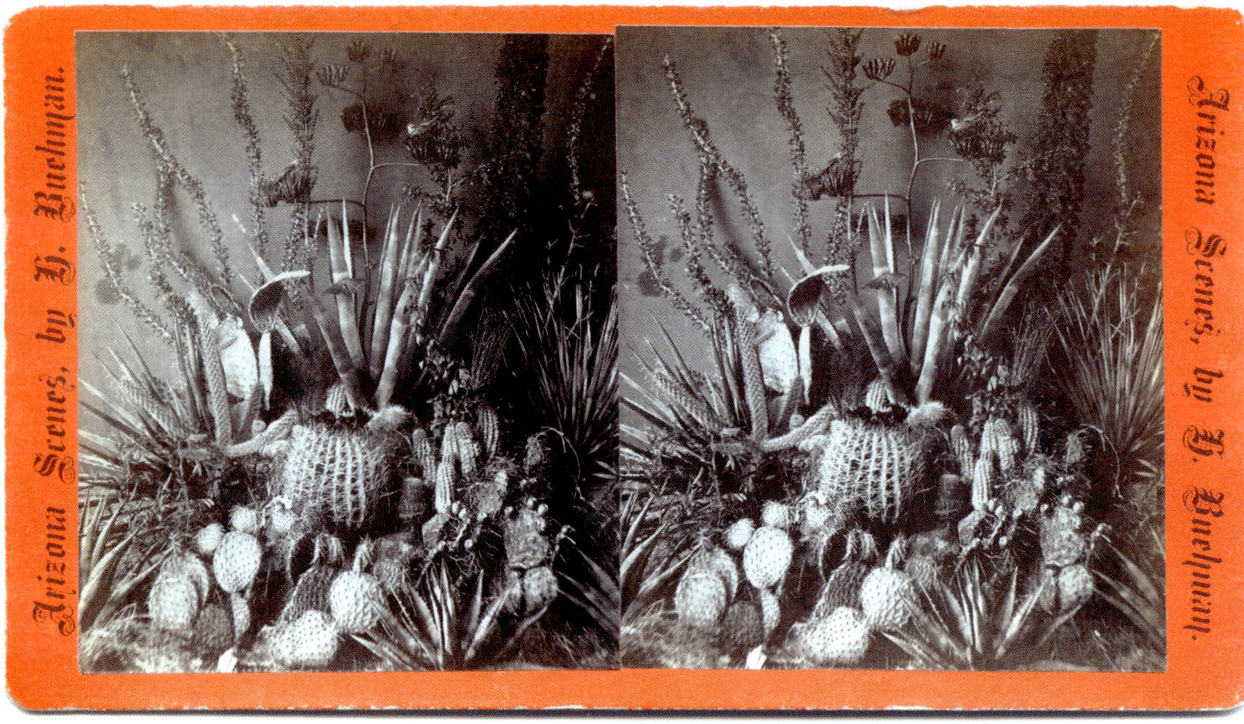

FIGURE 4.8. Stereoview. Arizona Scenes by Henry Buehman. 1878–1881. Arizona bouquet cactus and Spanish bayonet. Tucson, Arizona.

with portraits, cabinet cards, and cartes des visites, but for the close proximity to the newspaper editors he supplied with photographs to boost southern Arizona. Buehman's son Albert continued the business. His son Remick briefly assumed part ownership in 1949 before selling the studio to new owners who declared bankruptcy in 1950 (Arizona Daily Star 1880; Daniels 1968; Evans 1981; Spude 1989; Cooper 1995).

Photographer Carleton E. Watkins came about his Arizona views quite differently. He established a studio in San Francisco and traveled to build an inventory of large-format negatives for "mammoth prints" and stereoviews printed upon his return. Watkins set the standard for picturing the Southern Pacific right-of-way in descriptive passenger guides, brochures, and trackside photo portfolios of the kind that the railroad published well into the twentieth century. He arrived at Tucson shortly after the opening of the line from Los Angeles in the spring of 1880. A friend of Collis P. Huntington, Watkins traveled in a repurposed caboose coupled to a flat car carrying the horse-drawn wagon that he used as a field darkroom—an important consideration for processing the large wet-plate negatives he favored. Even with this elaborate set-up, he rued the complications of desert travel, especially its unrelieved dust. His "Arizona Views" featured San Xavier Mission, Casa

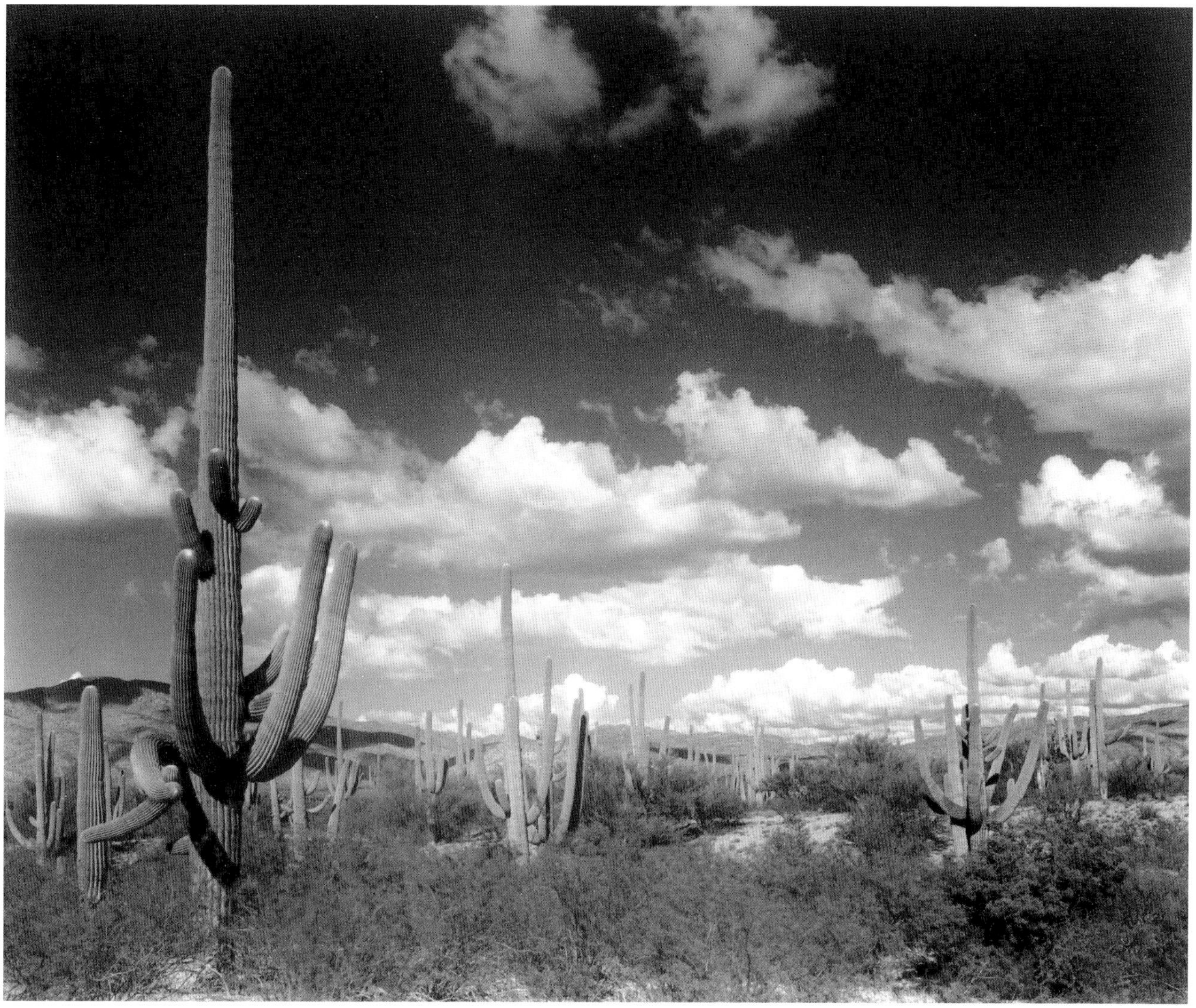

FIGURE 4.9. A copy photograph of saguaros attributed to Henry Buehman, ca. 1880.

Grande ruins, Tombstone mining camps, and desert plant life (Figure 4.10; Palmquist 1987:353–390; Green 2018).

The saguaro played a singular role as a foil in stereoviews documenting developments in irrigation. Stereoviews foregrounded saguaros like the columnar standout in a view of the Salt River Valley (also published as a lantern slide), and in another notable example, an alfalfa field in which all but the largest saguaros had been bladed (Figures 4.11 and 4.12).

Despite its contribution to the establishment of the Desert Laboratory in 1903, the publicity of the Tucson Chamber of Commerce spoke for a certain "it's here, but" ambivalence about native botany. Founded in 1896 as the Tucson Grocer's Association, the organization changed its name to its present title in 1898 to reflect

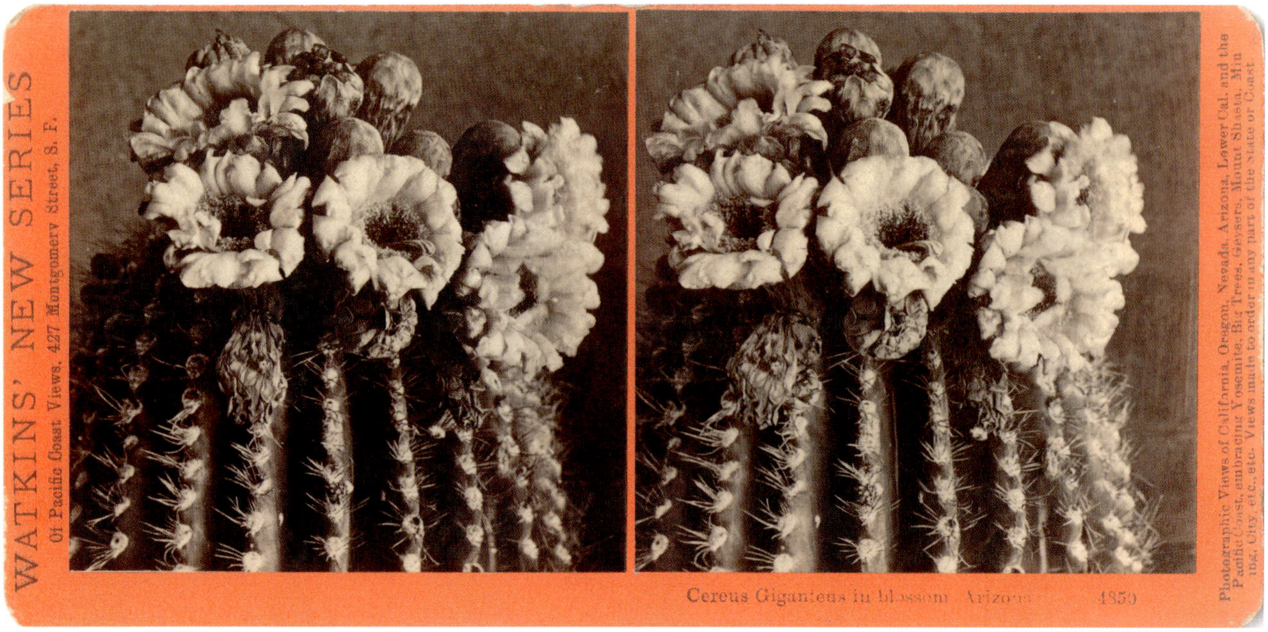

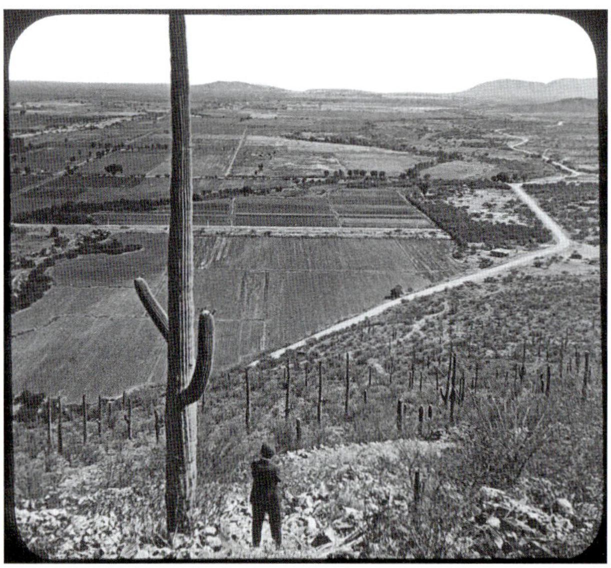

FIGURE 4.10. Stereoview. *Cereus giganteus* in blossom. Arizona. Carleton Watkins, ca. 1875.

FIGURE 4.11. Glass plate positive lantern slide. 1900–1910. The effects of irrigation on Arizona desert giant, Salt River Valley. Keystone View Company, Meadville, Pennsylvania. The Keystone View Company produced educational stereoviews and single glass lantern slides like this one that could be projected on a screen for use in the classroom or the lecture circuit.

a broader business constituency. In 1905, the organization formed a committee lead by J.M. Ormsby, a Western Union manager, and Emanuel Drachman, a theater owner, to combat false advertising in the city. That year Ormsby, the chairman of the Chamber's "subscription committee," raised a fund of several thousand dollars dedicated to the outward promotion of Tucson (Arizona Daily Star 1905, 1906; Kimmelman 1987). What appears to have been the beginning of an on-and-off ad campaign revolved around the 1904 publication of a chamber booklet entitled "Health Wealth Golden Opportunities" and national magazine advertisements soliciting interest in it (Figure 4.13; Tucson Chamber of Commerce 1904). The ads appeared infrequently over the next seven years in a diverse assortment of publications including the Southern Pacific Railroad's *Sunset* and *The Plant World*, edited by the Desert Laboratory's Volney M. Spalding (Figure 4.14; Tucson Chamber of Commerce 1908; Tucson Daily Citizen 1911; Arizona Daily Star 1912a, 1912b, 1914a, 1915a, 1915b).

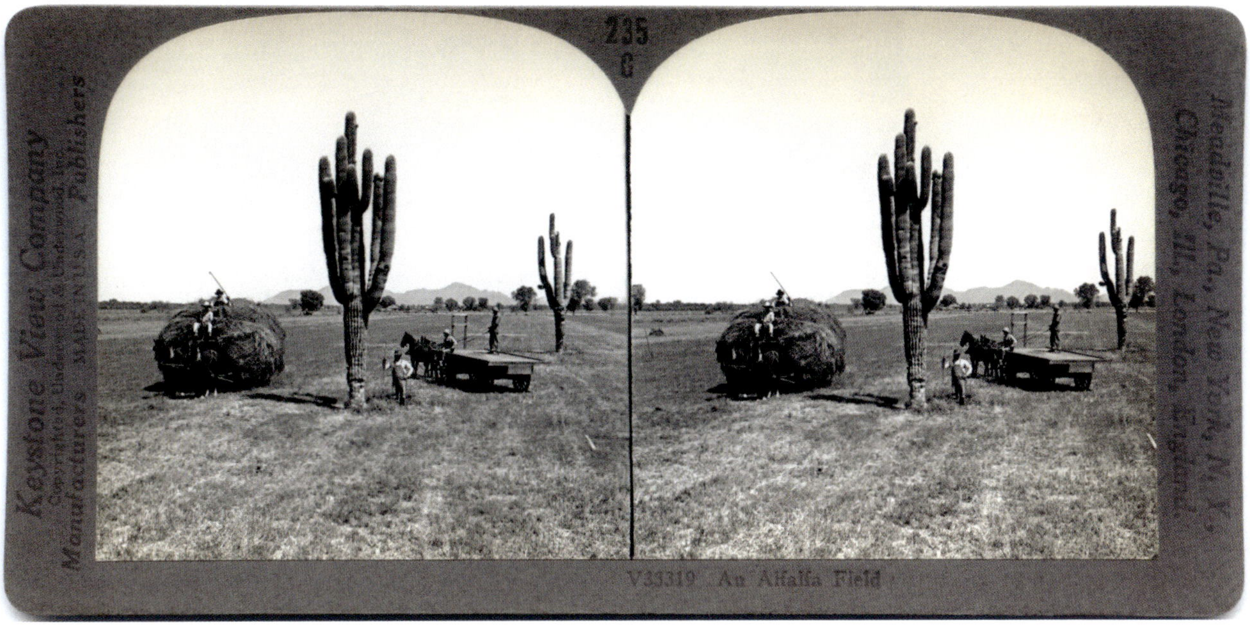

FIGURE 4.12. Stereoview. 1900–1910. Giant cacti of the desert in alfalfa field made productive by irrigation, Arizona. Keystone View Company, Meadville, Pennsylvania.

In the looming national emergency of the First World War, Tucson Chamber of Commerce Chamber secretary Ray B. Leach announced a postcard campaign conceived to promote its picture booklet. Instead of ads placed in magazines, postcards picturing Tucson and its environs would invite correspondence with the Chamber, and by return mail, its picture booklet. The campaign's justification lay in the closing of passenger ticket agencies that had freely displayed chamber literature. To compensate for the loss, Leach turned to Tucson's Red Cross Canteen. There, Red Cross workers would distribute complimentary photogravure postcards of Tucson scenes to troops as they passed through the city, or studied at the University of Arizona. Their postcards would be collected, sorted, and mailed to friends back home. Printed by the Albertype Company of Brooklyn, New York, the cards pictured Tucson homes, its business district, country club, Carnegie Library, YMCA, San Xavier del Bac Mission and the rippling waters of Sabino Canyon (Figure 4.15). With the exception of San Xavier and Sabino Canyon, the photographs might have been taken anywhere. But that was the point (Arizona Daily Star 1918a, 1918b, 1918c). By September 1918, the Red Cross had distributed 75,000 packets of 15 cards apiece featuring scenes imprinted with the Chamber's offer of a free booklet prepared by the photographers and writers of the Sunset Magazine Homeseekers' Bureau (Figure 4.16; Arizona Daily Star 1918d:6; Maulsby and Chamber of Commerce 1911[?]; Wilson 1913[?]).

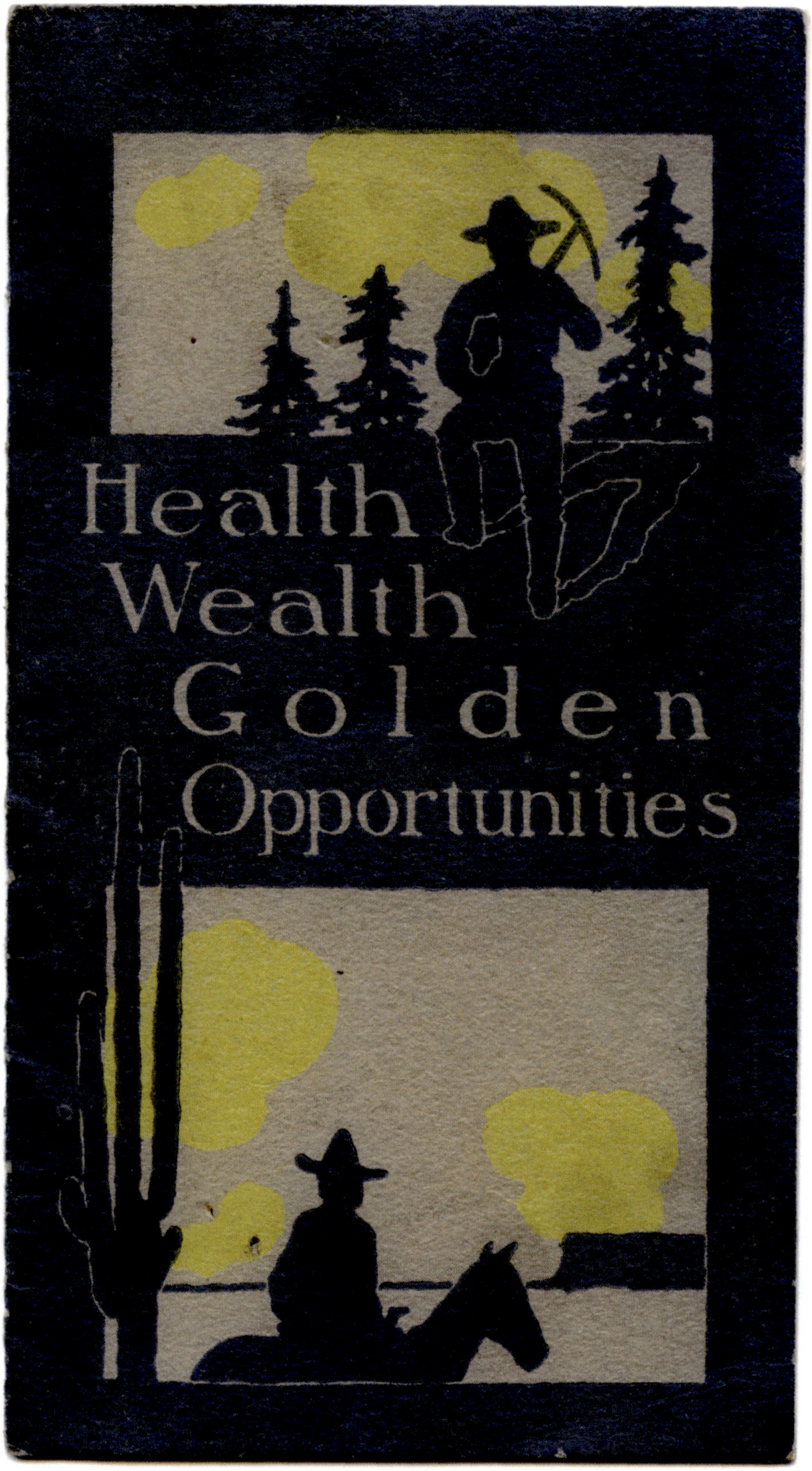

FIGURE 4.13. Booklet cover. "Health Wealth Golden Opportunities." Tucson Chamber of Commerce, ca. 1906.

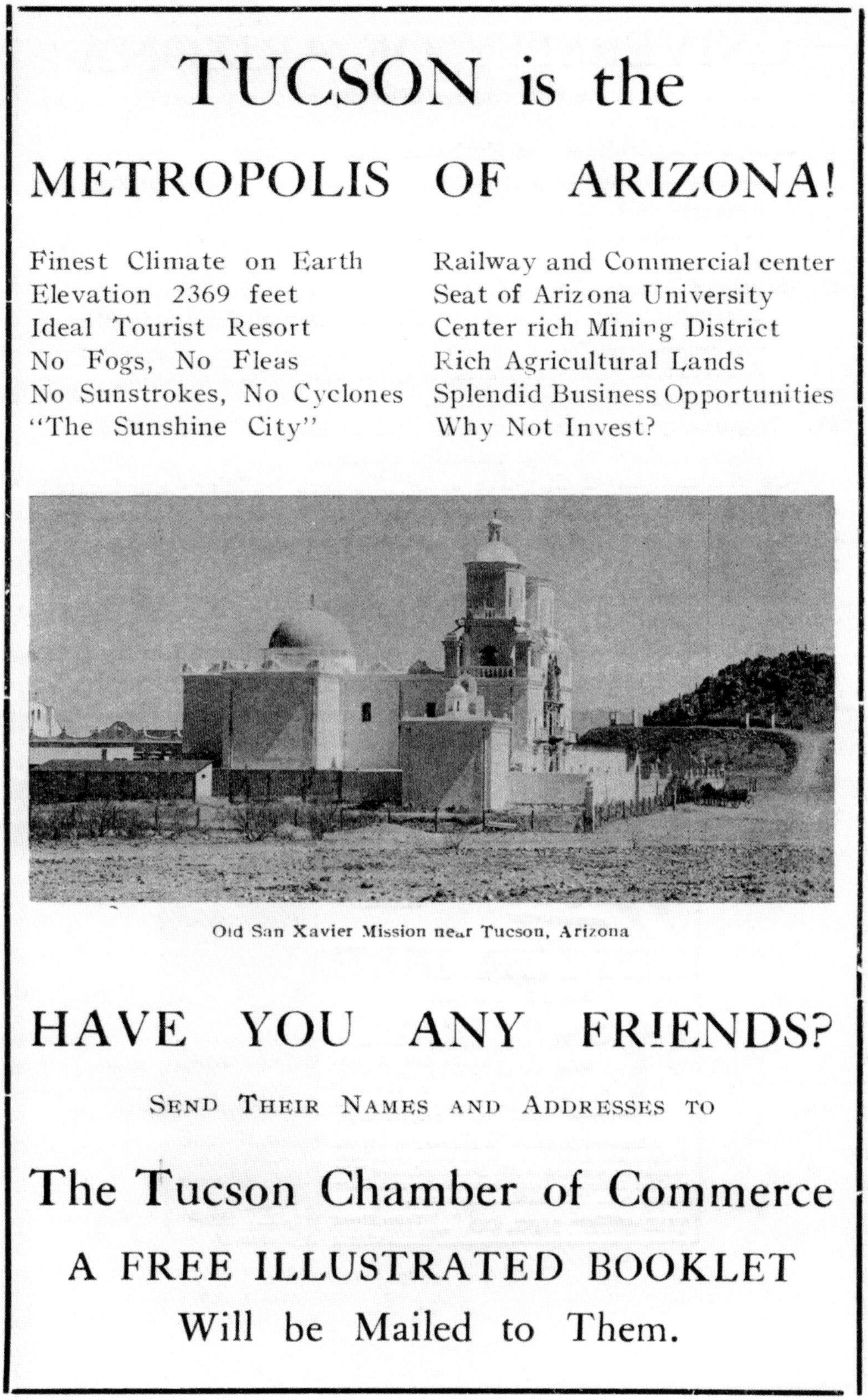

FIGURE 4.14. Advertisement. "Tucson is the metropolis of Arizona!" Tucson Chamber of Commerce (1911). *Plant World*, June 1911. The earliest Tucson Chamber of Commerce advertising wrapped itself around the continuing historical and architectural appearance of San Xavier del Bac Mission.

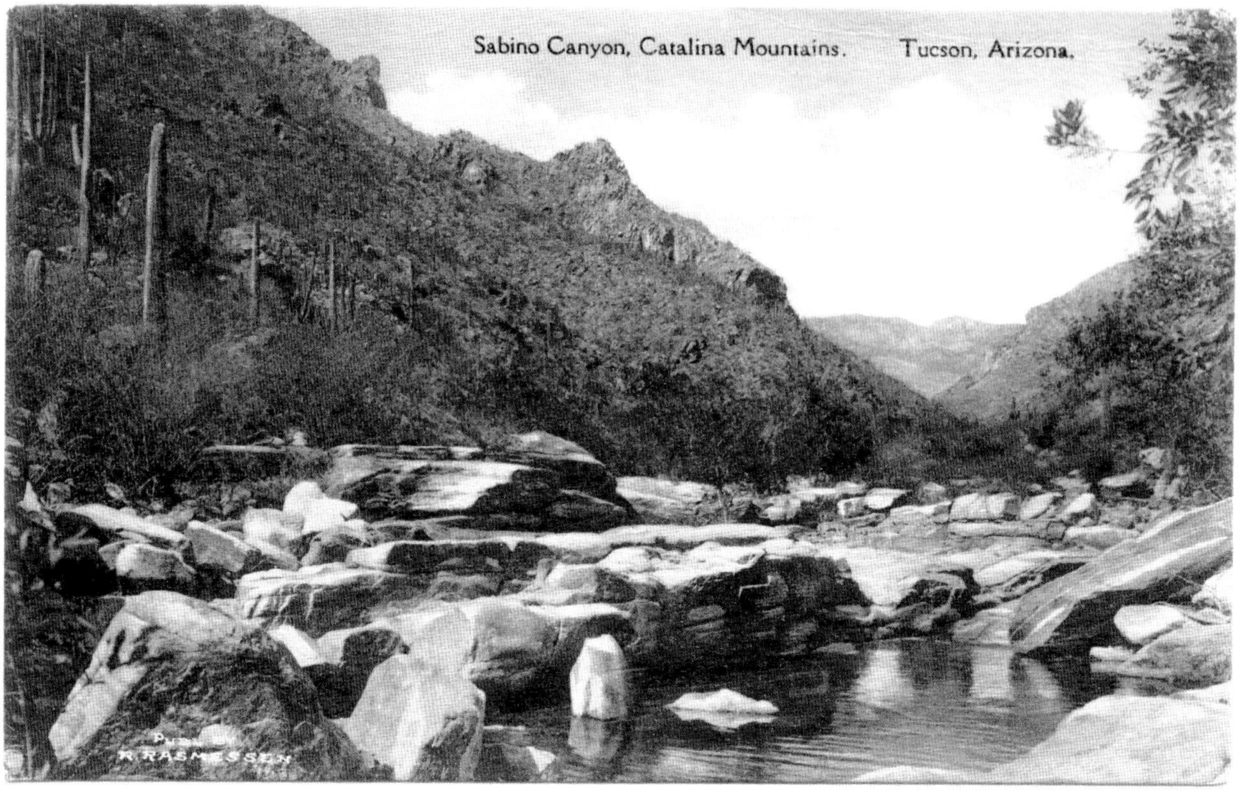

FIGURE 4.15. Postcard. Sabino Canyon, Catalina Mountains. 1921. Printed by Albertype Co., Brooklyn, New York. Published by Rudolph Rasmessen. The card's message side states: "For Tucson literature address Chamber of Commerce, Tucson, Arizona."

Distributed to servicemen with smokes and drinks, the postcards credited R. Rasmessen, a prominent chamber board member and downtown curio store owner who happened to be Tucson's most prolific postcard publisher. Born in Sweden in 1875, Rasmessen emigrated to Chicago with his parents at the age of nine. He moved to Tucson at 18 for his health and purchased a half-share in a downtown curio shop from Samuel F. Baumann, the son of a local Indian trader. Rasmessen soon bought out Baumann and took sole possession of the store in 1905. The civic-minded Rasmessen served on the University of Arizona Board of Regents and contributed his publishing expertise to the Chamber's postcard campaign. Like his predecessor Henry Buehman, Rasmessen became the Mayor of Tucson and served two terms. In 1921, the first year of Rasmessen's administration, the Chamber of Commerce distributed 30,000 Rasmessen postcards before ordering 10,000 more. Each invited the public to write for Tucson literature with the Chamber's new tag line: Climate, Copper, Cotton, Cattle (Arizona Daily Star 1921).

The Rasmessen Curio Store became a leading distributor of cultural treasures led by Navajo blankets, Pima and Hopi baskets, and ceramics and the hand-colored Albertype photogravure postcards and booklets that contributed to the building of

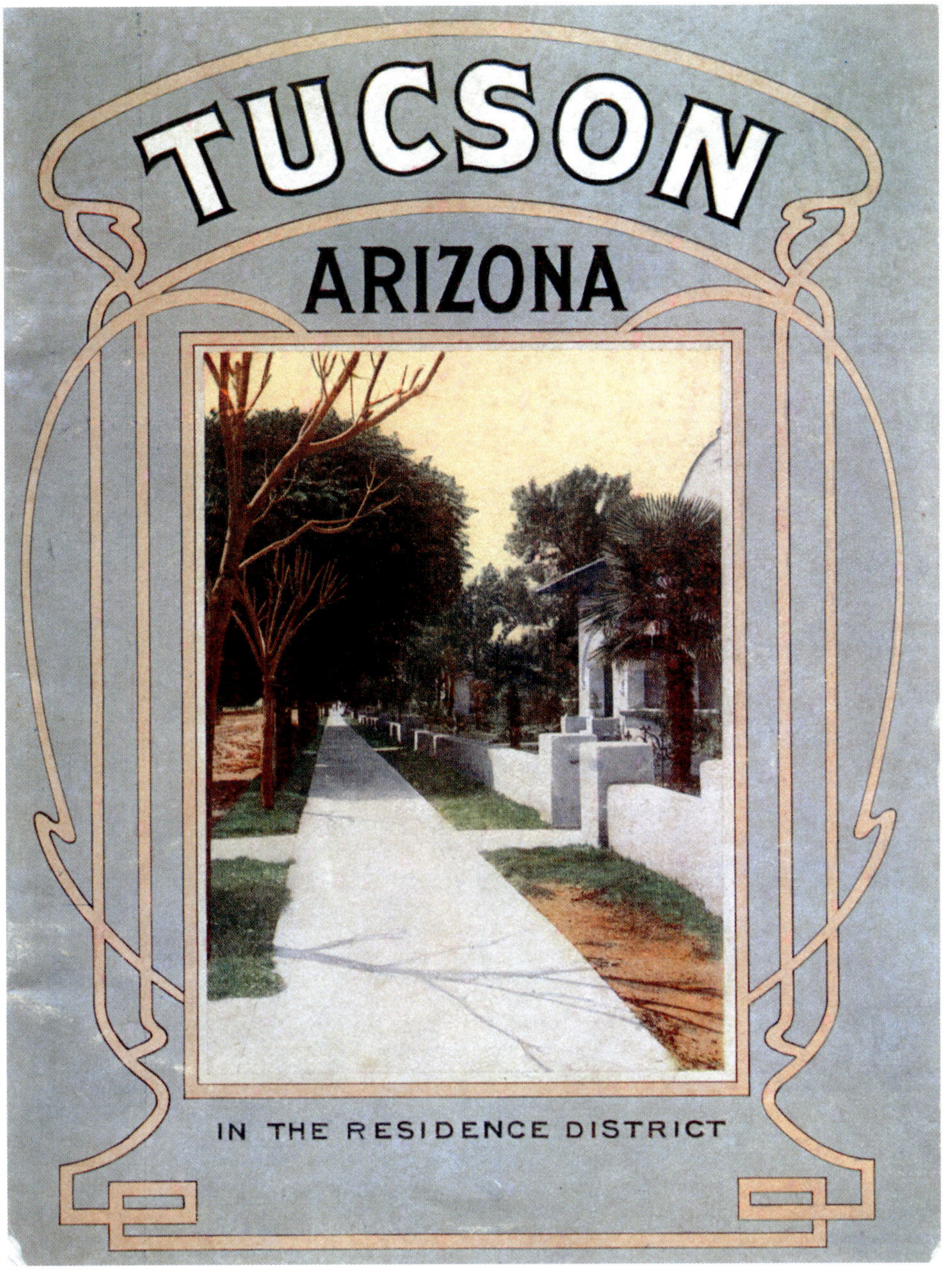

FIGURE 4.16. Booklet cover. "Tucson, Arizona, in the Residence District." Sunset Magazine Homeseekers' Bureau, San Francisco, California, ca. 1911.

southern Arizona's southwestern identity (Figure 4.17; Cooper 1989; Rowe 2006, 2010; Dittemore 2013; Frontz and Tackenburg 2014).

In their earliest days as business partners, Baumann and Rasmessen counted the Apache chief Geronimo among their customers. Late in life on a tour of New England, Geronimo sold a multicolored hair rope for $37.00 to an impressionable visitor who wondered aloud at the number of scalps the chief had taken. Geronimo excused himself and returned with the hair. The visitor could hardly believe his good fortune. Through his interpreter Geronimo later explained that the hair rope could be purchased for $1.50 at Baumann and Rasmessen (Arizona Daily Star 1901).

Rasmessen's functioned as a distribution point for monochrome and hand-colored postcards that pictured the plant life around Tucson without the editorial blinders of its Chamber of Commerce, whose postcard choices limited desert scenes (Figure 4.18; Rassmessen n.d. [1910?]). Unlike the customers perusing Rasmessen's postcard stock, the consumer of Tucson Chamber of Commerce literature had to look very carefully to spy a saguaro. Chamber brochures offset the desert giant—when it appeared at all—with an alternate something—a fir tree, a prospector, a pick axe—extractive symbols deployed in Arizona's state seal. The iterations of the Arizona territorial seal used in the years between 1864 and statehood in 1912 pictured the San Francisco Peaks of Flagstaff, foregrounded with a deer, coniferous trees, and an oddly out of place saguaro for good measure (Figure 4.19). The Arizona state seal adopted in 1912 does not picture cacti (Figure 4.20; Caldwell 1962:14–16; Department of State 2019).

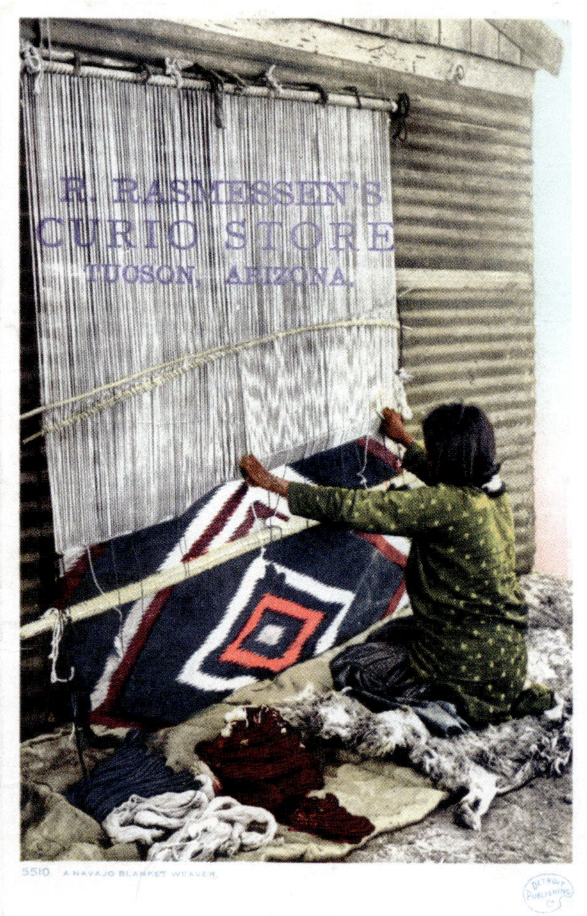

FIGURE 4.17. Postcard. A Navajo Blanket Weaver. 1901–1902. Detroit Publishing Co. stamped, R. Rasmessen's Curio Store Tucson, Arizona.

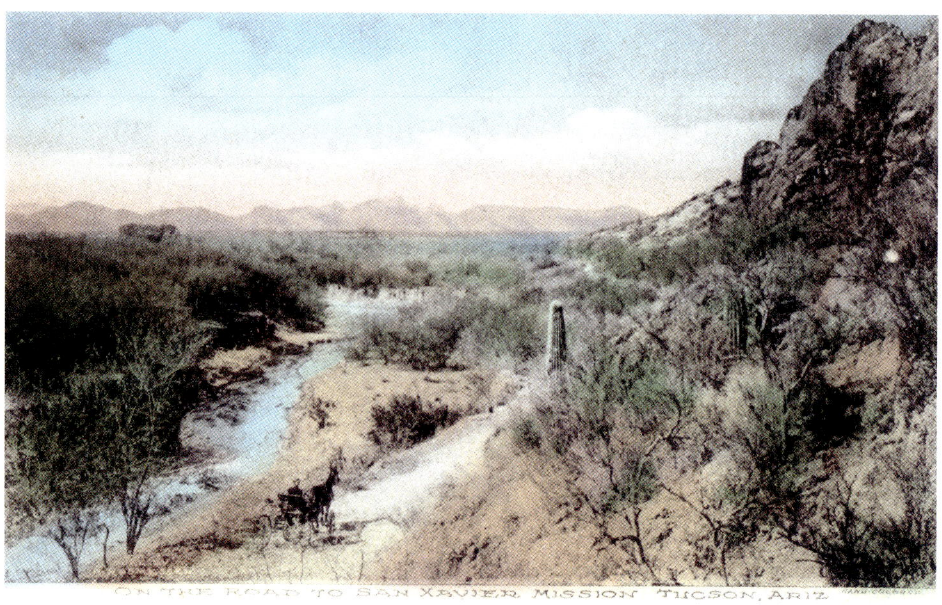

FIGURE 4.18. Hand-colored postcard. On the Road to San Xavier Mission, Tucson, Arizona. ca 1909. Published by R. Rasmessen.

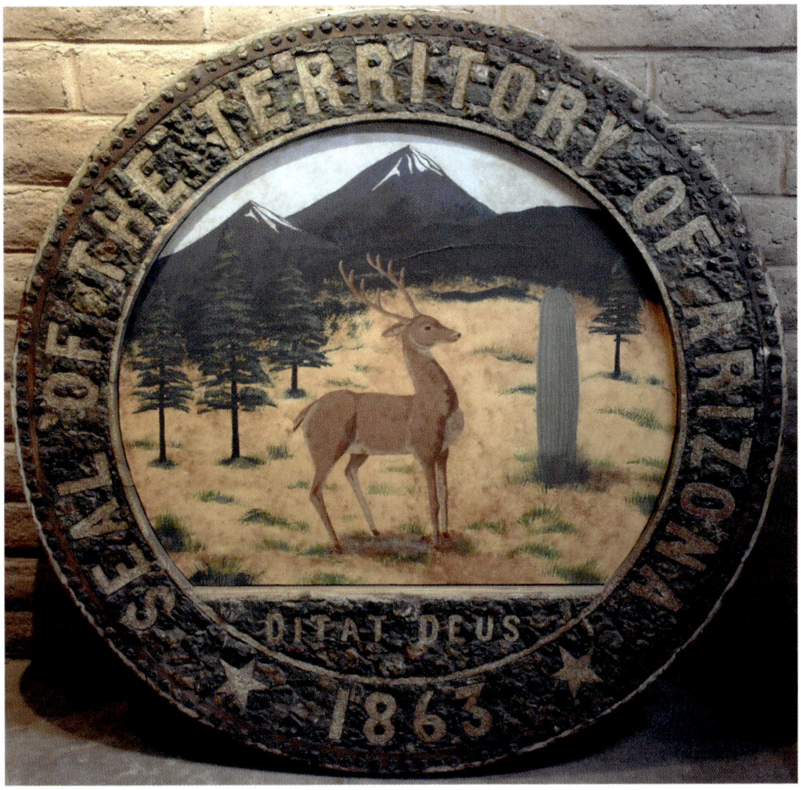

FIGURE 4.19. Contemporary recreation of the Arizona territorial seal picturing the San Francisco Peaks, a deer, conifers, and a saguaro. Tubac Presidio Museum, Tubac, Arizona.

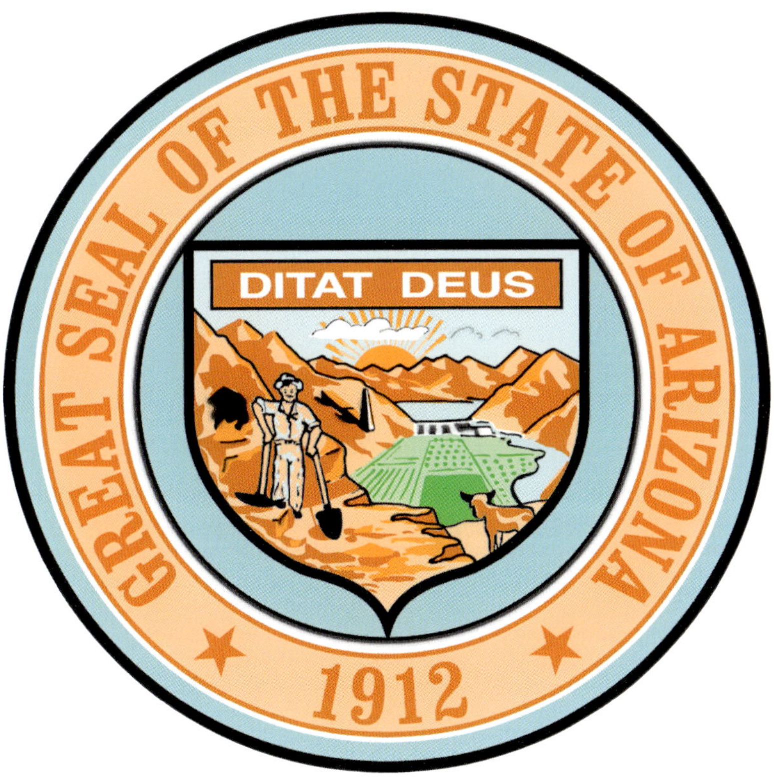

FIGURE 4.20. Great Seal of the State of Arizona. 1912. The iconography of the state seal adopted in 1911 includes a mountain range with the sun rising behind the peaks, a storage reservoir, and a dam, below which are irrigated fields and orchards with cattle grazing to the right. To the left, a mountainside with a quartz mill, a miner with a pick and a shovel. Above, the motto *Ditat Deus*, "God enriches." Arizona was admitted to the Union in 1912.

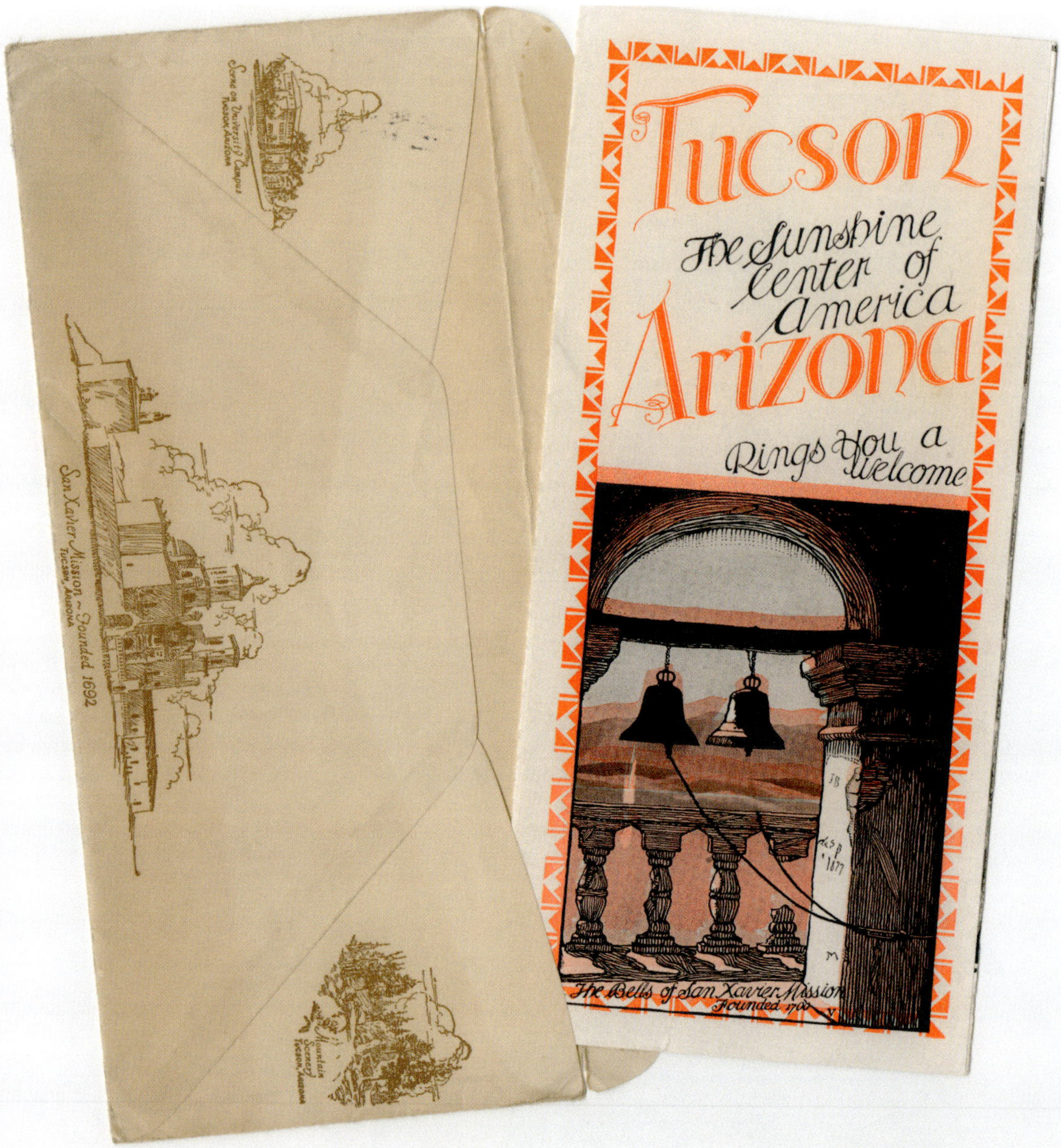

FIGURE 4.21. Brochure with envelope. "Tucson the sunshine center of America rings you a welcome." Tucson Chamber of Commerce, 1930. This brochure unfolds to reveal thumbnail photogravure scenes around Tucson and Pima County. The accompanying envelope is imprinted with line drawings picturing the University of Arizona, mountain scenery, and San Xavier del Bac Mission.

The cover of a Tucson Chamber of Commerce brochure entitled, "The Sunshine Center of America," featured San Xavier del Bac Mission (Figure 4.21). The open brochure presented, "Scenes around Tucson," a panel of thumbnail photogravures that pictured two saguaros (Figure 4.22). Widened to the entirety of Pima County, the following panel pictured none.

An uptick in the saguaro's quasi-official relevance occurred in 1922 when the Tucson Sunshine Climate Club (SCC) organized to advertise the city's winter climate. But even among the city's most ardent tourism boosters the saguaro's value was not immediately apparent. The saguaro played a limited role in the pam-

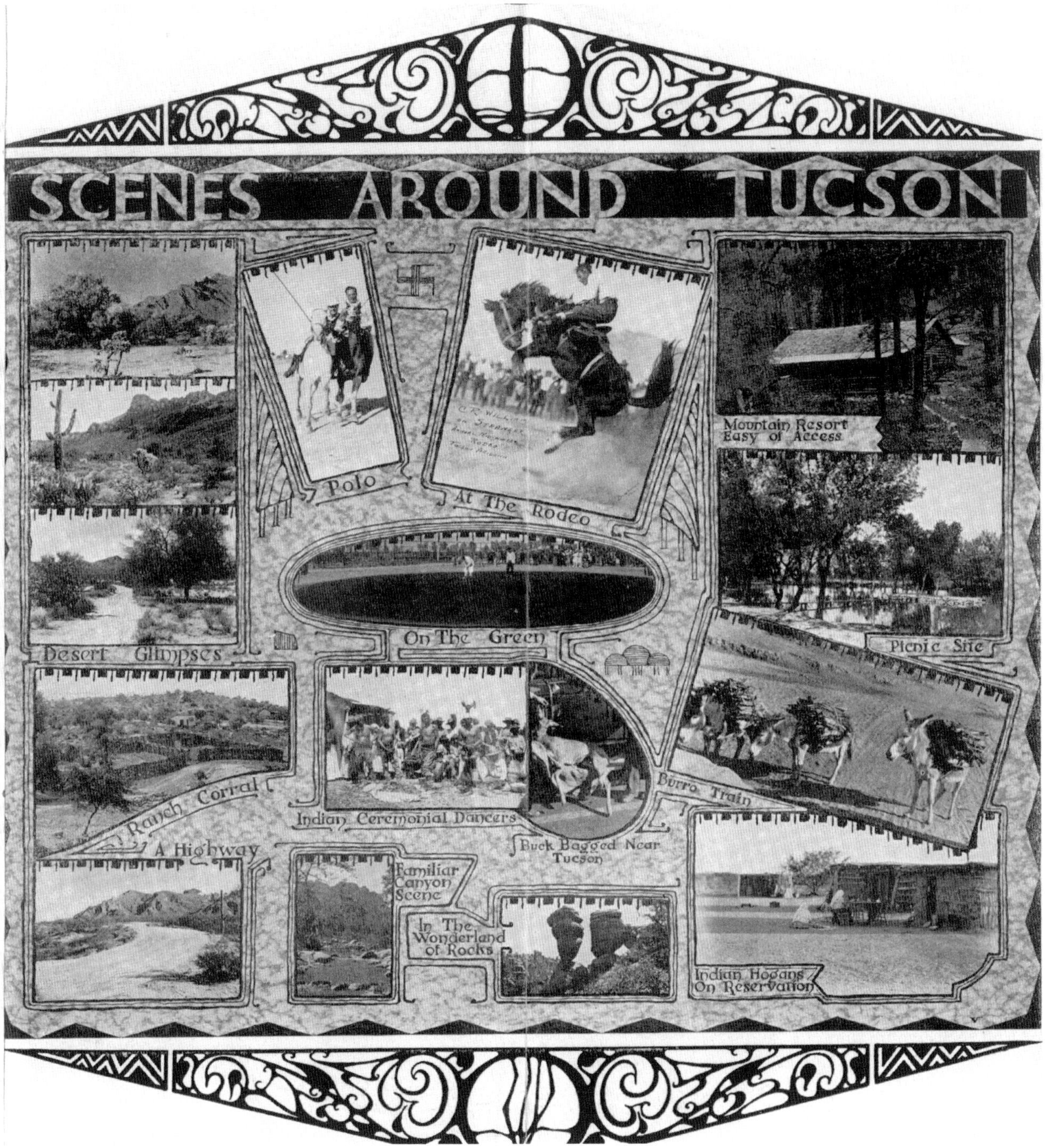

FIGURE 4.22. Photoengraving. "Scenes around Tucson." Tucson Chamber of Commerce, 1930.

phlets of the Sunshine Climate Club and its sibling rival the Phoenix-Arizona Club. Each advertised their cities as attractive winter destinations for sunshine and golf (Sonnichsen 1982:214–215, 263, 280–281; Kimmelman 1987; Luckingham 1982:57–67, 1993:84–91; Logan 1995:19–30, 70–77, 2006:89–92, 104–106; Sheridan 1995:240–241; Otero 2010:67–70; VanderMeer 2010:76–79).

Begun with the broadcast distribution of photogravure postcards and brochures, the marketing of southern Arizona as a tourist destination took on the sophisticated techniques of what the advertising trade described as "community advertising." A national survey of community-based advertising undertaken at the height of the leisure boom in 1924 noted the campaigns of southern California's All-Year Club; the Atlantic City Hotel Men's Association; the Salt Lake City Chamber of Commerce; the Denver, Colorado Publicity Bureau; the cities of El Paso, Texas, Phoenix and Tucson, Arizona; and the states of Maine and Florida (Evening Star 1924). These campaigns targeted specific audiences with small ads placed in Sunday newspaper travel sections and the backs of quality magazines to collect the names and addresses of well-heeled travel prospects. A subtle letter keyed the coupons to the publications in which they appeared, allowing advertising specialists to tabulate the return on each ad by publication, an important factor when managing the expectations of clients who questioned the value of their own advertising. Each individual who filled out a coupon with a name and address received an illustrated booklet by return mail.

The Sunshine Climate Club's campaign subsumed the anecdotal promotion work of the Tucson Chamber of Commerce. The two groups enjoyed an interlocking leadership. The SCC's sole function was to fund national advertising directed to the leisure traveler while skirting the political issues that preoccupied the Chamber of Commerce (Arizona Daily Star 1933, 1946; Tucson Daily Citizen 1960). Tucson's political economy revolved around the operations of miners, farmers, small businessmen, respiratory victims seeking dry desert air, and World War I veterans. Many staked claims on the saguaro-studded land west of the city that later became Tucson Mountain Park (Ascarza 2010; Negri 2010:6).

Officials began to weigh the necessity of a tourism campaign when the population of Phoenix passed Tucson in 1922 during Mayor Rasmessen's first year in office. On the SCC's thirtieth anniversary in 1952, Roy Drachman, Emanuel Drachman's son, who became SCC president in the late 1930s, recalled that Chamber officials sought the counsel of the H.K. McCann advertising agency of Los Angeles. A pioneering firm in the community advertising field, McCann advised Tucson officials to advertise the naturally occurring benefits of "sunshine and scenery" (Arizona Daily Star 1952).

The initial series of Sunshine Climate Club advertisements began with the all-important coupon that offered an illustrated booklet picturing Tucson's natural features, hotels, and guest ranches. The ads ran in the *New York Times*, *Outlook*, *National*

Geographic, *Literary Digest*, and other up-market magazines in the fall and winter of 1922. The headline of the first implored, "Unlock the Best That's in You." In successive weeks the copy unspooled as a conversation among friends. "Spring Days Now! Come to Tucson," "Are You Paying the Price of Health—And not Getting It?", "Golf on New Year's Day," and the clincher, "Must Hearts Break for Want of Sunshine?" (Figure 4.23; Sunshine Climate Club 1922a, 1922b, 1922c, 1922d, 1922e). The conversation broke with the tearing of a coupon and resumed with an illustrated booklet entitled, "Man-Building in the Sunshine Climate" (Figures 4.24 and 4.25).

Tucson's community advertising campaign attracted the attention of the Phoenix Chamber of Commerce, which had recently organized the Phoenix-Arizona Club for similar effect. In naming their club, its leaders specifically rejected references to the Salt River Valley for its untoward association with salt. The group soon met with Tucson auto dealer and SCC President Monte Mansfield, who explained the sunny premise of the Tucson campaign and the intricacies of the coupon. In short order the Phoenix-Arizona Club hired McCann to roll out a comparable community advertising campaign headlined, "Where Winter Never Comes" (Figure 4.26; Arizona Republic 1924a, 1924b, 1924c; Phoenix-Arizona Club 1924a, 1924b).

As the common creation of H.K. McCann, the Tucson and Phoenix campaigns developed what have since become the familiar tropes of travel and leisure iconography: golf, swimming, horseback rides, and families a world apart. Like the large-format Phoe-

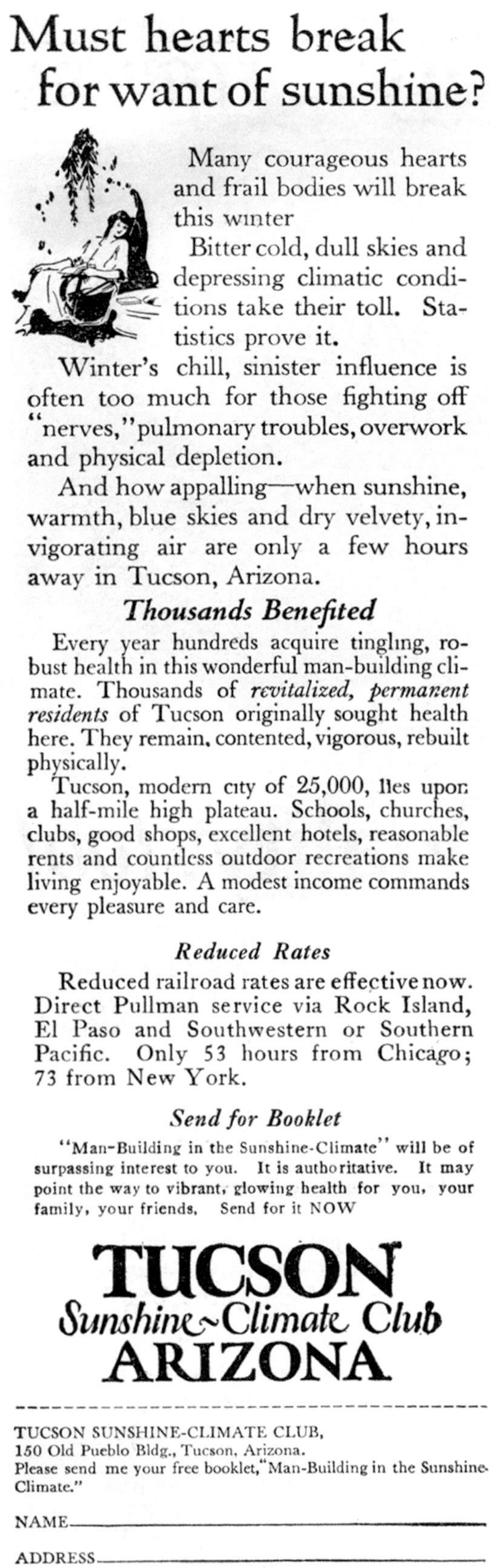

FIGURE 4.23. Advertisement. "Must hearts break for want of sunshine?" Tucson Sunshine Climate Club, 1922 (Sunshine Climate Club 1922:44).

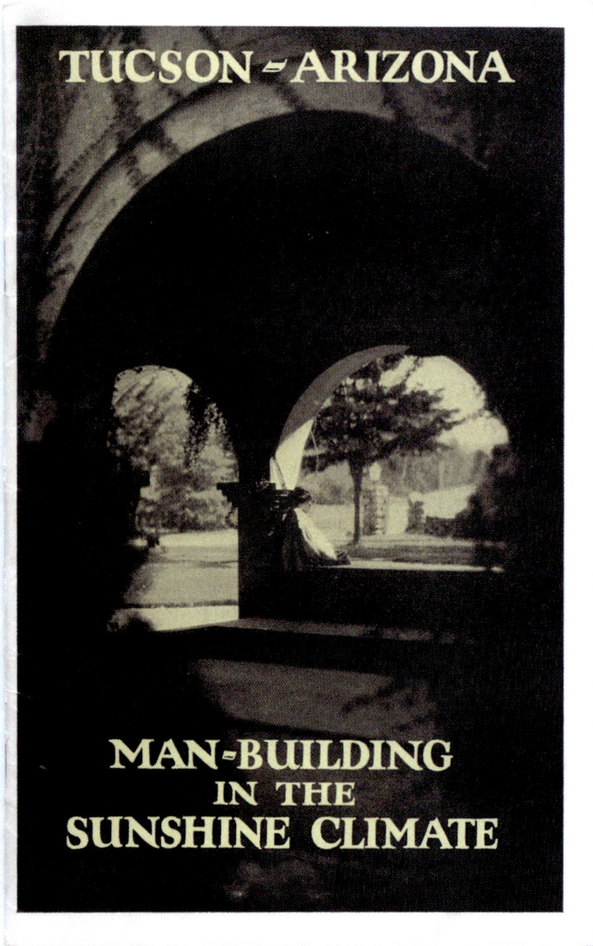

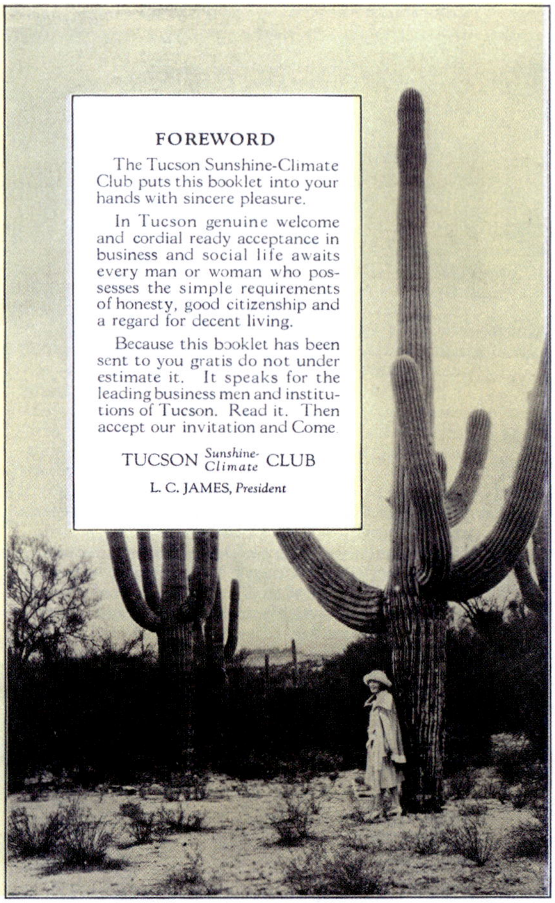

FIGURE 4.24. Booklet cover. "Man-Building in the Sunshine Climate." Tucson Sunshine Climate Club. 1922–1924.

RIGHT: FIGURE 4.25. Booklet foreword. "Man-Building in the Sunshine Climate." Tucson Sunshine Climate Club. 1922–1924.

OPPOSITE PAGE: FIGURE 4.26. Booklet cover. "Phoenix Arizona Where Winter Never Comes." Phoenix-Arizona Club, 1924.

nix booklet, the smaller Tucson booklet pictured exceptional homes as the norm, nodding to the saguaro forest when away from lush lawns, gardens, and deciduous trees. More so than the publications of the Phoenix group, Tucson's pictured saguaros, but after the foreword, in which a well-dressed woman posed for a saguaro moment, said little about them. After all, the booklet's purpose lay in dramatizing the amenities of a city.

Had its leadership thought about it, the Sunshine Climate Club might have published Buehman's pictures of the beauty spots in the saguaro forest east and west of the city. Enjoying success through the late 1920s measured in an increasing rate of reader response quantified by McCann, local support for the campaign faltered in the deepening depression of the early 1930s. Not until the 1940s did the saguaro come into its own as a free-standing symbol, unencumbered by the indifferent editorial direction of postcards, travel, and tourism booklets that consigned its appearance to the margins of the metropolis.

WHERE WINTER NEVER COMES

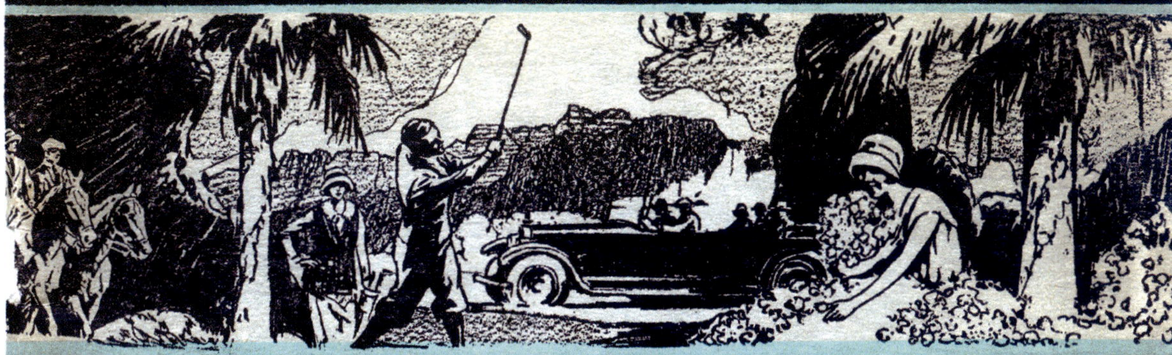

Phoenix

ARIZONA

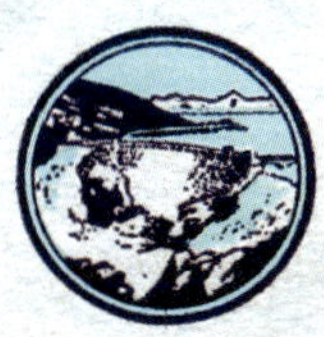

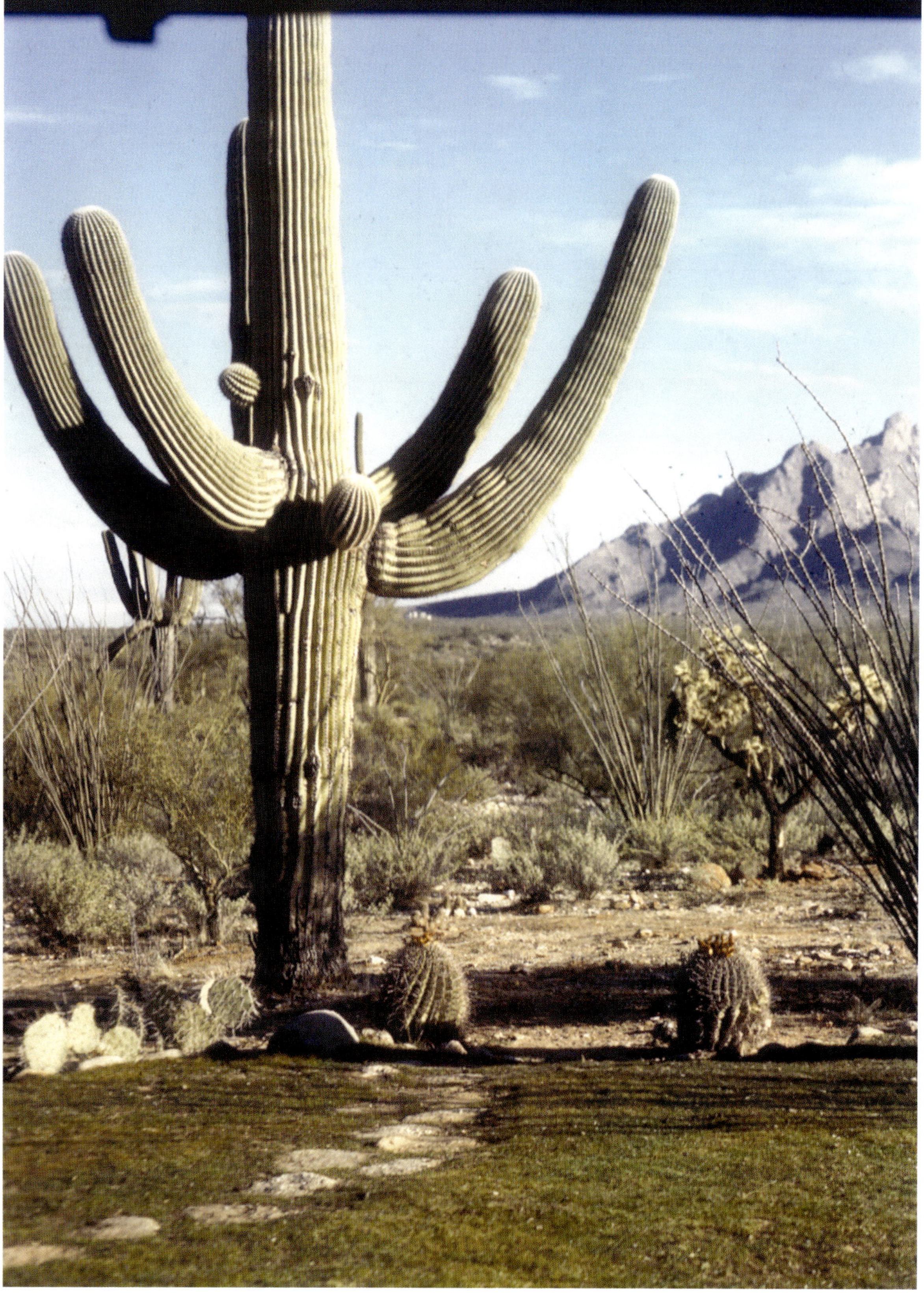

CHAPTER FIVE
FRAMING UP

If the frame of advertising consigned the desert's natural features to the margins of pamphlets and brochures—the framing of desert views for homes, parks, and gardens anticipated the saguaro's iconic status. From the picture window to the cactus forest, there wasn't one person or event that shaped a counter-narrative in the headlong rush of development—with the possible exceptions of novelist Harold Bell Wright, who set the standard for a desert home landscaped with native plants, and University of Arizona president Homer LeRoy Shantz, who purchased the saguaro forest that Wright and others suggested be set aside and protected as a national monument (Shantz 1937; Martin 1960:150, 160, 236–237; Burtner 2011:34–35, 46–53; National Park Service 2016). Wright's success in framing the desert for domestic consumption—and Shantz's campaign to conserve it—elevated the saguaro's habitat from the shadows of non-native trees and sunken lawns.

Novelist and Hollywood screenwriter Harold Bell Wright moved to Tucson for his health in 1915. With more resources and more intimate acquaintanceship with the desert than the average tubercular, Wright set up a serviceable outdoor camp for himself with two cooks in the Catalina foothills. Dressed in white, he chose to live outdoors at a desk of his own design with a shroud for writing in direct sunlight. He famously told about his convalescence in a 1924 *American Magazine* article entitled, "Why I Did Not Die" (Wright 1924; Arizona Daily Star 1968; Langdon 1975; Grubb 1984:19–25, 260–261, 271).

Wright purchased land for a home on a low mesa in the desert east of Tucson. Dominated by creosote, ocotillos, chollas, and saguaros, the area had been written off by local residents as distant and unattractive. Wright purchased 80 acres, constructed a Pueblo revival house with a detached writing studio and a garden of transplanted yuccas, barrels, prickly pears, and saguaros (Figure 5.1). Wright selectively deployed the plants along his electrically illuminated driveway leading to the mesa. Like the Desert Laboratory's Daniel T. MacDougal, Wright cautioned his construction crew to leave existing vegetation unharmed. When

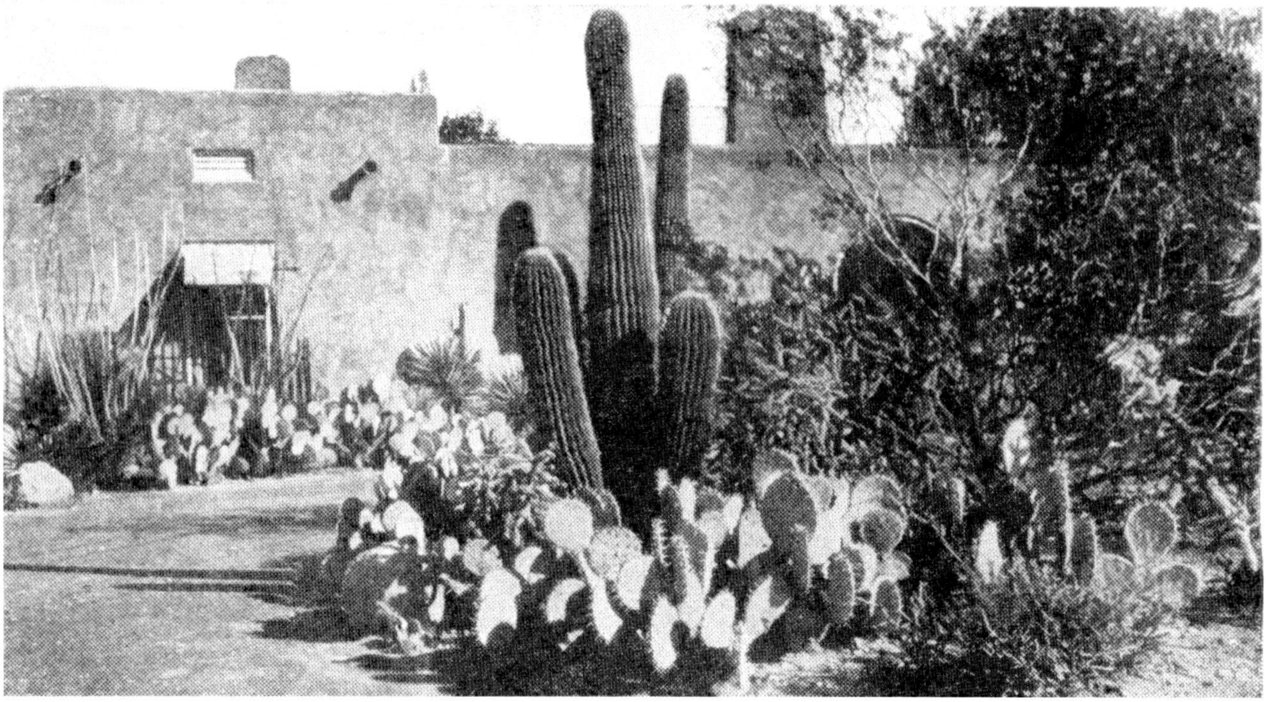

FIGURE 5.1. Harold Bell Wright's home and garden, Tucson, Arizona, ca. 1922.

completed, Wright's home offered a view of the Rincon Mountains to the east and the Santa Catalinas to the north in sight of his first camp. Wright enthused about the mountain ranges visible from the mesa. "Where else in Arizona or the world could I find a site for this home of mine with eleven such ranges of mountains in the plainest view and in sight of the country I love best?" (Tucson Daily Citizen 1922)

Wright's investment set off a land scramble initiated by Tucson's winter visitors, many of whom like Wright had come west for their health. Tucson realtor Herbert Drachman noted that "the beautiful desert flora that surrounded the Wright home for miles in every direction opened the eyes not only of Tucson people, but those visitors fortunate enough to see it" (Tucson Daily Citizen 1929). Of course it had been there all along. What changed was the introduction of a palatial home and the careful curation of plants that left the appearance of a natural effect.

A conservationist for whom the desert had become personal, Wright devised a comparable landscaping plan for the Desert Sanatorium that broke ground on a 160-acre tract west of his estate in 1924. Wright joined the sanatorium's first board of directors. A sequence of photographs taken after the period of Wright's greatest activity in landscaping his estate documents the excavation of large saguaros from his property by the sanatorium grounds crew (Figures 5.2–5.4).

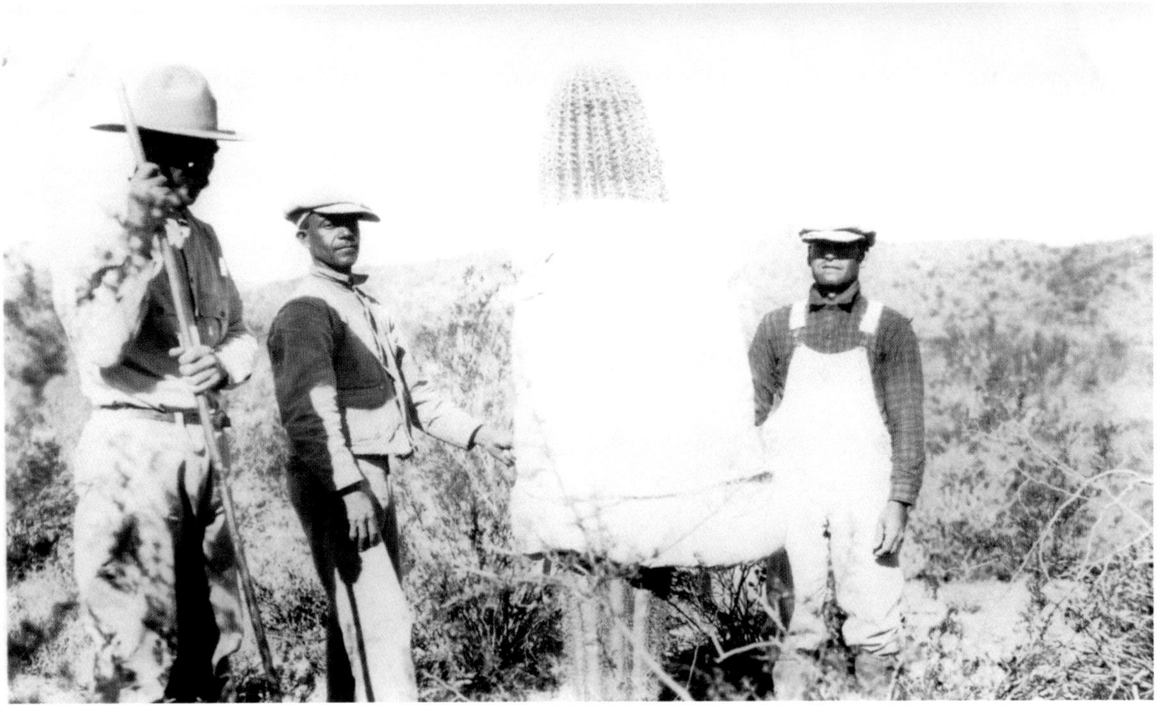

ABOVE, FIGURE 5.2. Desert Sanatorium buildings and grounds crew members pose with a padded saguaro ready for excavation from the estate of author Harold Bell Wright. 1927. The mattress pad protects both the excavators and the plant's spines: once broken, spines cannot grow back.

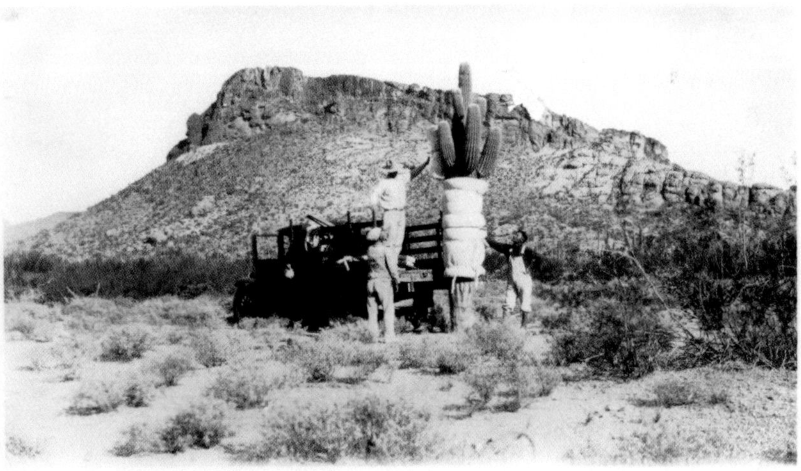

FIGURE 5.3. Desert Sanatorium buildings and grounds crew preparing to load a padded saguaro from the Wright estate for planting on the Desert Sanatorium grounds. 1927.

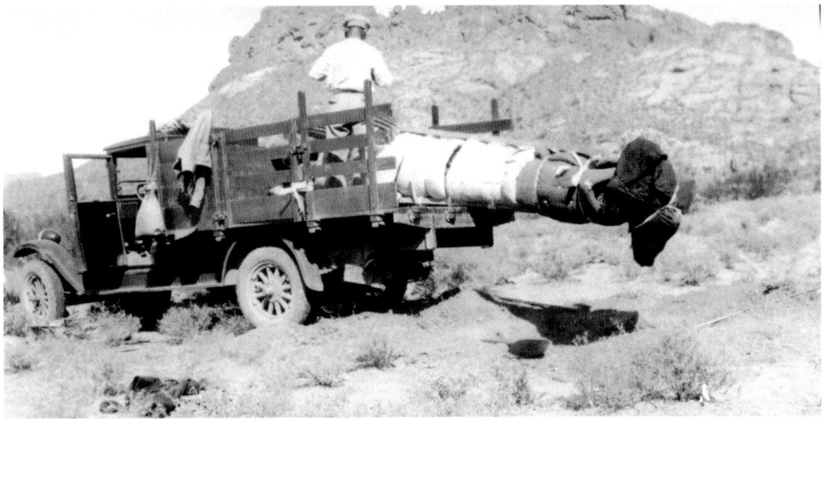

FIGURE 5.4. Desert Sanatorium buildings and grounds crew with loaded saguaro. 1927.

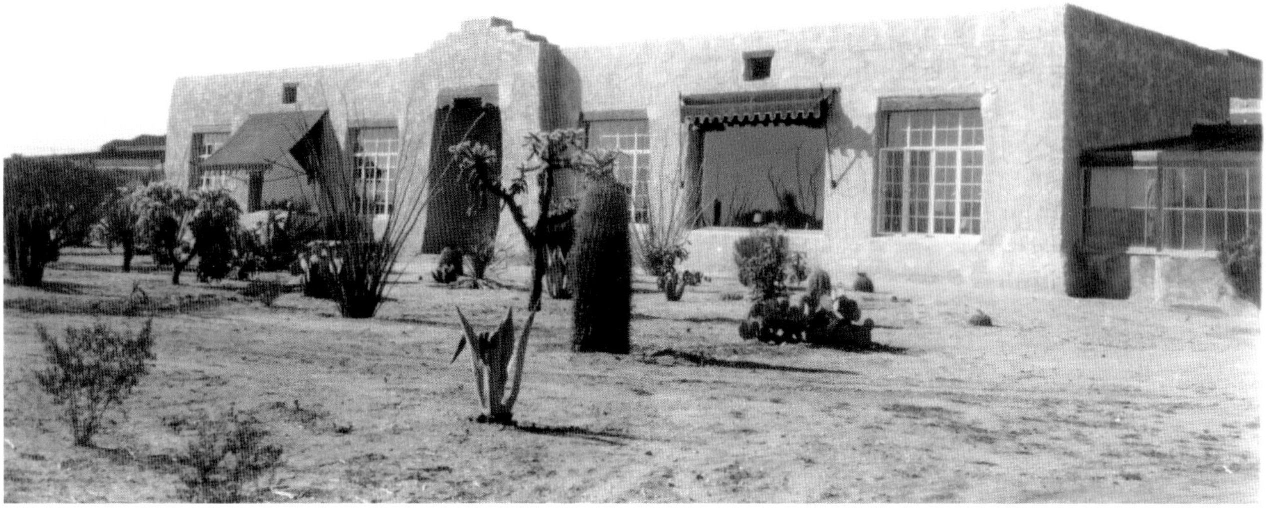

FIGURE 5.5. Zuni Dining room landscaping, Desert Sanatorium, Tucson, Arizona. 1930. The Zuni Dining Room picture window looked out on a landscape enhanced with native plants.

The Desert Sanatorium enjoyed substantial local support, drawing top medical professionals to the promising new field of heliotherapy for the treatment of tuberculosis and chronic lung disease. Opened in 1926, a campus of charming casitas in the Pueblo revival style designed by architect Henry O. Jastaad featured sun-drenched patios and a Native American themed food service, the Zuni Dining Room (Durbin 1965; Tucson Historic Preservation Foundation 2012:12). The restaurant's picture window took full advantage of Wright's landscaping plan (Figures 5.5 and 5.6). Rutger Porter, the nephew of Wright's New York publisher, carried out the plan after his recuperation from a bout with pneumonia (Figure 5.7). Porter massed saguaros, prickly pears, and barrels at the sanatorium entrance and along the driveway into the property, creating a pleasing effect where there had been none (Figures 5.8 and 5.9). Porter went on to become a professional landscape designer and to found a cactus nursery. As the city of Tucson grew and enveloped his home and business, the nursery became today's Tucson Botanical Gardens (Brown 1975; Tucson Botanical Gardens 2020).

By the mid-1920s a trip to see the Desert Sanatorium or Wright's home might include a visit to the impressive stand of saguaros at the base of the Rincon Mountains some seventeen miles east of Tucson. Though one can only speculate about its discovery, Harold Bell Wright had come under its spell and showed the tract to University of Arizona entomologist Charles T. Vorhies shortly after Vorhies joined the university's College of Agriculture in 1915 (Burtner 2011:34).

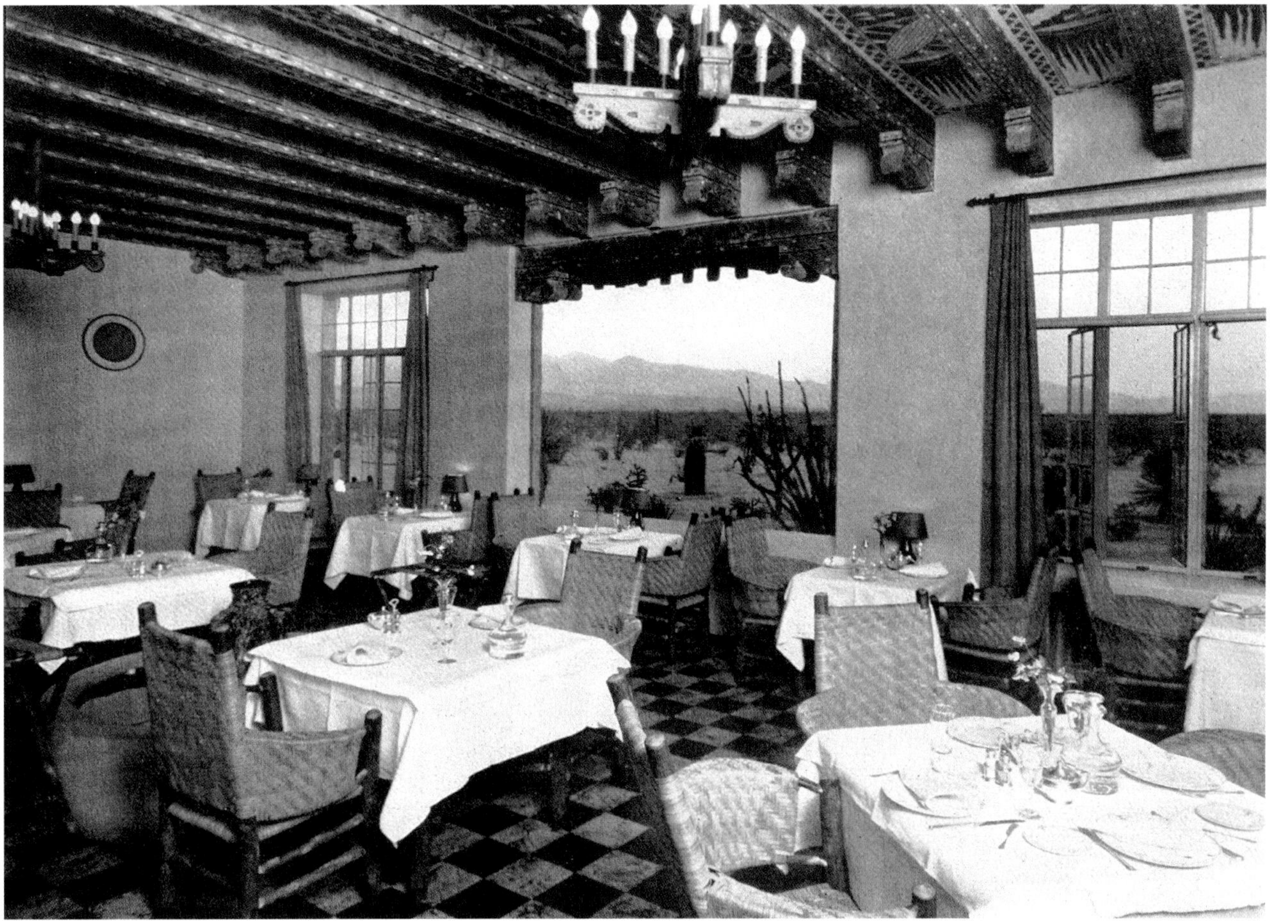

FIGURE 5.6. Zuni Dining Room, Desert Sanatorium, Tucson, Arizona. 1928.

A visit to the saguaro-filled tract, locally known as the Tanque Verde Cactus Forest, became something of a right of initiation among the university's leading academic figures, most notably botanist Homer LeRoy Shantz, who became the President of the University of Arizona in 1928. Shantz thus joined the individuals who not only had become aware of the stand of massive saguaros but increasingly protective of it as the finest of its kind.

Born on a Michigan farm in 1876, Shantz attended Colorado College and received a PhD degree from the University of Nebraska in 1905. After a brief stint teaching biology and botany, he joined the U.S. Department of Agriculture's Bureau of Plant Industry specializing in the study of arid lands and grasses. He signed on to the Smithsonian African Expedition of 1919–1920 arranged for former President Theodore Roosevelt. Shantz visited Africa a second time in 1923–1924. Adept at field photography, Shantz returned from Africa with some 5,000 negatives documenting plant geography. From his position as a federal plant pathologist, Shantz joined the faculty of the University of Illinois Department of Botany in 1926 and

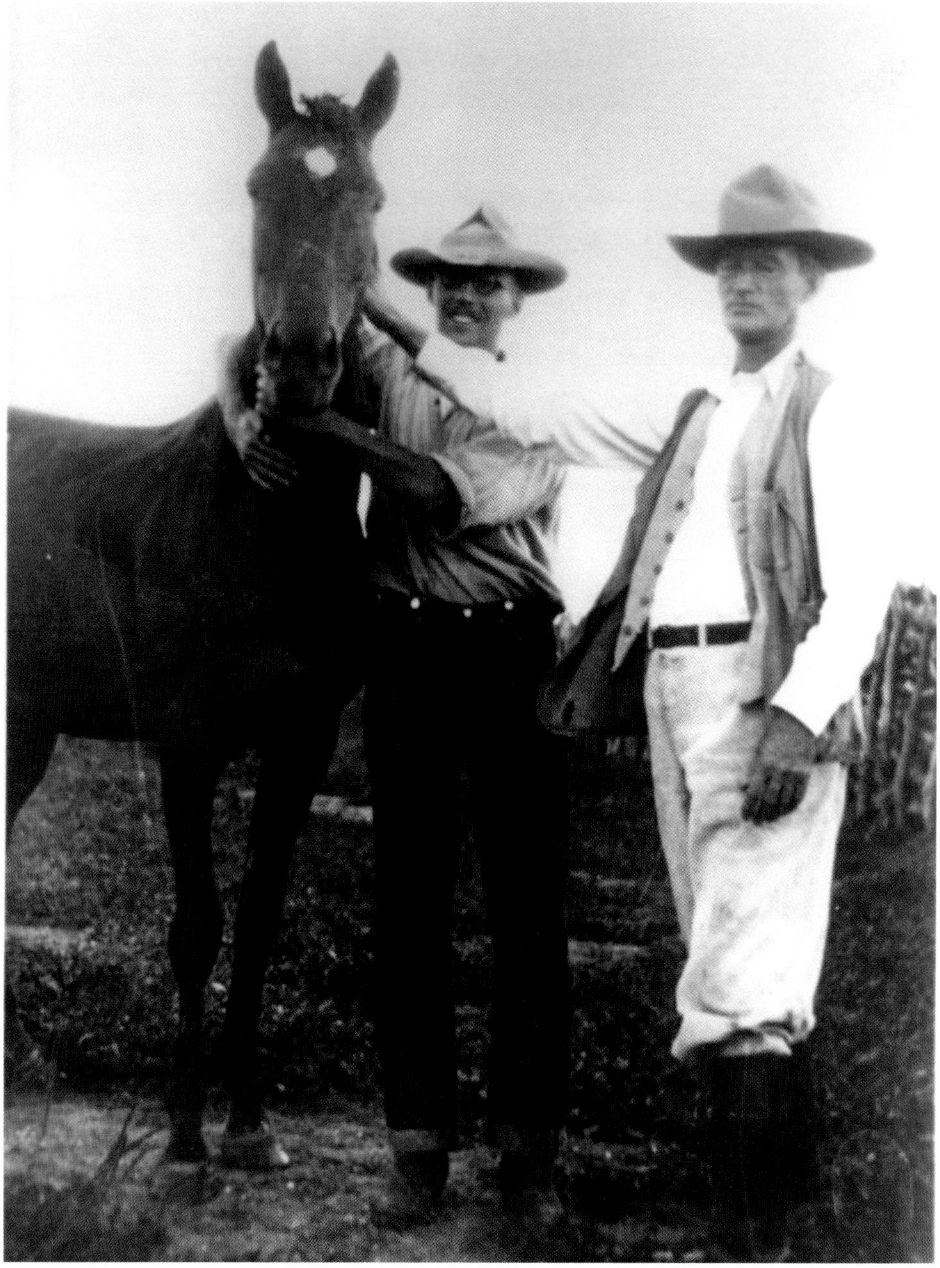

FIGURE 5.7. Rutger Porter and Harold Bell Wright. 1927.

OPPOSITE PAGE, TOP: FIGURE 5.8. Main entrance construction, Desert Sanatorium, Tucson, Arizona.

BOTTOM: FIGURE 5.9. Main entrance with landscaping, Desert Sanatorium, Tucson, Arizona, published by Frasher's Fotos, Pomona California, ca. 1928.

from there became the President of the University of Arizona. From 1928 to 1936, Shantz lead the university through the depression with a round of pay cuts, a policy that included the university's academic departments and his own salary. Through it all Shantz remained a popular and respected figure on campus. With federal assistance he pursued a modest building program, made improvements to the campus' physical plant, completed the football stadium and expanded the campus cactus garden (University of Arizona Library Special Collections 1942, 2020a; Geographical Review 1959).

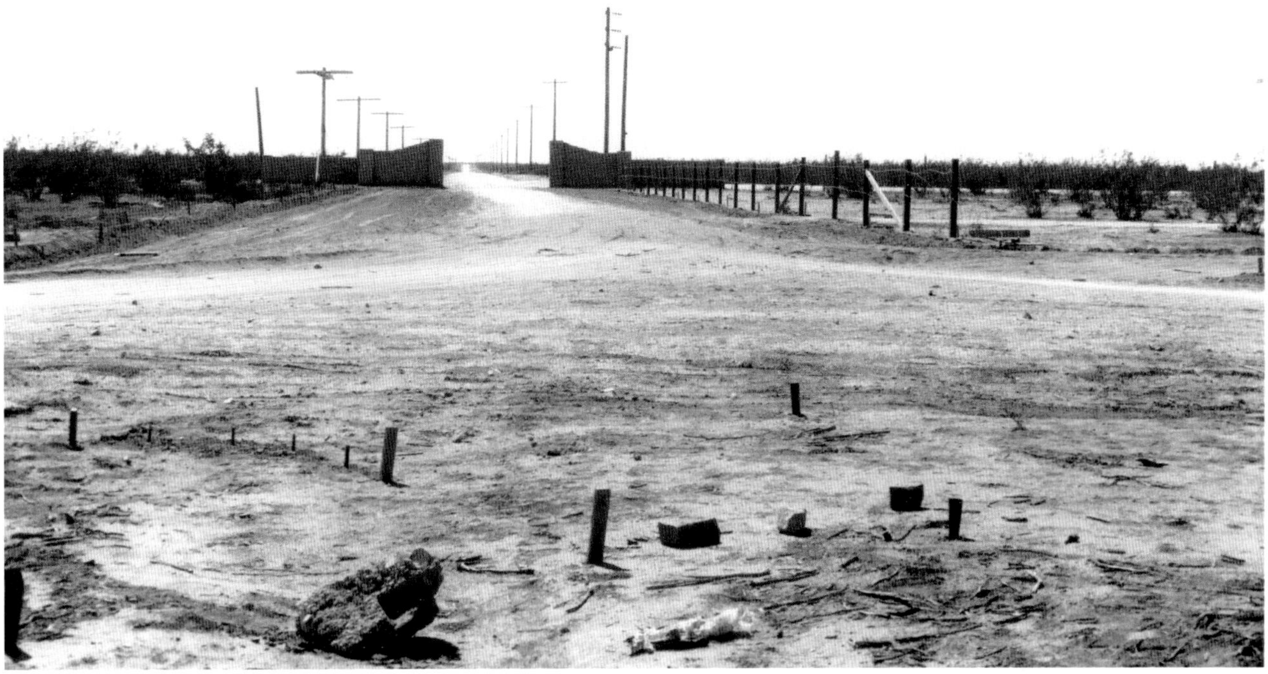

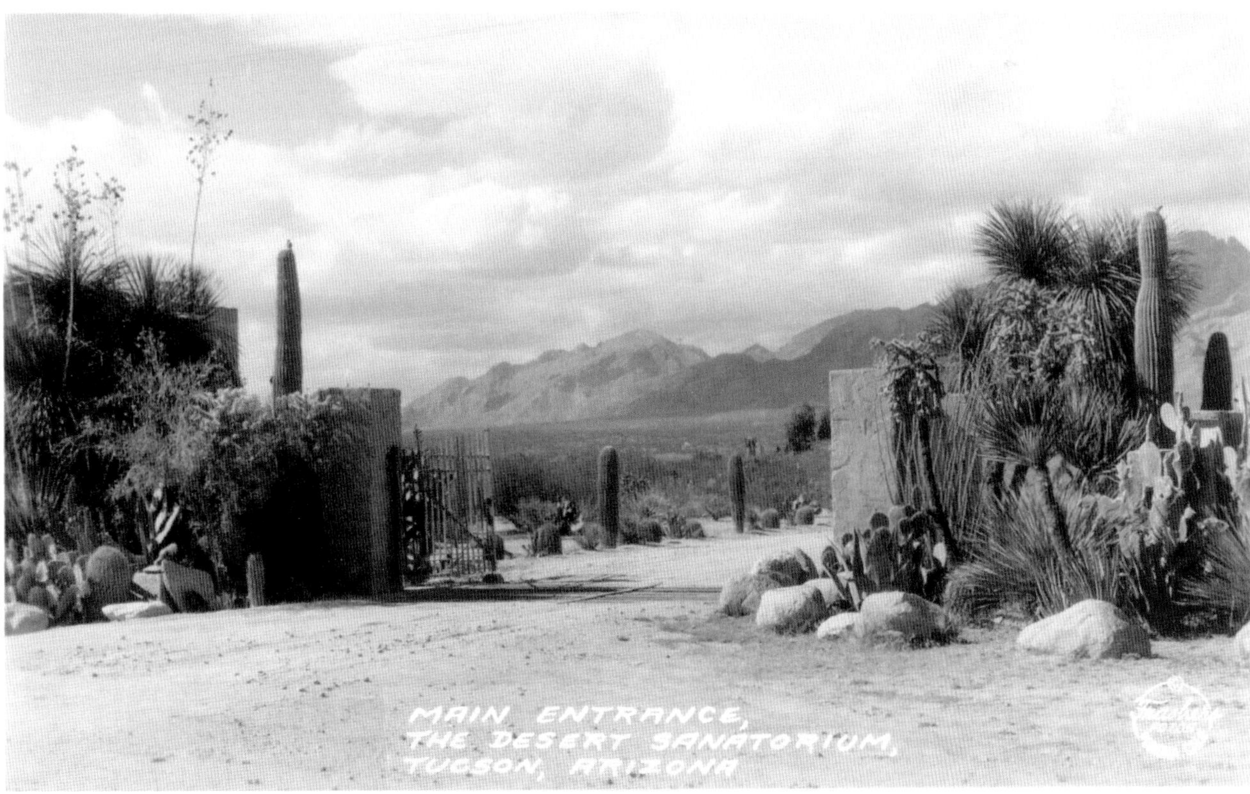
MAIN ENTRANCE,
THE DESERT SANATORIUM,
TUCSON, ARIZONA

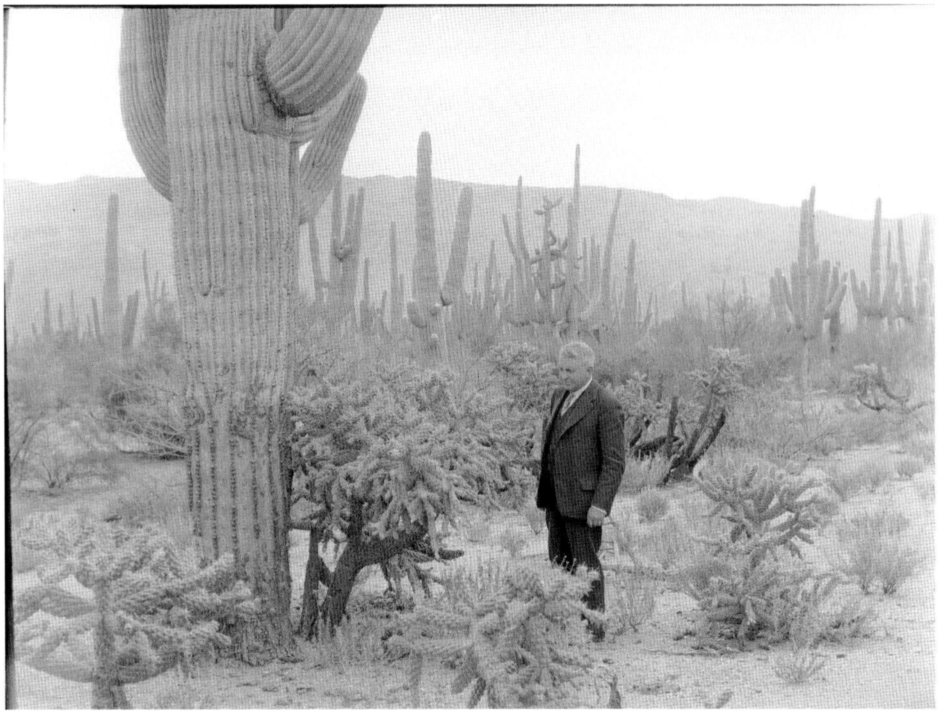

FIGURE 5.10. University of Arizona President Homer LeRoy Shantz, a frequent visitor to the Tanque Verde Cactus Forest, the future Saguaro National Monument. February 22, 1930.Photograph by R.B. Streets.

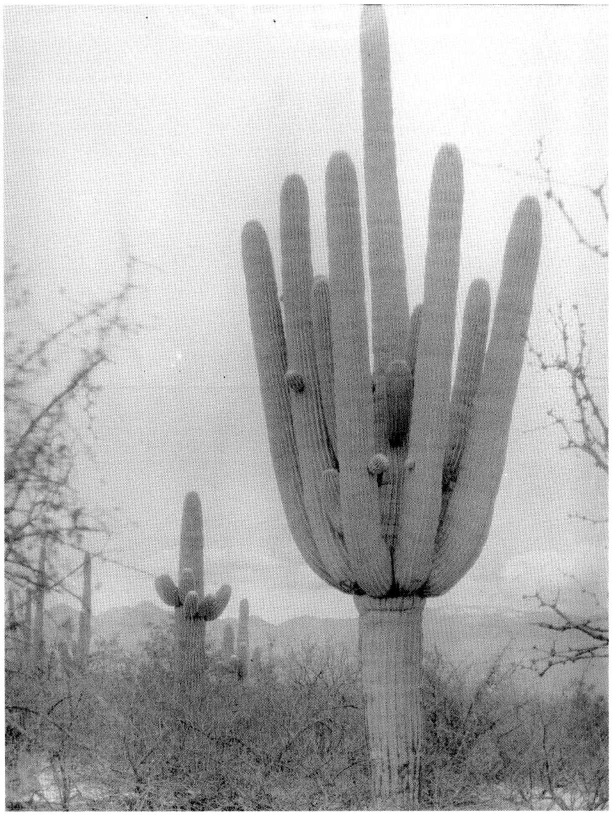

FIGURE 5.11. Chandelier-type saguaro in the Tanque Verde Cactus Forest. February 22, 1931. Photograph by Homer Shantz.

Shantz availed himself of every opportunity to visit the Tanque Verde Cactus Forest, posing for photographs with its massive saguaros for the first time in 1930. R.B. Streets, professor of plant pathology in the Department of Botany, took the photo (Figure 5.10). Shantz soon made the acquaintance of Tucson realtor and eastside resident John Elles Harrison. Shantz purchased the Tanque Verde tract with University monies augmented by privately donated funds that Harrison solicited in the name of the University Cactus Forest Association. Not without irony, Harold Bell Wright declined Harrison's solicitation to participate in the new association (Wright 1930; Arizona Daily Star 1957; Burtner 2011:42).

Most of what we know about the stand of saguaros that inspired the purchase of the Tanque Verde tract comes from Shantz's photographs, which focused upon the tract's most massive specimens. Shantz documented the area in nine visits from 1929 to 1935, often

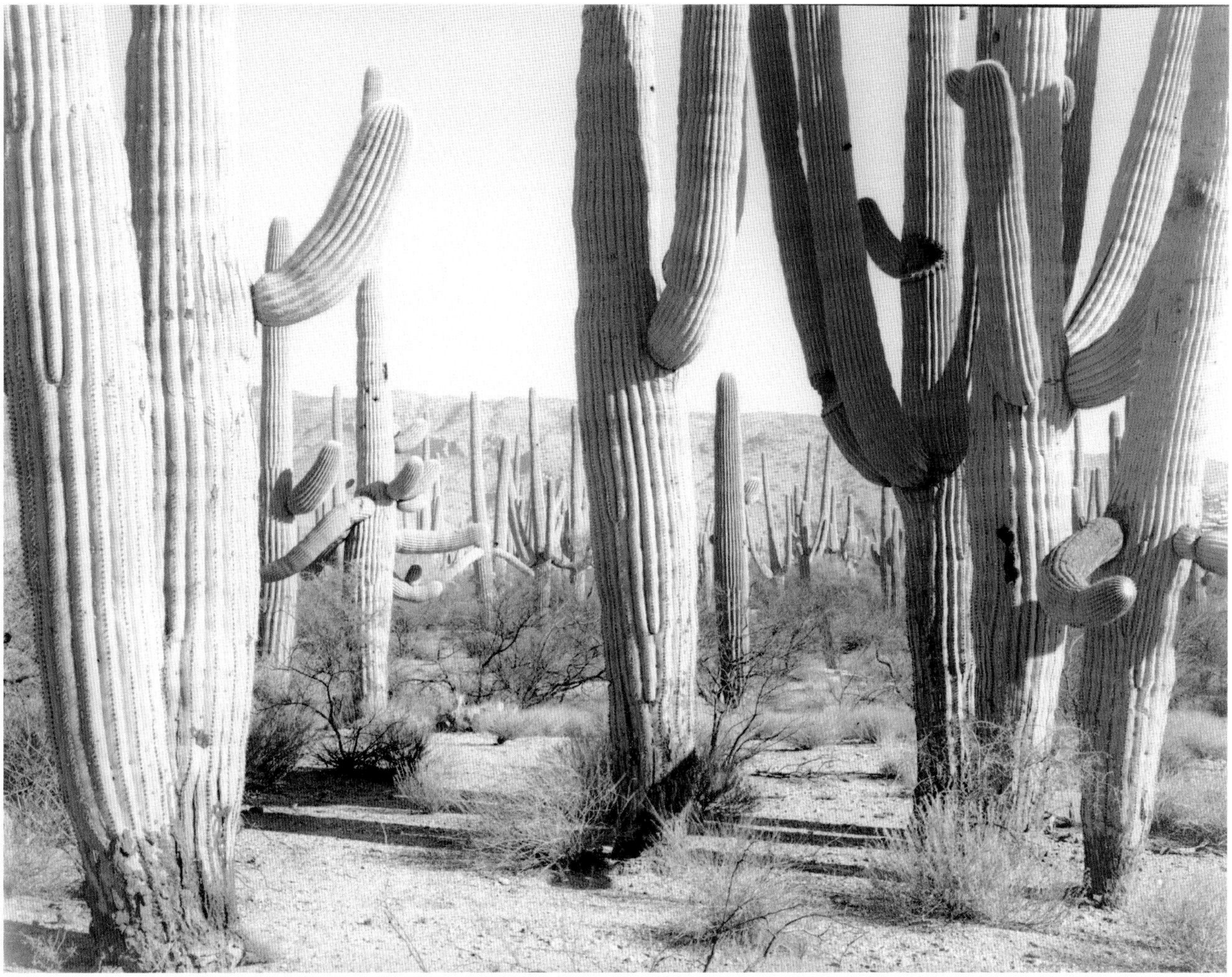

FIGURE 5.12. Saguaro stand in the Tanque Verde Cactus Forest. February 20, 1933. Photograph by Homer Shantz.

taking advantage of long holiday weekends to make the day trip from Tucson, sometimes in the company of the visiting botanists that he entertained. The remarkably dense arrangement of the saguaros held his attention and was a subject to which he often returned (Figures 5.11–5.13; Shantz 1937).

More to the point of Shantz's daily attention, the University of Arizona campus cactus garden enjoyed a new and greater prominence in the lives of students, faculty, and the public at large. With Shantz's installation as president, workers set to transplanting saguaros, barrels, ocotillos, and prickly pears on the east side of the Old Main building, the first to be built by the university. The territorial-style stone and wood frame structure with a wrap-around porch housed university offices and classroom space. Professor James W. Toumey established a cactus garden northeast of the building in 1891. The following year, the cactus garden moved to the building's west front, where it grew from 1892 to 1929. The move

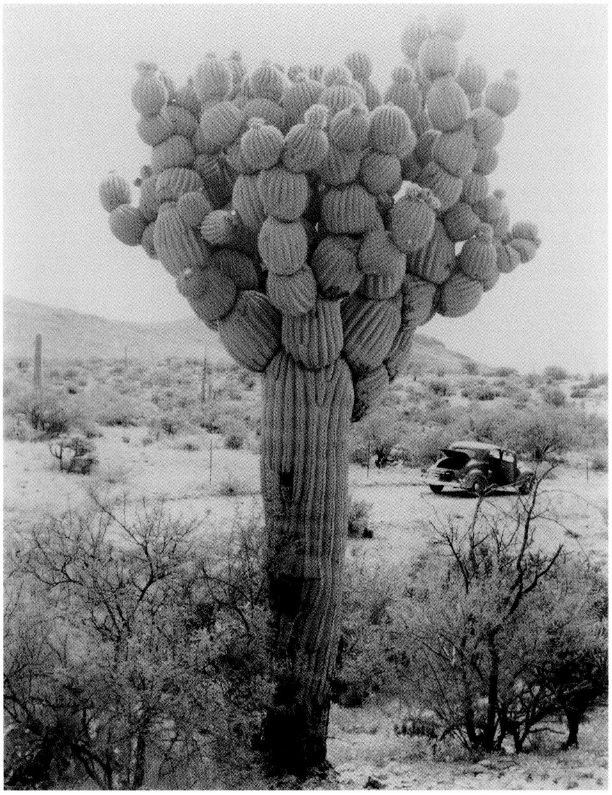

FIGURE 5.13. The locally known "watermelon" saguaro and waiting car on the Old Spanish Trail just outside the boundary of Saguaro National Monument. September 22, 1935. Photograph by Homer Shantz.

incorporated cacti left over from the local dig that supplied the saguaros for the 1893 Chicago World's Columbian Exposition. The plants were placed on the west front of Old Main in the center of the heart-shaped drive leading to and from the university's main gate (Rowley 1984). By 1920, the arrangement included a fountain. At the time of Shantz's arrival in 1928–1929, the garden featured native flowering queen-of-the-night plants (Figure 5.14). Under Shantz's administration a new and more ambitious cactus garden was created on the east front of Old Main in a 50-foot-wide trench extending some 1,000 feet to the university gymnasium (Ball 1987:28, 179, 220). An aerial view postcard postmarked in 1943 pictures the garden from Old Main to the gymnasium, with the two buildings circled by the sender, a Navy ensign who identified them as classroom and dormitory space, respectively (Figure 5.15). Landscaped with rocks and boulders, the trench functioned as a water collecting arroyo. The below-grade paths and shoulder-high boulders created an environmental showcase for the plants of the Sonoran Desert (Figures 5.16 and 5.17). Near the gymnasium end of the display, an island-like break that crossed the trench at grade level encircled a "Mexican Garden" of boojum trees (Figure 5.18).

President Shantz's second successor, Professor Alfred Atkinson, found little use for the university cactus garden, or for that matter cacti of any kind. President from 1936 to 1945, Atkinson, according to his official university biography, was notably "short on sentimentality" (University of Arizona Library Special Collections 2020b). In the practical manner of finding a reason to remove the garden, its upkeep was allowed to slip in proportion to Atkinson's complaint that it cost too much to maintain—an issue that Shantz never faced, even when taking a self-imposed salary cut in the depths of the Depression. Post Shantz, Atkinson's vision for the campus revolved around landscaped lawns with vistas of new buildings, not a 1,000-foot environmental rockwork and cactus-filled trench along the east axis of Old Main. Atkinson concurred with Old Main's condemnation as a campus eyesore in 1938 with feigned regret, not unlike the cactus garden, the victim of

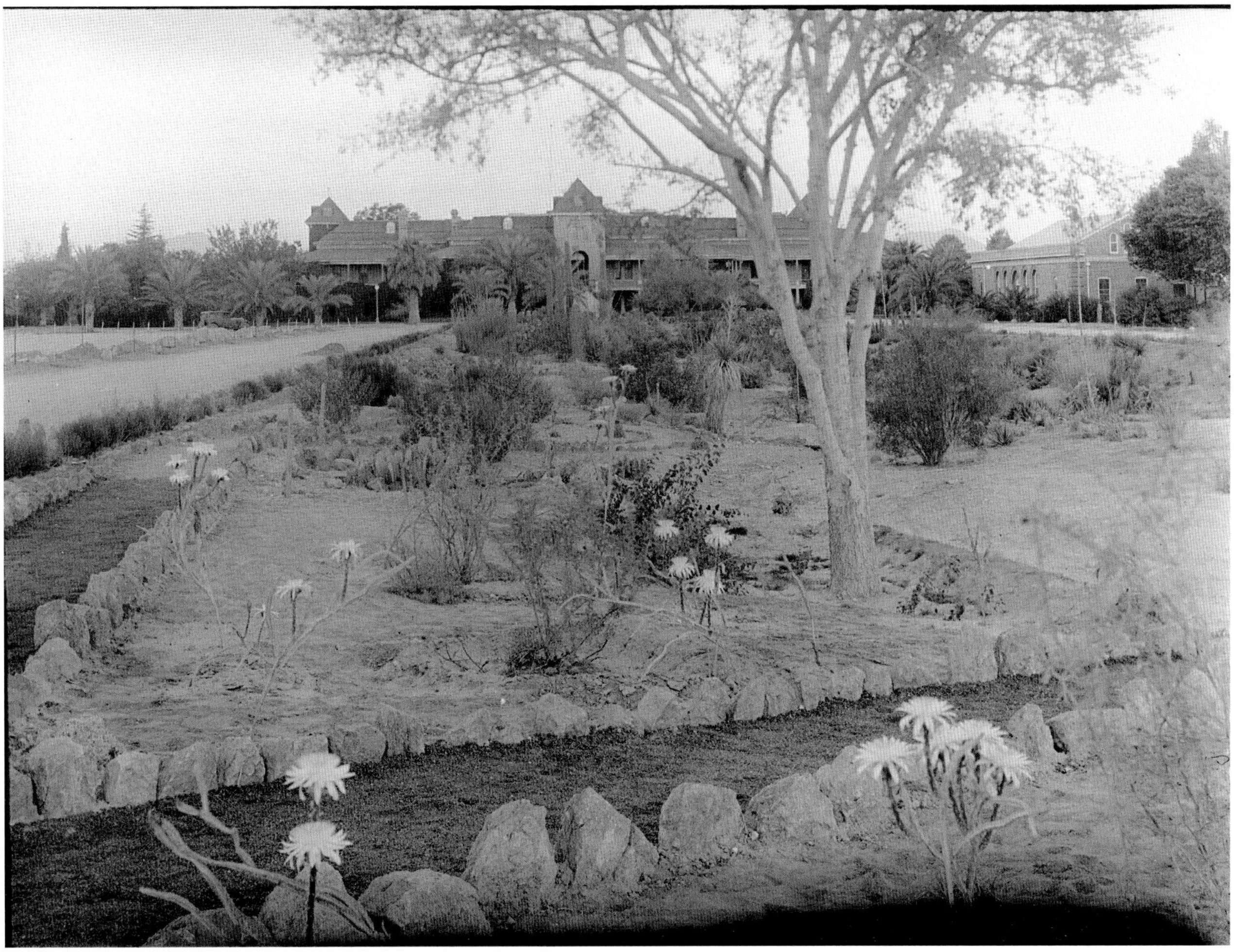

FIGURE 5.14. Queen of the Night in bloom in the cactus garden west of Old Main building, University of Arizona. June 15, 1932. Photograph by Homer Shantz.

deferred maintenance. In the war emergency following Pearl Harbor the U.S. Navy, in desperate need of classroom space, restored Old Main and returned it to service in 1943–1944 (Tucson Daily Citizen 1938, 1944; University of Arizona Library Special Collections 2020a).

Atkinson's interest in campus greenscape betrayed the cactus-or-not tensions displayed in Arizona's travel literature and state seal. A proposal made by the campus grounds committee that Atkinson tasked with figuring out the cactus conundrum recommended doing away with the plants entirely. Seeking an alternative for the west side of Old Main, for example, his committee proposed installing a large lily pool in front of the Physics-Chemistry building. The inground pool would be flush with the lawn. Its maintenance would be minimal (Atkinson 1939; University Campus Committee 1939; Streets and Finch 1940).

Lily Starkweather, co-chairman of the Statewide Beautification Program of the Arizona Federation of Gardens, was having none of it. In February 1939, she

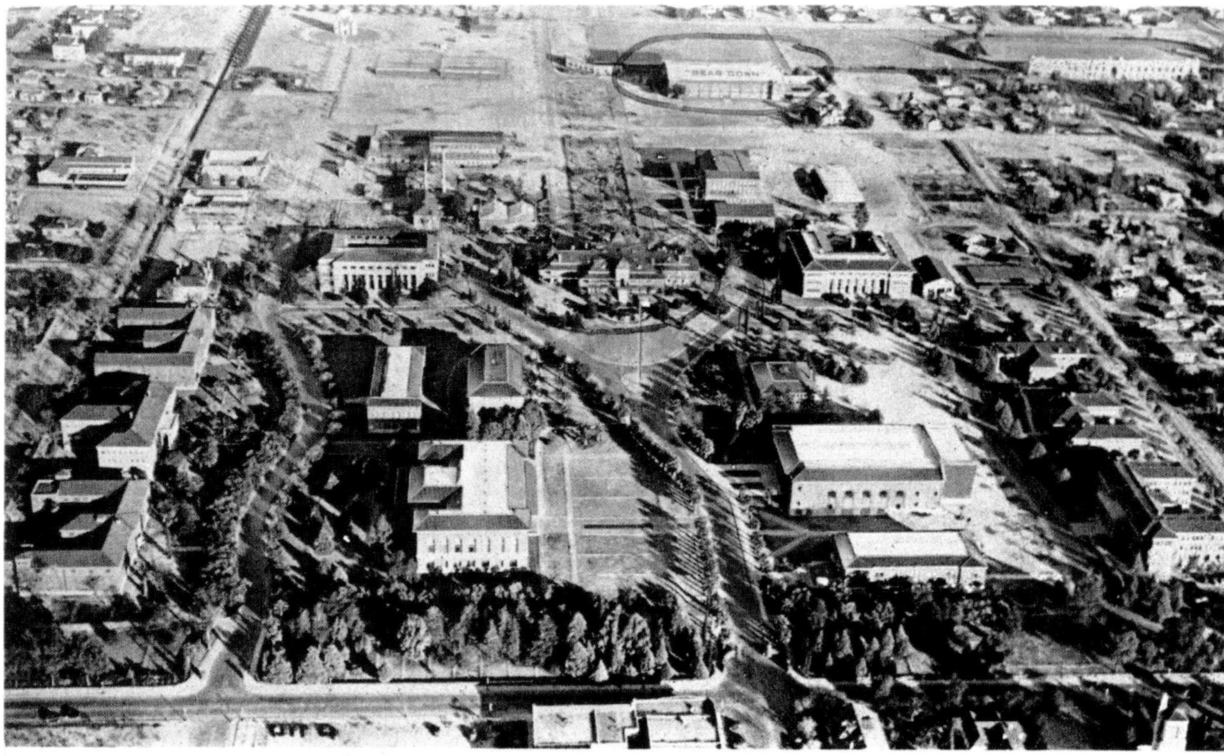

FIGURE 5.15. Postcard. Aerial View, University of Arizona, Tucson, Arizona. Postmarked 1943. Photograph by Ben Gross. The postcard's sender, a U.S. Navy ensign at the University during the Second World War, circled the campus locations where he spent most of his time: his gymnasium dormitory and the Old Main building recently restored as a classroom space. The 1,000-foot campus cactus garden connects the buildings in a line.

telephoned Atkinson and followed up with a note protesting the proposed destruction of "our native cacti and desert growth that has been planted on the campus" (Starkweather 1939). Perhaps to better meet the challenge, she and other members of the Tucson Garden Club organized the Tucson *Desert* Garden Club on the heels of Atkinson's campus committee's stated conclusion that the cactus garden "was not less expensive to maintain than a grass-shrub-tree" landscape. In what would become a familiar litany of complaint justifying the cactus garden's removal, the committee stated that the irrigation of cacti allowed grass and weeds to germinate; the constant replacement of the cactus plants that died or became unsightly was more trouble than it was worth; and smaller and rarer cactus specimens planted in the garden were routinely stolen. Lastly, Atkinson's campus committee tendered its justification for the garden's removal with the disingenuous statement that the completion of the road to Saguaro National Monument "made it unnecessary to maintain a separate cactus garden for our students and the public" (University Campus Committee 1939).

The issue of campus cactus-or-not stalled through the Second World War, when the university's priorities lay elsewhere. Revisiting the issue in late 1945, Atkinson

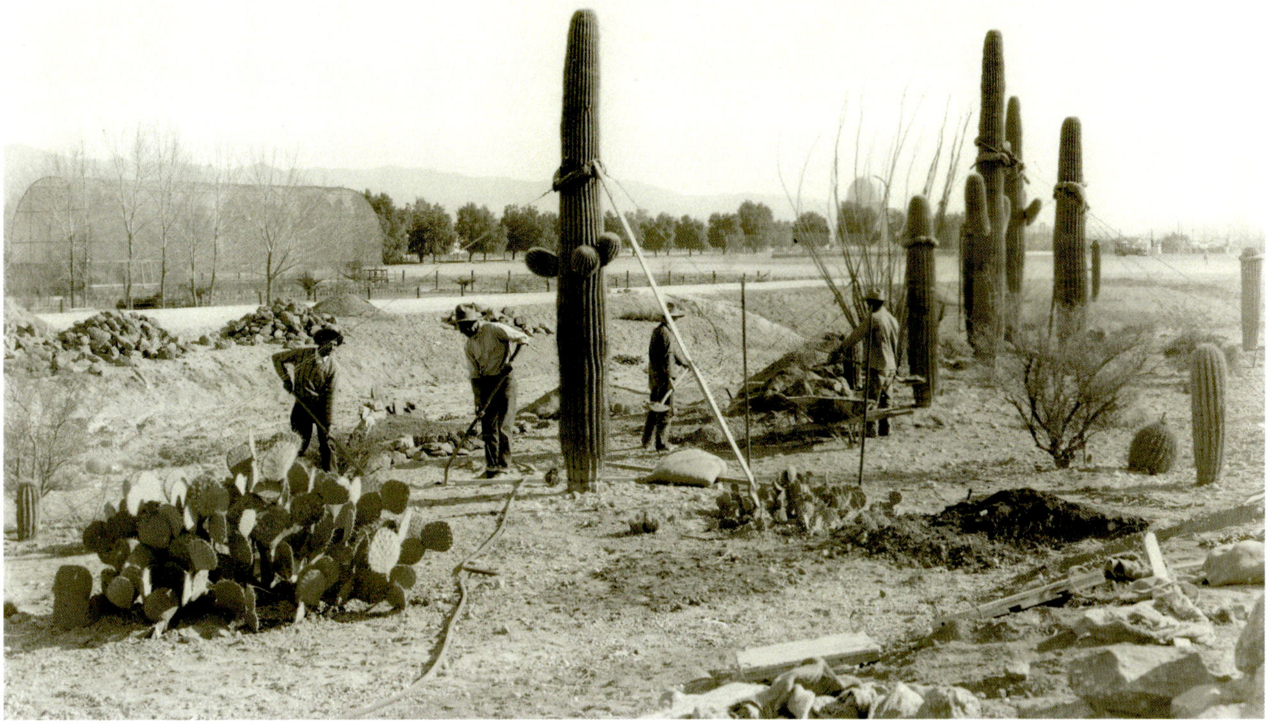

FIGURE 5.16. University buildings and grounds crew members planting the cactus garden arroyo east of the Main building, University of Arizona, ca. 1929. Photograph by R.B. Streets.

wrote to Starkweather reprising the shortcomings of maintaining a cactus garden: lack of labor and outright theft. Starkweather, addressing her continuing concerns up the chain of command to Governor Sidney P. Osborn, suggested that the University clean up its moribund garden and secure smaller cactus specimens in a hothouse. Keeping up the pressure, Starkweather noted that the cactus garden "could be the center attraction, instead of an eyesore on our otherwise attractive campus" (Atkinson 1945; Starkweather 1945, 1946).

Among enthusiasts of the compositional frame of cactus specimens and gardens, Charles and Lucile Herbert built their desert home with a picture window that looked out on a massive chandelier-type saguaro. The Herberts moved to Tucson to open the Western Ways Features Service in 1939. Charles Herbert had recently retired as a *March of Time* cinematographer. Lucile Herbert, a photographer in her own right, contributed many of the feature stories published by their news publicity service that simply became known as Western Ways (Figures 5.19 and 5.20). The Herberts' oft-repeated dictum to their stable of contract photographers and freelance writers: "Photographs must accompany articles" (Keatley 1949; Balestrero 1972; Jenkins 2016:1–14).

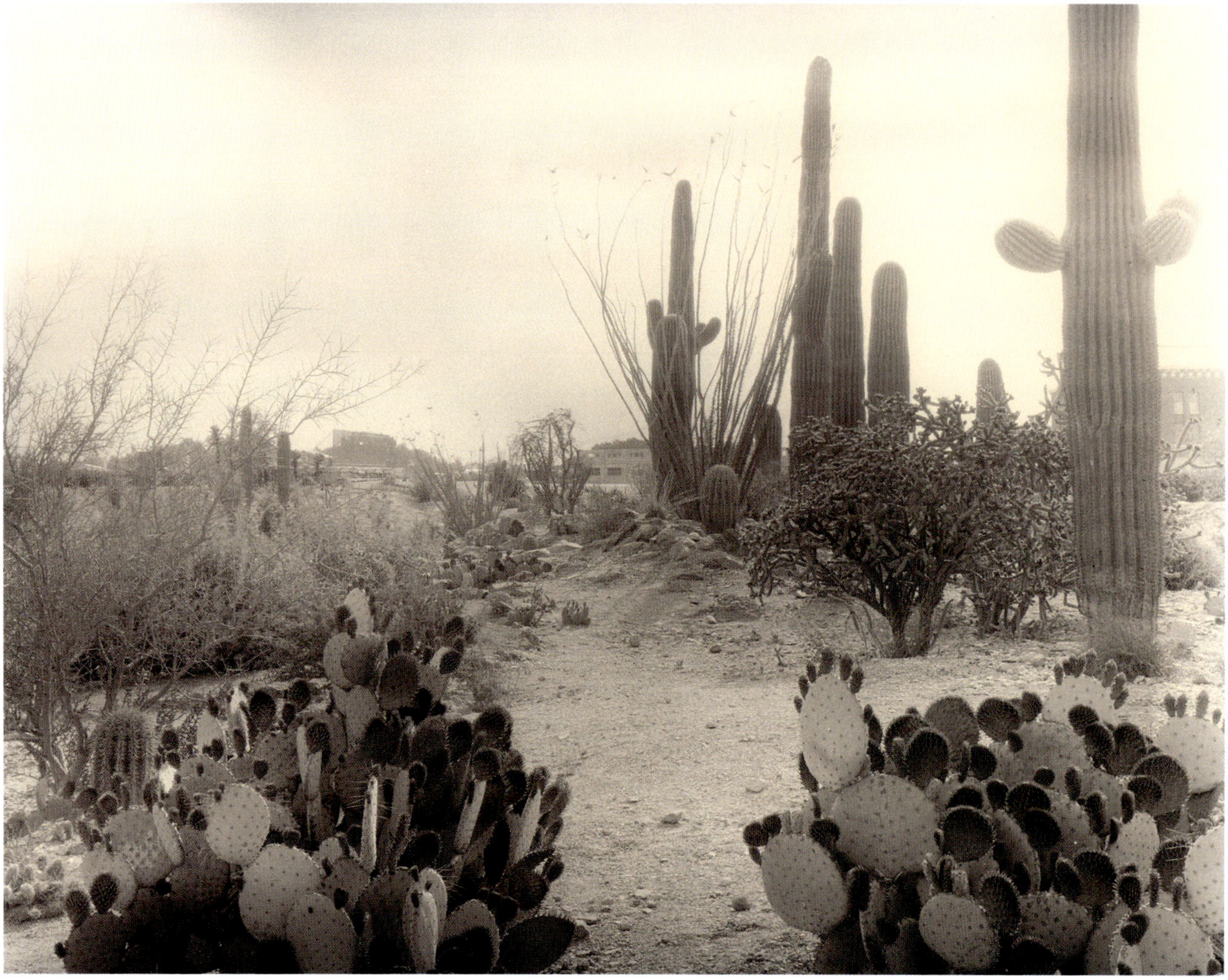

FIGURE 5.17. Cactus garden arroyo east of Main building, University of Arizona. April 1931. Photograph by Homer Shantz.

Described in perennial syndicated features as "The View with a Room," the Herberts' living room window framed the most carefully curated saguaro moment in Arizona (Figure 5.21). As Herbert (n.d. [1940?]) explained to Sunshine Climate Club president Roy Drachman, who welcomed them with open arms, "Our picture window is going to be one made by a photographer instead of a builder who only deals in the mechanics of four sides of a window." A carpenter tacked up a temporary window frame conforming to a viewfinder to test the sightlines. Lucile Herbert stood on a wooden box approximating the level of the finished room floor to confirm the view (Hall n.d. [1940?]; Herbert n.d. [1941?]; Winsor n.d. [1962?]; Jenkins 2016:9–10).

Though siting a home with a scenic view was not new, in the Herberts' understanding of what was important, the saguaro was the critical element. Over the years they amortized the cost of their home's construction with illustrated articles

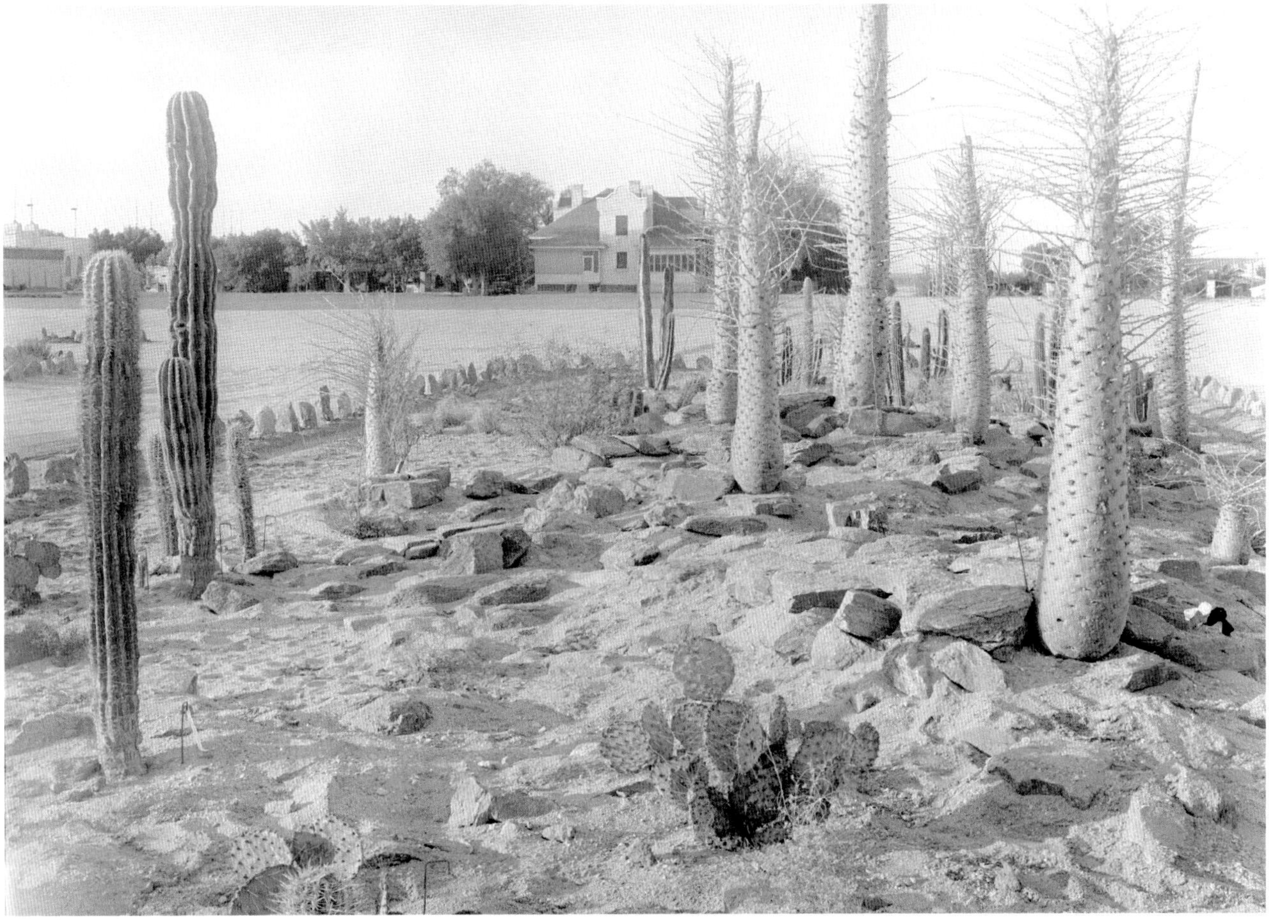

FIGURE 5.18. Boojum trees in the Mexican garden, University of Arizona. July 7, 1932. Photograph by Homer Shantz.

about it, syndicating feature stories picturing the "view" as late as 1947 (Figure 5.22; Detroit News Pictorial 1947).

For those with unlimited resources to create a view from scratch, the transplantation of saguaros once beyond the means of even the most zealous botanical collector yielded to the diesel-powered crane. In 1939, a landmark in the history of cactus transplantation culminated with the relocation of massive saguaros to heiress Louise N. Grace's Catalina foothills home, Eleven Arches. The placement of the mattress-clad specimens at Eleven Arches became the subject of Western Ways magazine, newspaper features, and a short subject film later recut for NBC Television (Figures 5.23–5.25; National Geographic 1953; Desert Magazine 1960; Jenkins 2016:19–22).

Louise Grace was the daughter of the late, two-term mayor of New York, William R. Grace, who founded the W.R. Grace trading, investment, and shipping company. Used to doing things in a big way, she lined the living room of Eleven Arches with the monochromatic floor-to-ceiling murals of desert wildlife and cacti that she

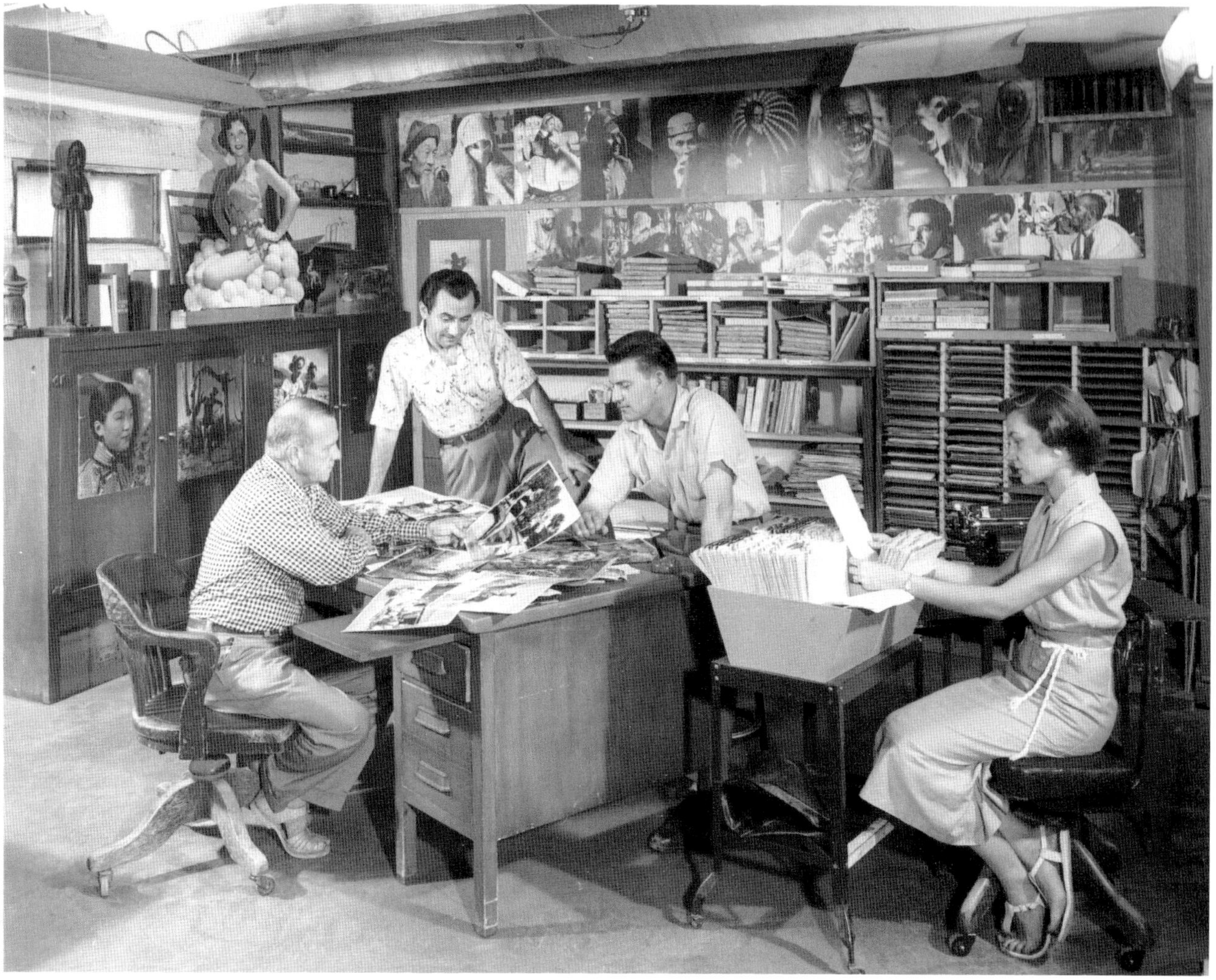

FIGURE 5.19. Charles Herbert (left) with Western Ways Features staff. Tucson, Arizona. Visible are some of the products of Herbert's life in pictures: from left to right, a cardboard cut-out of a model posing with Florida oranges, a cactus ballet photo mounted to the cabinet below, and a wall-length collage of portraits reflecting Herbert's international experience as a cinematographer and photographer.

painted. Grace exhibited the murals at Rockefeller Center before installing them in her Tucson home. Artist Cleve Gray helped complete their installation at Eleven Arches, painting in outlined forms and other unfinished areas under her direction (Figure 5.26; New York Times 1937; New York Daily News 1954; Weber 1998).

Reaction to the story of the saguaros' removal to the grassy forecourt of Grace's foothills mansion came from as far away as England (Arizona Daily Star 1948). Despite the extensive precautions taken for the protection of the plants, the limits of saguaro transplantation had been reached. Within a matter of years the plants died in place. Arguably they died the moment the crane lifted each one of them out of the ground. The film documenting the excavation of the last and largest of the saguaros—a replacement for one that had already died in place—shows its tap root breaking as the crane lifts the plant. Today it is known among film historians as "the cactus snuff film" (Jenkins 2016:xv–xvi).

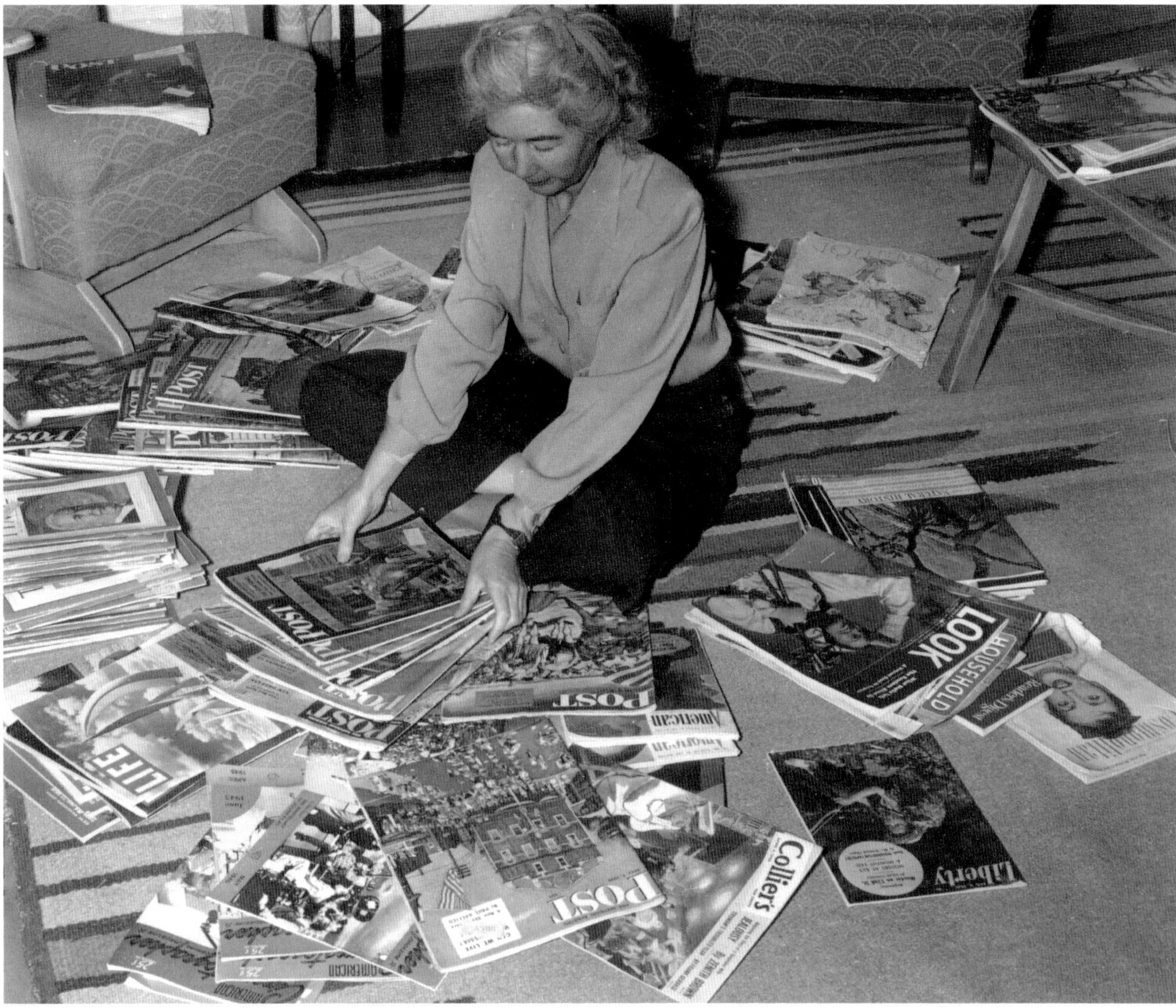

FIGURE 5.20. Writer and photographer Lucile Herbert with some of the magazines in which their photos and stories appeared.

The advertising agency partnership of Ernest Cabat and Norval Gill (Tucson Historic Preservation Foundation 2020) interpreted the saguaro as a free-standing artwork. Their Tucson-based Cabat-Gill Advertising favored the frame technique in graphic designs for a long list of desert hospitality clients whose brochures folded the viewer's attention right into the subject at hand. Their saguaro logo exemplified the technique. The logo pictured an artist who held a framed saguaro picture at arm's length, comparing it to the full-sized plant, itself in a frame (Figure 5.27). The firm's artwork deployed window and picture frames to similar effect. For Tucson's Lazy K Bar Guest Ranch, for example, Cabat-Gill created brochures whose largest graphic element was the picture window in the ranch's main lounge. Pictured in successive brochures over the years, the room's furnishings changed along with the ranch's owners. But the saguaro in the window never did (Figures 5.28–5.30).

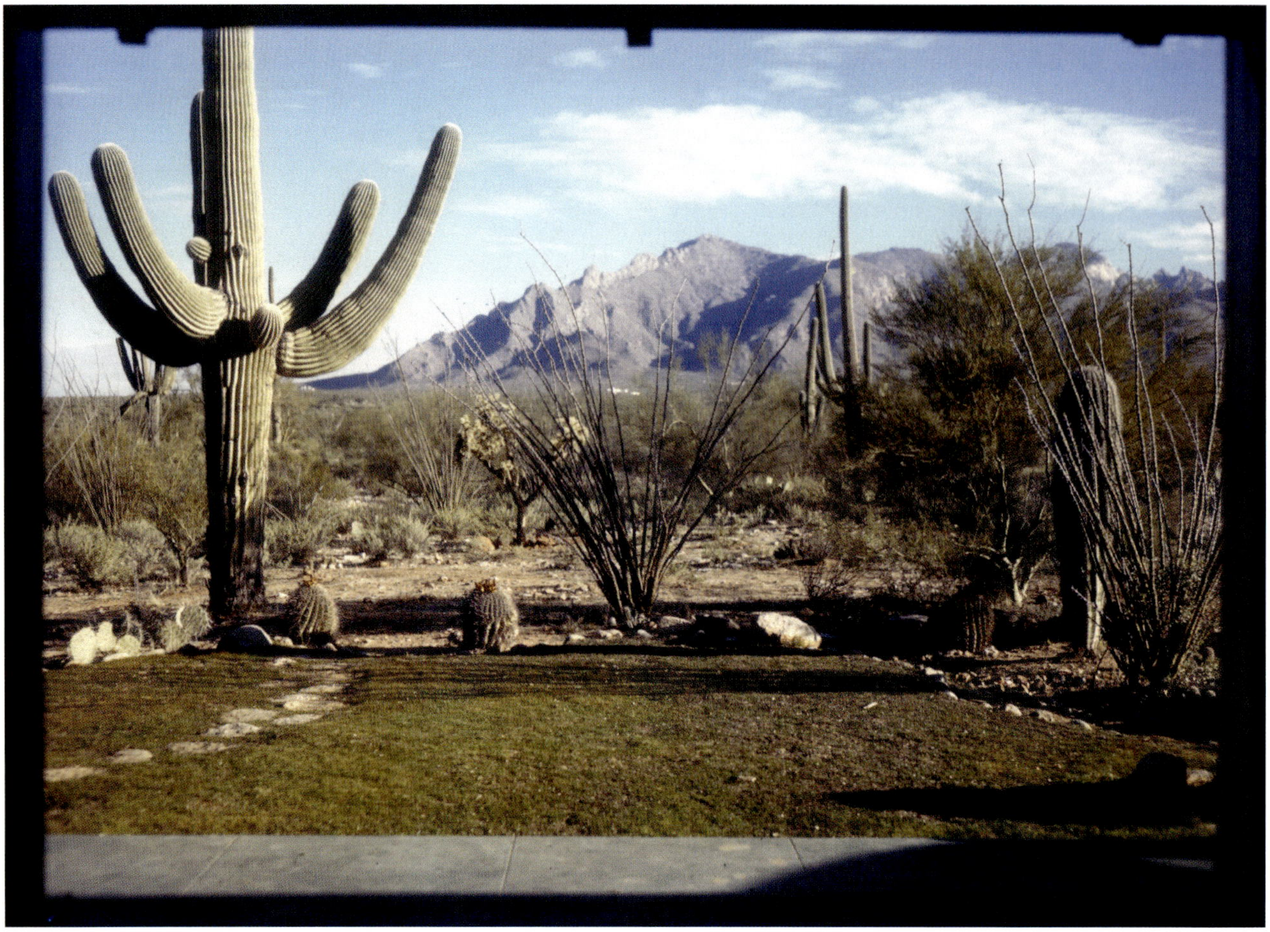

FIGURE 5.21. Color transparency of the Herberts' picture window, ca. 1947. Photograph Charles Herbert.

OPPOSITE PAGE: FIGURE 5.22. Dream House Located with Camera Eye. *Detroit News*, August 3, 1947.

Dressing faux windows with a certain Sonoran timelessness, the Mural Room restaurant in Tucson's El Presidio Hotel showcased the Shantz stand at Saguaro National Monument with backlit color transparencies recessed in the dining room's walls (Figure 5.31). Continuing the window theme on the menu cover, Cabat-Gill framed a saguaro in the spirit of the agency's familiar logo (Figure 5.32).

As William Temple Hornaday had foreseen, a commercial use for the saguaro cactus failed to materialize beyond its ribs' use as a traditional fencing and roofing material. Yet Hornaday and others did not anticipate the ribs' use as a decorative element, much less the cottage industry of rib crafters who furnished the lamps, picture frames, and desert dioramas that carried the wood and the likeness of the plant far from its native habitat.

Increasingly saguaro ribs became a signature decorative element in roadside décor, like the Cactus Bar in Salome, Arizona. Varnished saguaro ribs graced the bar's walls from floor to ceiling. Two saguaro arms attached to the wall above the bar cradled an impressive souvenir rib cage (Figure 5.33).

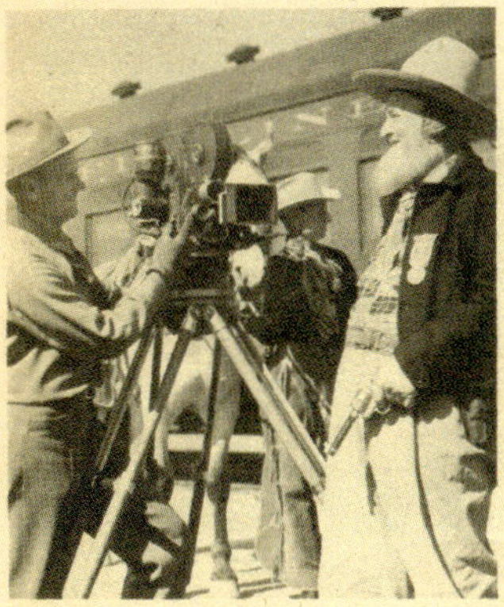

Dream House Located with Camera Eye

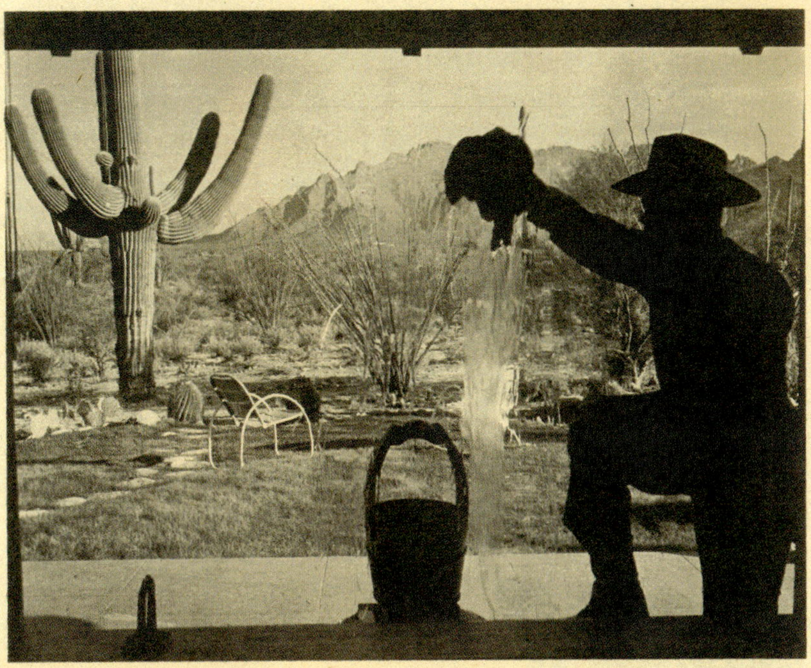

Herbert washes the picture window with a view like an 8x10 photograph around which he built his whole house.

CHARLES W. HERBERT, of Tucson (above left), special correspondent of The Detroit News, many of whose photographs have appeared in this magazine with the credit line, "Western Ways," built his house with a camera in more ways than one. The first part of the house to be constructed was the picture window. Herbert discovered a view in the finder of his camera. He nailed together a wooden frame in the proportions of an 8 by 10 picture. He placed it in the spot where it framed the very same scene his camera had shown him. He called the contractor and told him: "Build my house around that."

For 20 years, Herbert had been traveling the globe making newsreels for major moving picture companies. He got fed up on travel. Deciding to settle down, he selected southern Arizona as the general area. Tramping the foothills of the Catalina Mountains near Tucson, he found a site. A reflex camera showed him how to orient the house. His yard became an outdoor studio for his pictures.

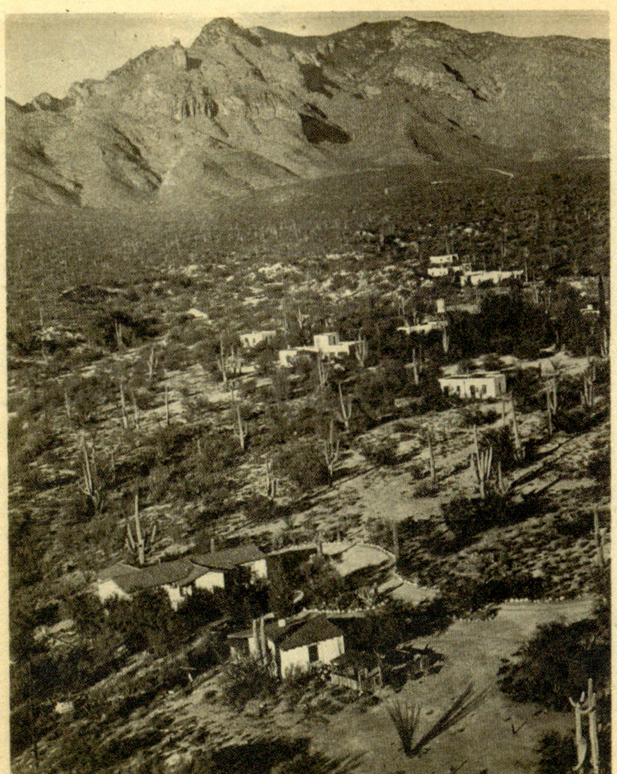

A stone-lined lane leads to the house that was located and laid out with the help of a camera lens and the principles of pictorial photography. Herbert's front yard has been the setting of many of his photographs published here. The mountains beyond and the desert landscape were selected as a backdrop for picture features of the southwest desert country. The sweeping shots were made elsewhere, closeups here.

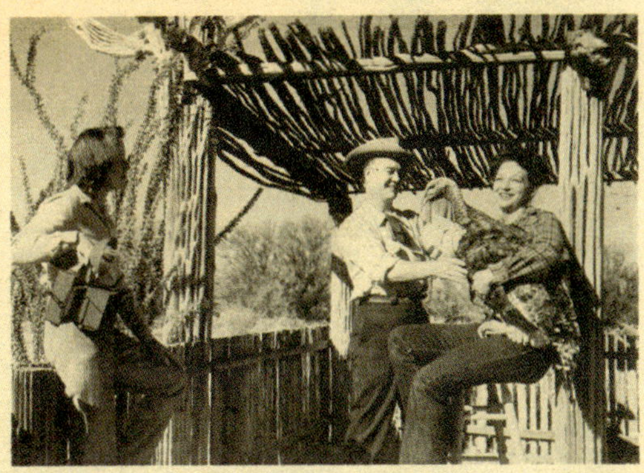

John Hicks, former Detroiter, moved to Tucson and evolved a turkey processing plant. Some of the closeups were made in Herbert's own yard.

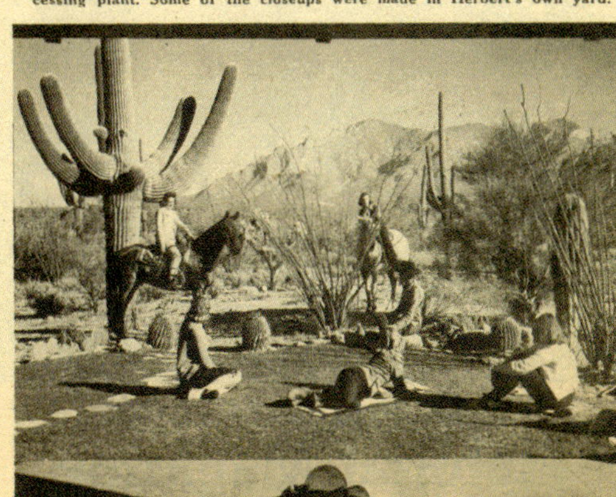

A picture story on five school teachers and their vacation activities called for a shot of girls on horseback and relaxing in the sun. The site? The photographer chose his own yard and area in front of picture window.

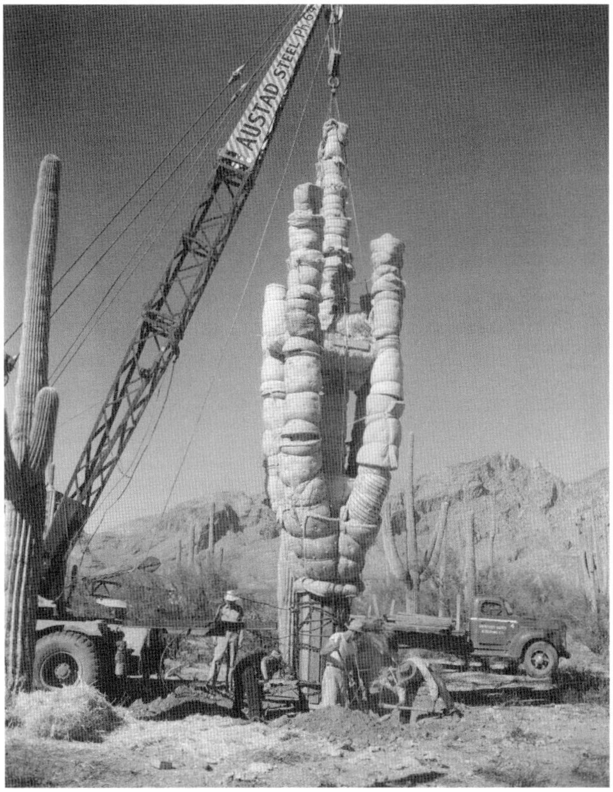

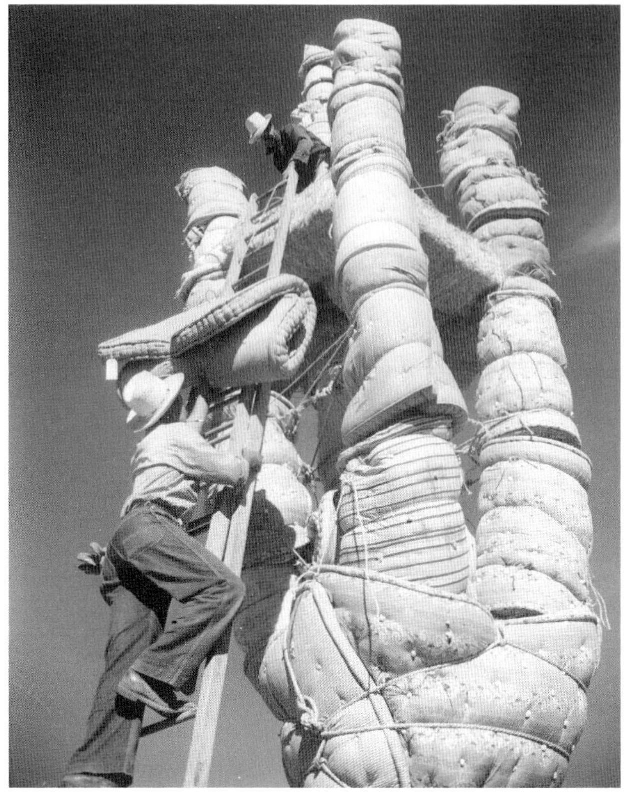

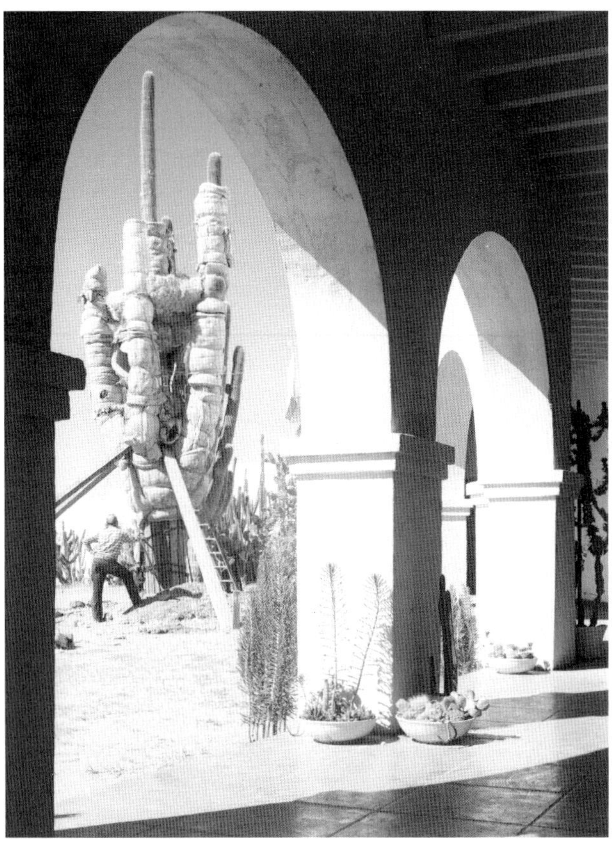

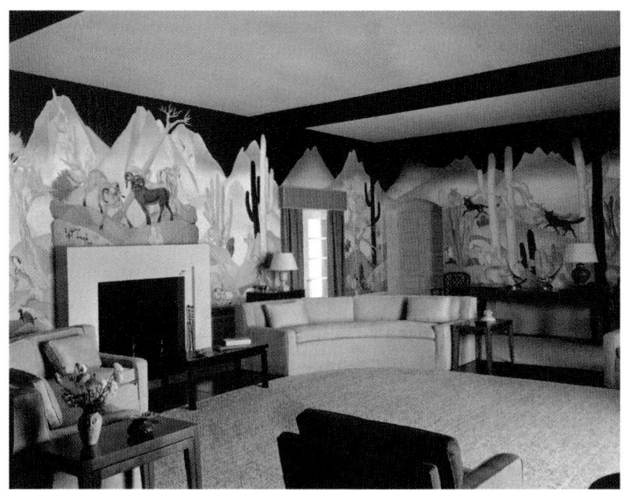

FROM TOP LEFT: FIGURE 5.23. Crane with mattress-wrapped saguaro ready to be transplanted to the home of Louise Grace, Tucson, Arizona. 1939. Photograph by Charles Herbert.

FIGURE 5.24. Workmen wrapping saguaro in mattress pads.

FIGURE 5.25. Saguaro viewed through arch at Grace's home, Eleven Arches.

FIGURE 5.26. Louise Grace's desert mural installed in the living room at Eleven Arches. 1939. Photograph by Maynard Parker.

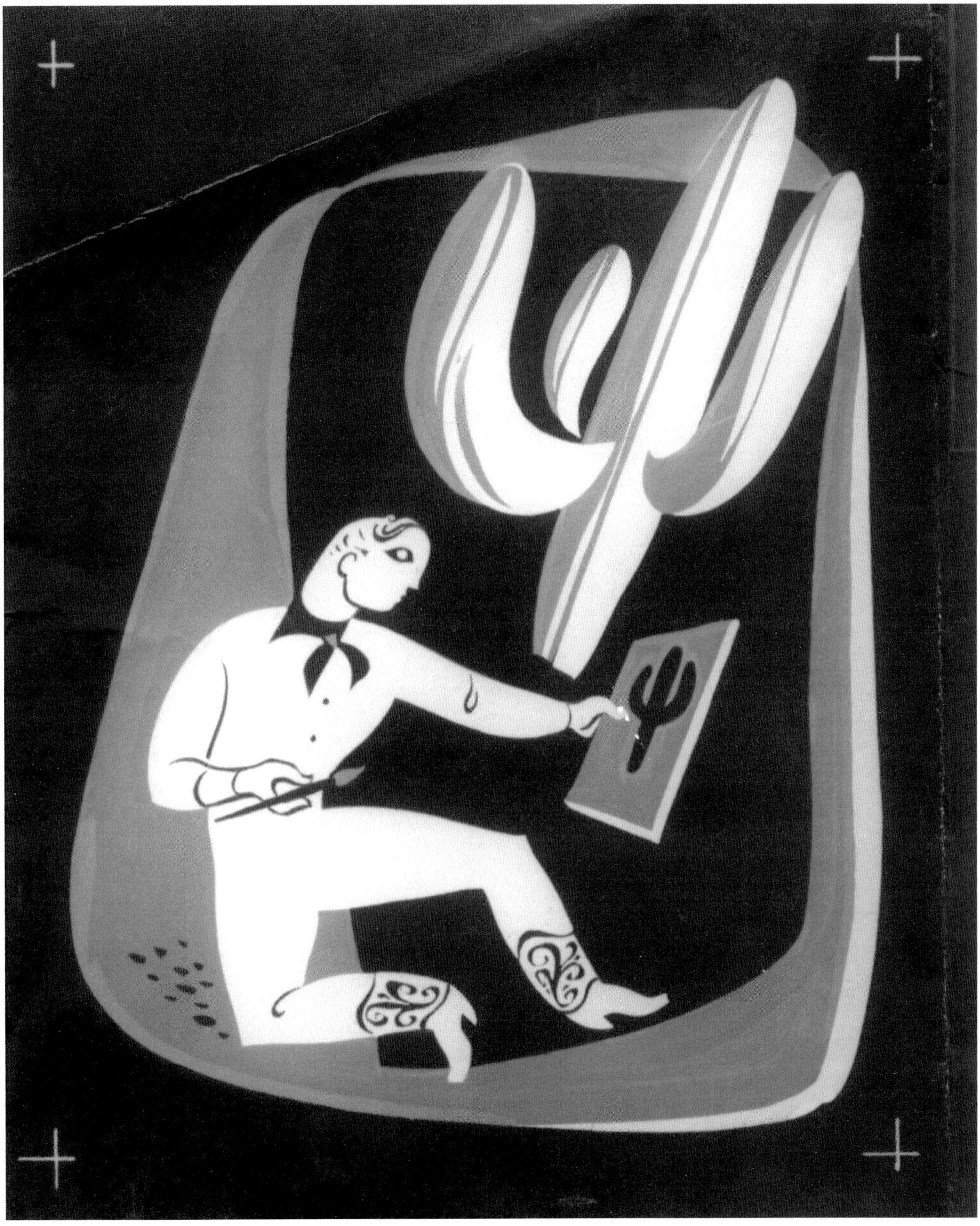

FIGURE 5.27. Photo offset negative, Cabat-Gill Advertising Agency company logo.

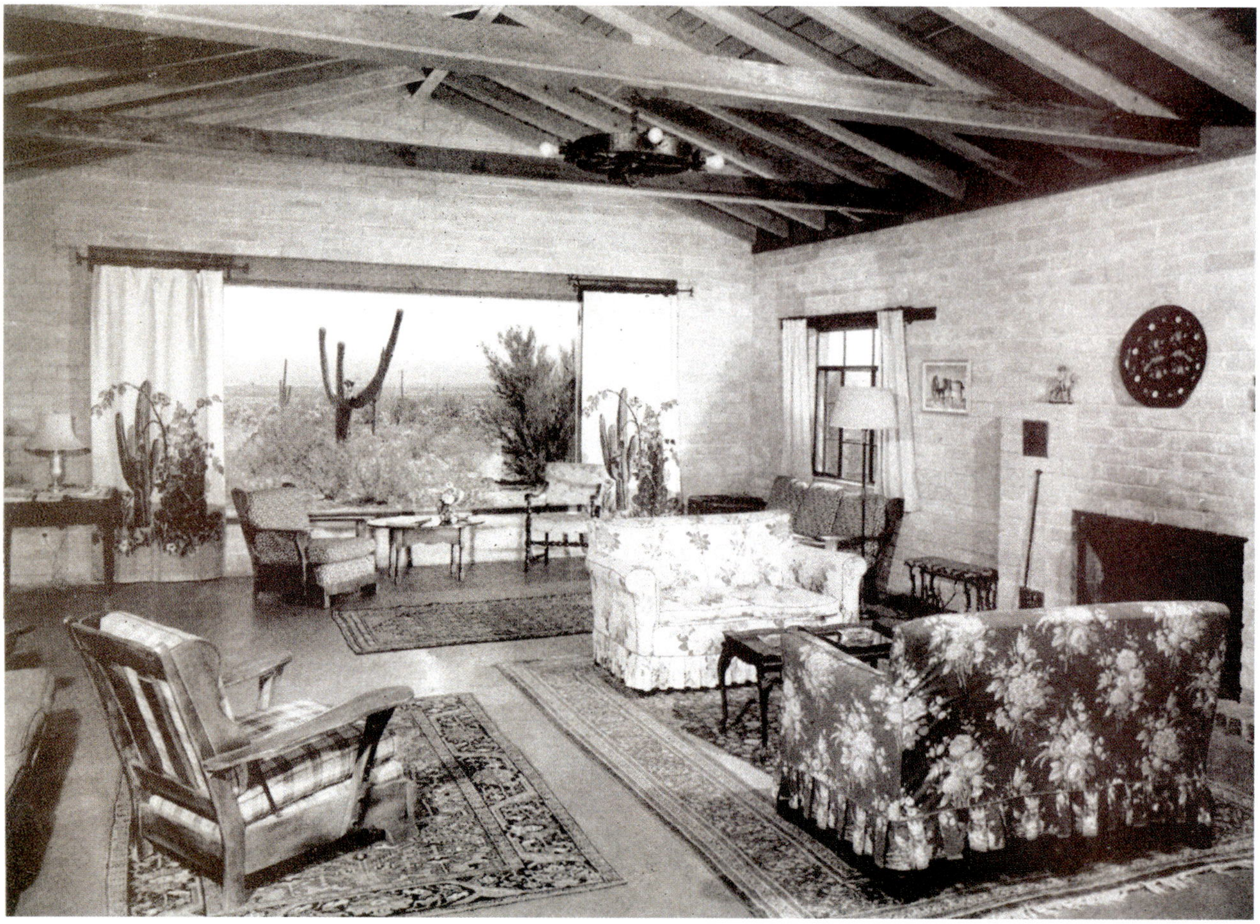

FIGURE 5.28. Brochure detail. Lazy K Bar Ranch monochrome living room with picture window frames a saguaro view.

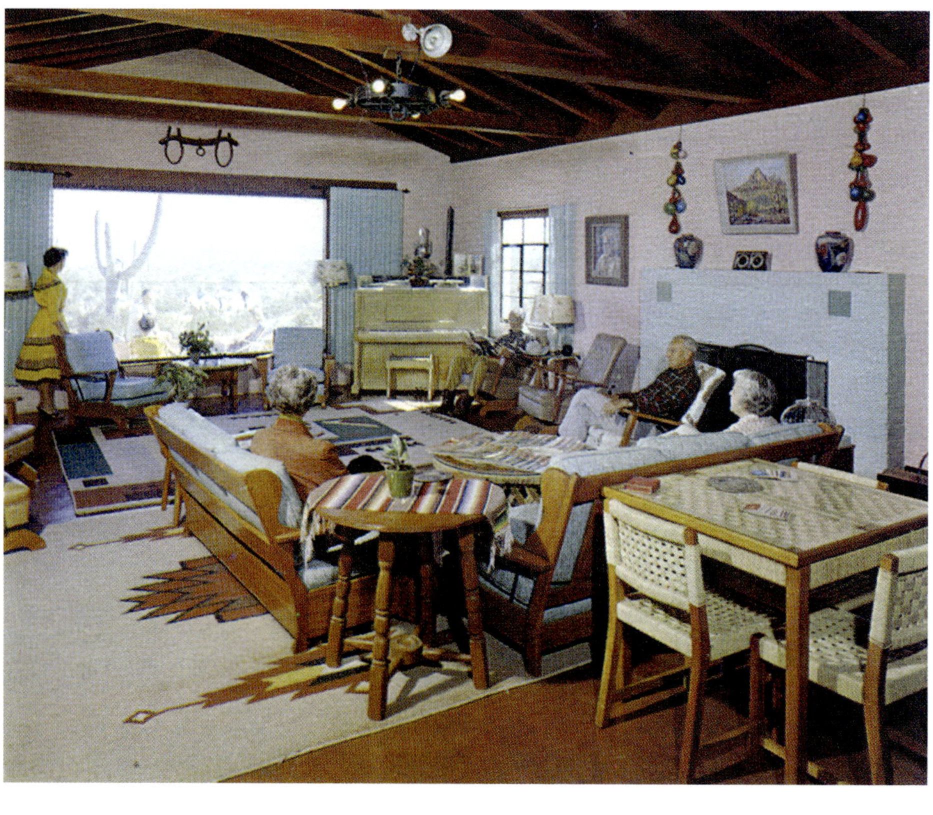

FIGURE 5.29. Brochure detail. Lazy K Bar Ranch living room (blue drapes) with picture window.

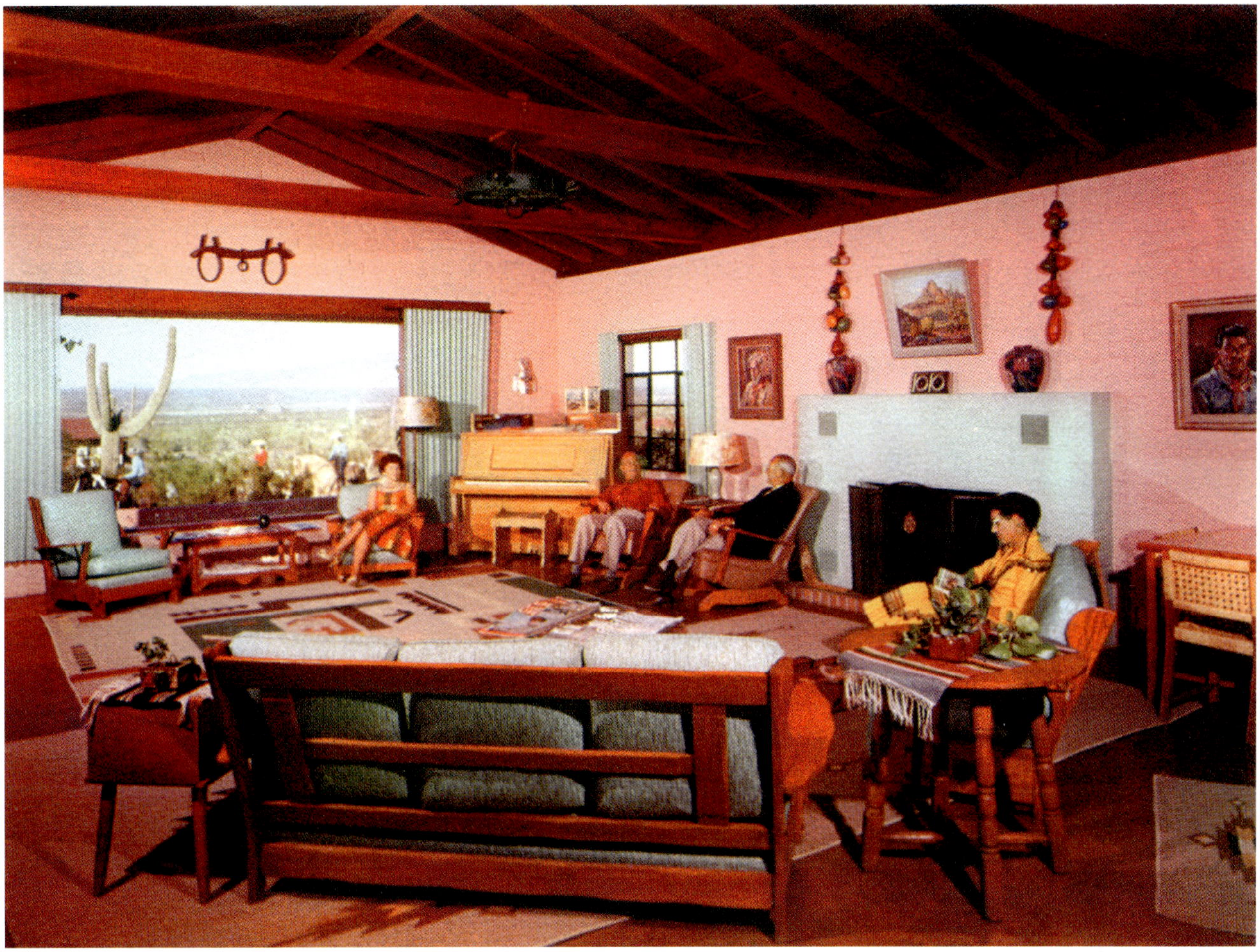

FIGURE 5.30. Brochure detail. Lazy K Bar Ranch, living room (pink walls) with picture window.

"Cactus" Herb Wood, a Pennsylvania stone mason, retreated to Tucson in 1931 after a years-long recuperation from a back injury. He purchased a rock-strewn lot where he built a house of stone. With little money left after its completion, Wood improvised furniture from saguaro and cholla wood. Wood's creations attracted requests from friends and acquaintances. His wife christened the furniture, "Cacticraft." They opened a roadside stand with a workshop on the side (Arizona Highways 1946; Family Circle n.d. [1947?]; Holiday 1947a). Wood stockpiled an impressive collection of saguaro ribs with which to work (Figure 5.34).

Many rib crafters became quite adept at plumbing the taproots of the dried plants for ever-widening widths of saguaro wood with which to work. William F. Bucher, a Washington, D.C., cabinetmaker, compiled a collection of tree photographs framed with the wood so pictured (Figure 5.35). Bucher's collection entitled, "Our Trees and Their Woods," attracted the attention of the U.S. Forest Service, from whom Bucher obtained photos, and officials of the Smithsonian

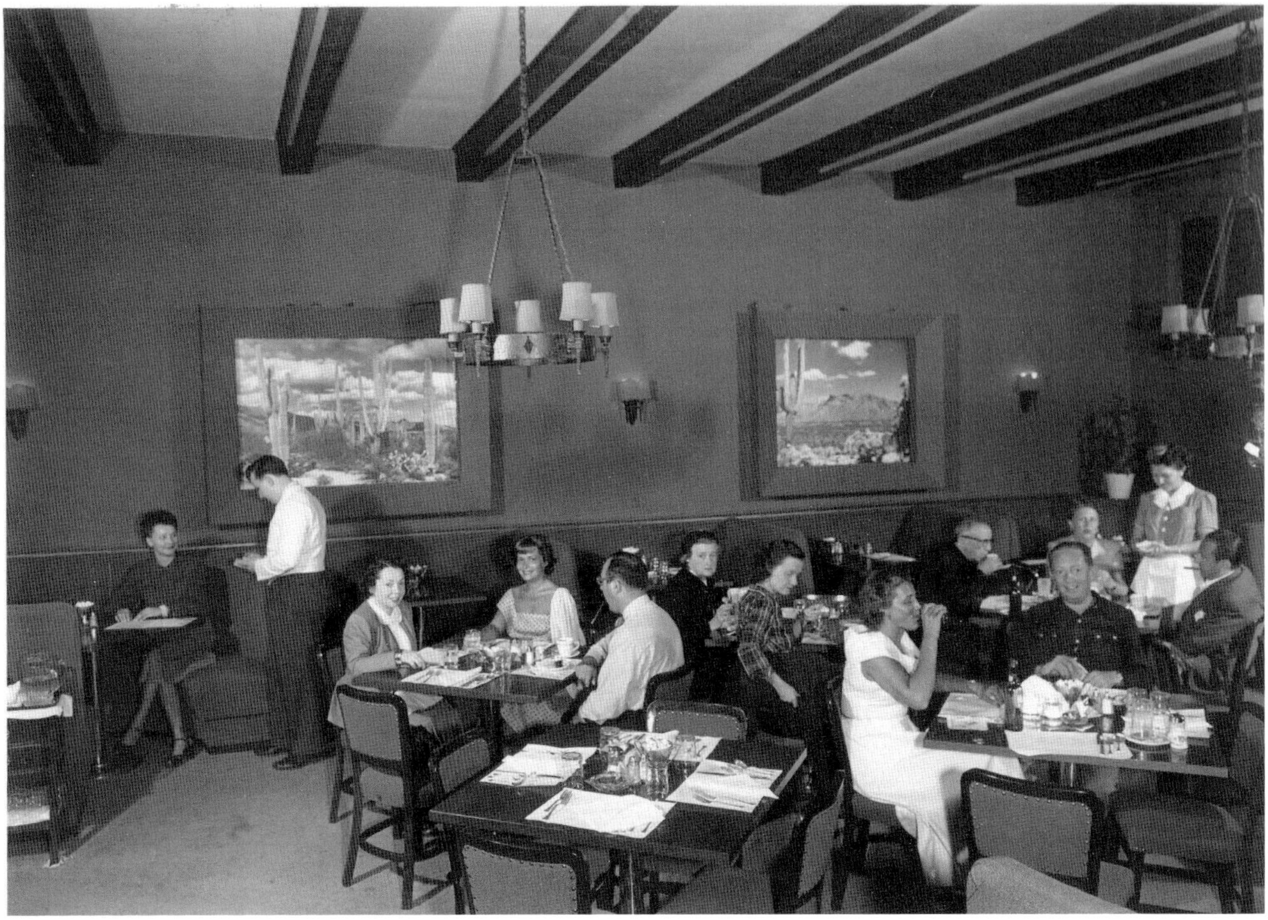

FIGURE 5.31. El Presidio Hotel Mural Room Restaurant, Tucson, Arizona.

Institution, who exhibited his exquisitely framed collection of 59 tree photographs. The collection featured the "Giant Cactus," "Cholla," "Mesquite," and "Monterey Cypress." The goal, Bucher stated, was to display the "beauty and utility of trees." The Forest Service contributed a photo of the world's largest saguaro at Saguaro National Monument. The ribs for the frame were as remarkable for their width as the beauty of their natural finish. Bucher started the collection in 1915 and placed it on loan for a long term exhibition at the United States National Museum in 1932 (Watkins 1911; Bucher n.d. [1932?]; United States National Museum 1933:88).

The cactus craft of the Guy Wellington Knauff family both capitalized upon and furthered saguaro iconography in miniature desert dioramas. Merchandised as a decorative souvenir for the home, each diorama rendered a three-dimensional saguaro moment under glass framed with saguaro ribs.

Charles Herbert and camera followed Knauff and his family into the desert to document the production process that began with a rib harvest. Herbert photo-

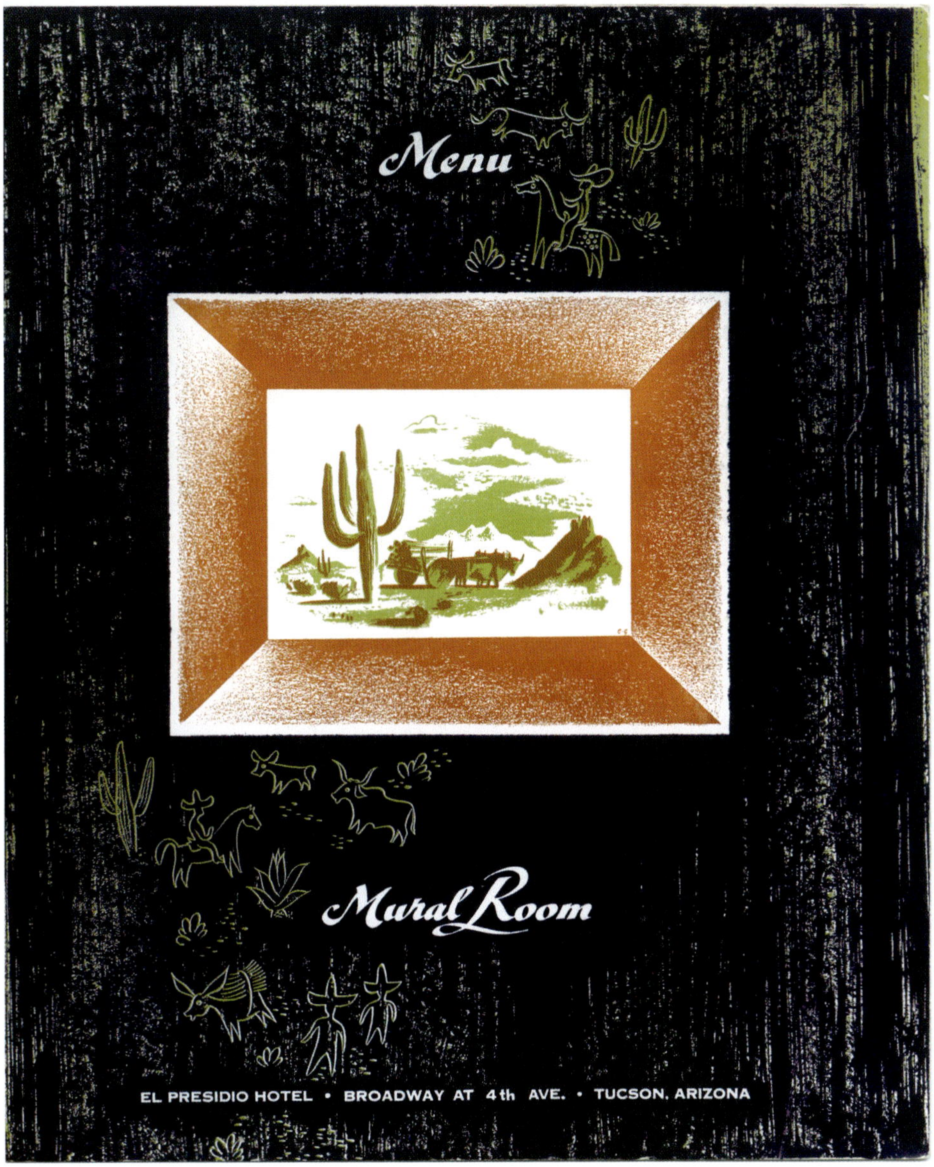

FIGURE 5.32. Menu cover, El Presidio Hotel Mural Room.

graphed the final stages of production in his backyard saguaro rib *ramada*, where he posed the Knauffs painting Santa Catalina mountain backdrops on concave cardboard shells brought to life with natural grasses and tiny clay saguaros wrapped around wire frames (Figure 5.36). The forced perspective of the road receding into the Catalinas rendered a pleasing facsimile of the Sonoran Desert (Figure 5.37). The Knauff's dioramas achieved their effect despite the fact that all but one family member was color blind. The Knauffs sold their dioramas to department stores and gift shops under the name of the Landscape View Company. At the time of Guy Knauff's death in 1947, the family had completed its 50,000th diorama. In 1956, his widow, Mary A. Knauff, sold the company to Homecrafters of Tucson, which employed the physically disabled. In the course of Mary Knauff's life, she, her husband, and family oversaw the creation of 80,000 desert dioramas that hung on walls in homes from Maine to California and beyond (Arizona Daily Star 1947, 1956; Tucson Daily Citizen 1947a; Rydell n.d. [1951?]).

To appreciate the take-off point of the saguaro's iconic legacy, one need only drive a short distance from Tucson to Saguaro National Park, in reality two parks that bookend the city to the east and west. To the west, over Gates Pass, lies Old

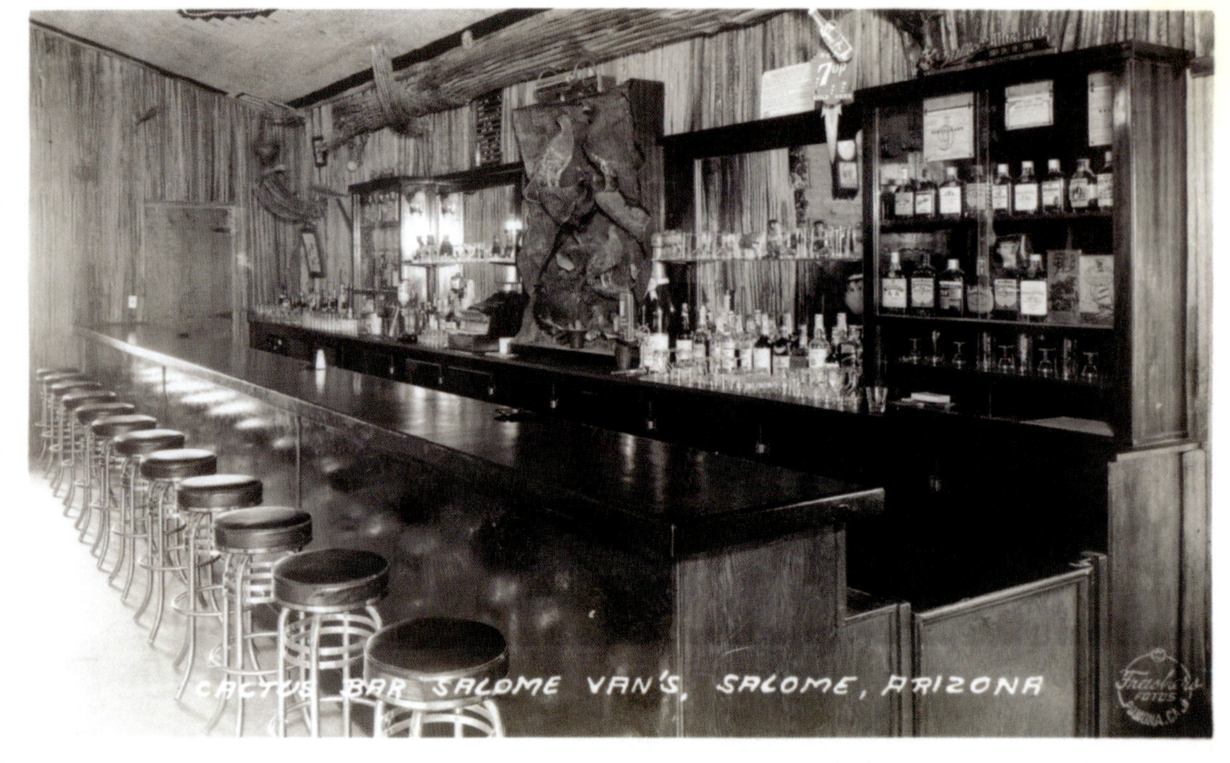

FIGURE 5.33. Postcard. Cactus Bar, Salome Van's, Salome, Arizona. Published by Frasher's Fotos, Pomona, California, ca. 1939.

Tucson, the western theme park cum stage set built for the 1939 theatrical film "Arizona," and just beyond that, the state-of-the-art environmental exhibits of the Arizona-Sonora Desert Museum that began as a modest trailside museum in 1952. On the east end of town Harold Bell Wright's home still stands on the mesa much as he left it. After the Second World War, the surrounding acreage was subdivided into Harold Bell Wright Estates. The Zuni Dining Room at the Desert Sanatorium was enveloped by the sprawling Tucson Medical Center. Modest decorative elements from the Zuni Dining Room may be seen today in the hospital's art-filled halls. The dining room's wooden beams live on above the drop ceiling of the outpatient pharmacy.

The march of development cashiered the view from the lounge of the Lazy K Bar Guest Ranch. The saguaro pictured in successive versions of the ranch's brochures passed long before the building did in 2019. In a similar instance of now-you-see-it, now-you-don't, the chandelier saguaro around which Charles and Lucile Herbert framed their living room window toppled over and died in 1995. The Herberts' saguaro rib *ramada*, the setting for feature news and photo stories including the Knauff's, fell into disrepair and scattered to the winds. Only the concrete footings remain.

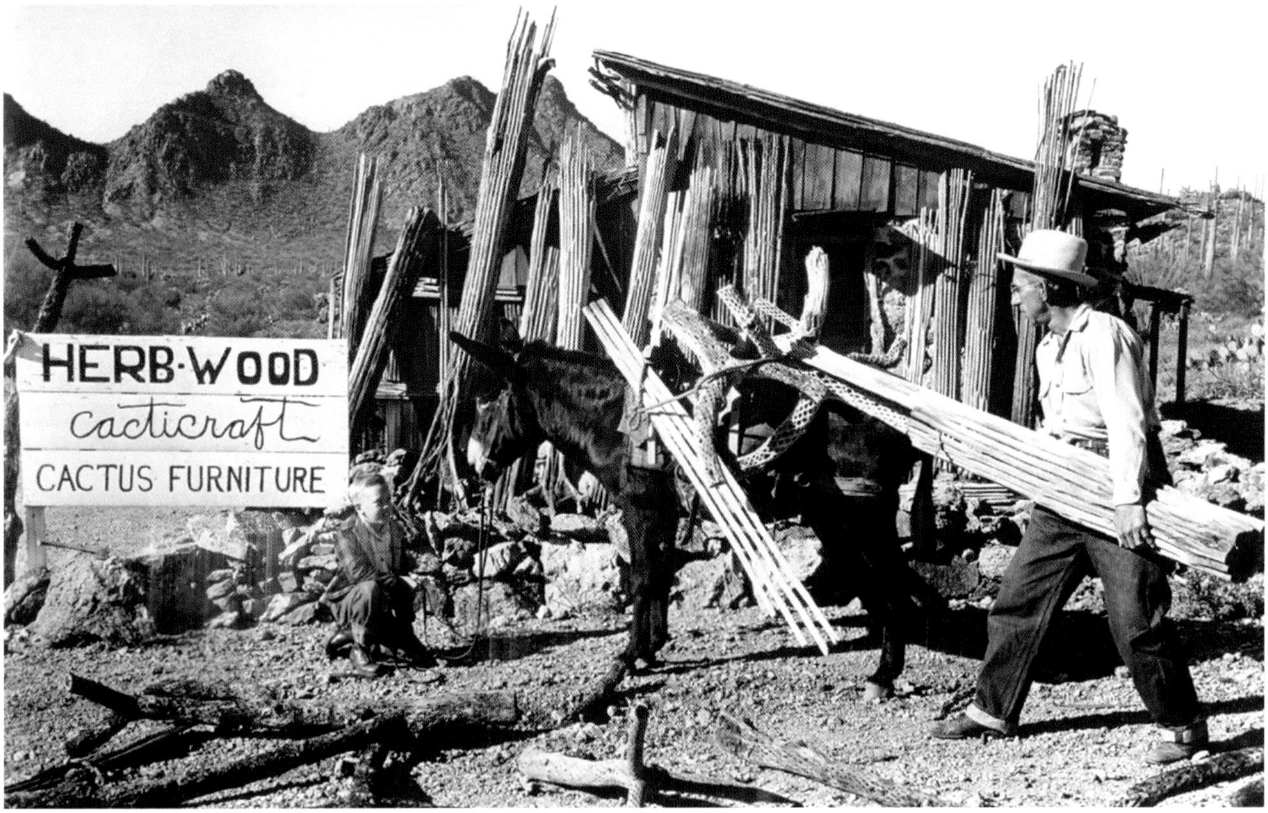

FIGURE 5.34. Realphoto advertising card. Herb Wood Cacticraft Cactus Furniture, Tucson, Arizona, ca. 1947.

On the campus of the University of Arizona, a token cactus garden surrounded by a greenscape with vistas of buildings is all that's left of the era in which cactus-or-not was a question and the framing of life with saguaros was an ongoing project. Leaving office in 1945, university president Alfred Atkinson's vision of a cactus-free campus had largely failed. Yet the pressures of development remained. With the construction of a new student union building in the early 1950s, the cactus garden in front of it—in the trench on the east side of Old Main—became a lawn. The easternmost portion of the arroyo remained for the time being (Figure 5.38). In 1952, a wanton act of vandalism cut down three of the garden's largest saguaros and uprooted other specimens (Figure 5.39). The perpetrator or perpetrators were never found (Tucson Daily Citizen 1952; Rowley 1984). In later years, university officials reduced the garden's footprint to its present configuration as an island across from the university's administration building. In 1980, University President John Shaeffer dedicated the garden to the memory of author and naturalist Joseph Wood Krutch.

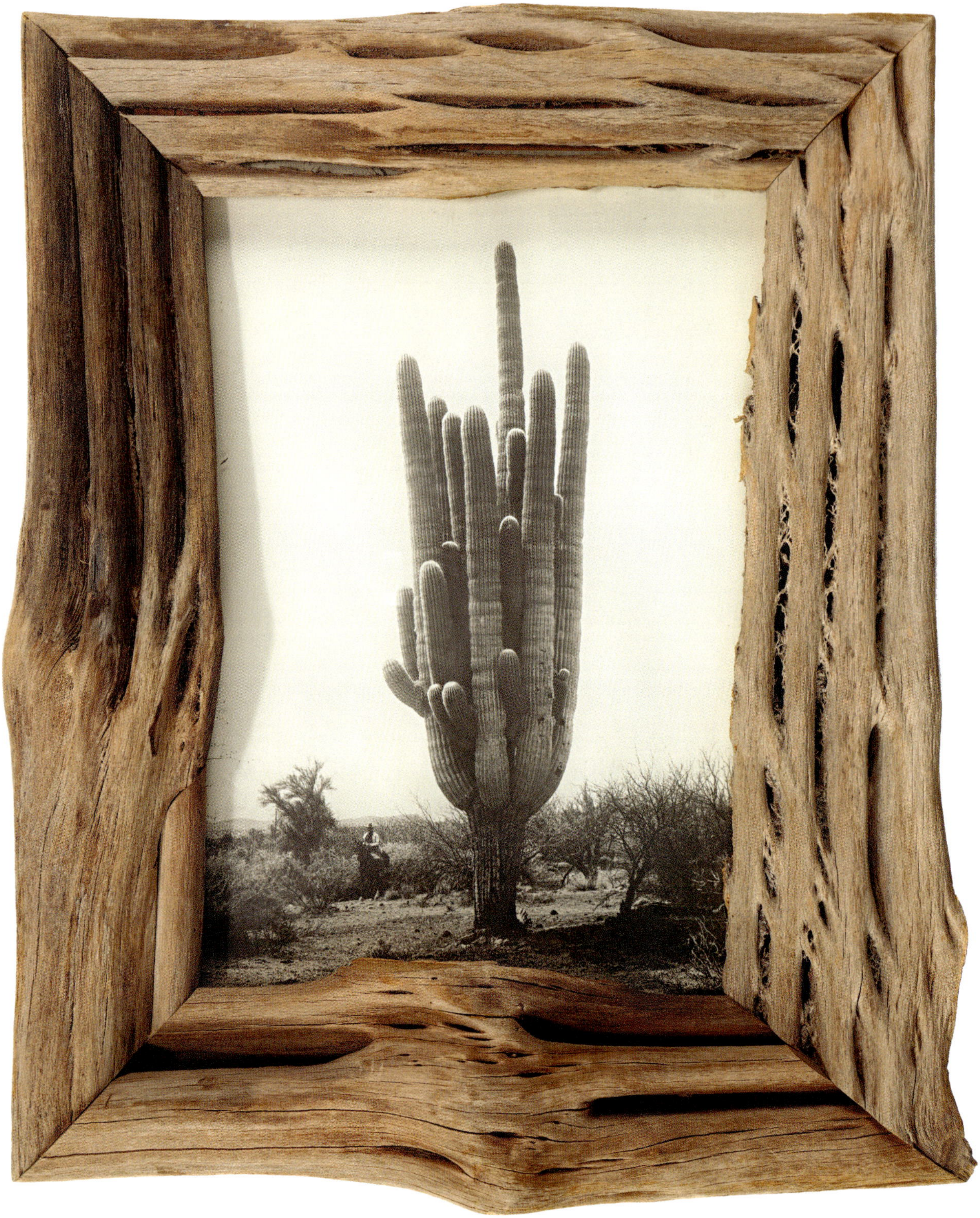

FIGURE 5.35. Photograph mounted in saguaro rib frame. Photograph by U.S. Forest Service, ca. 1930. Frame by William Bucher, 1932.

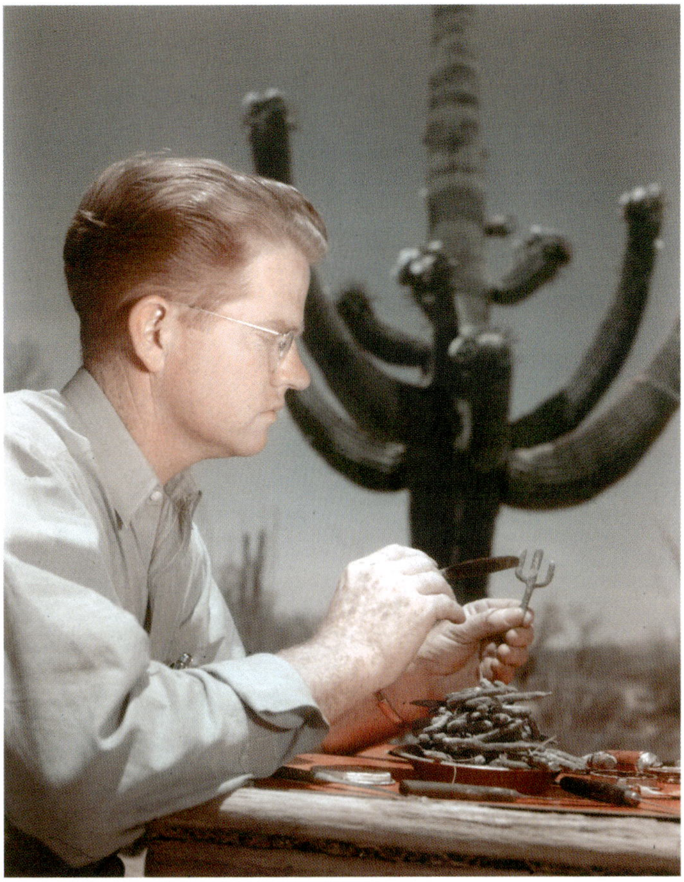

FIGURE 5.36. Color transparency. Knauff man painting miniature saguaro. 1947. Photograph by Charles Herbert.

FIGURE 5.37. Photograph. Completed Knauff saguaro diorama under glass. The Knauff's shipped desert dioramas to Scotland, London, Denmark, France, Sweden, Hawaii, Alaska, and Canada.

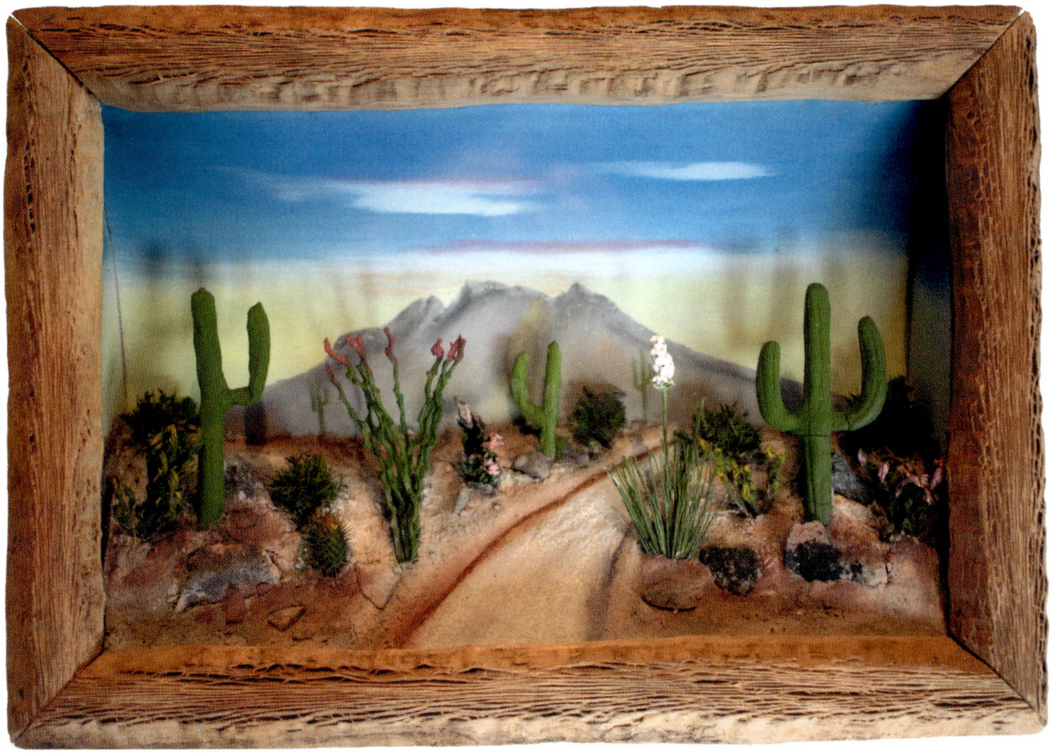

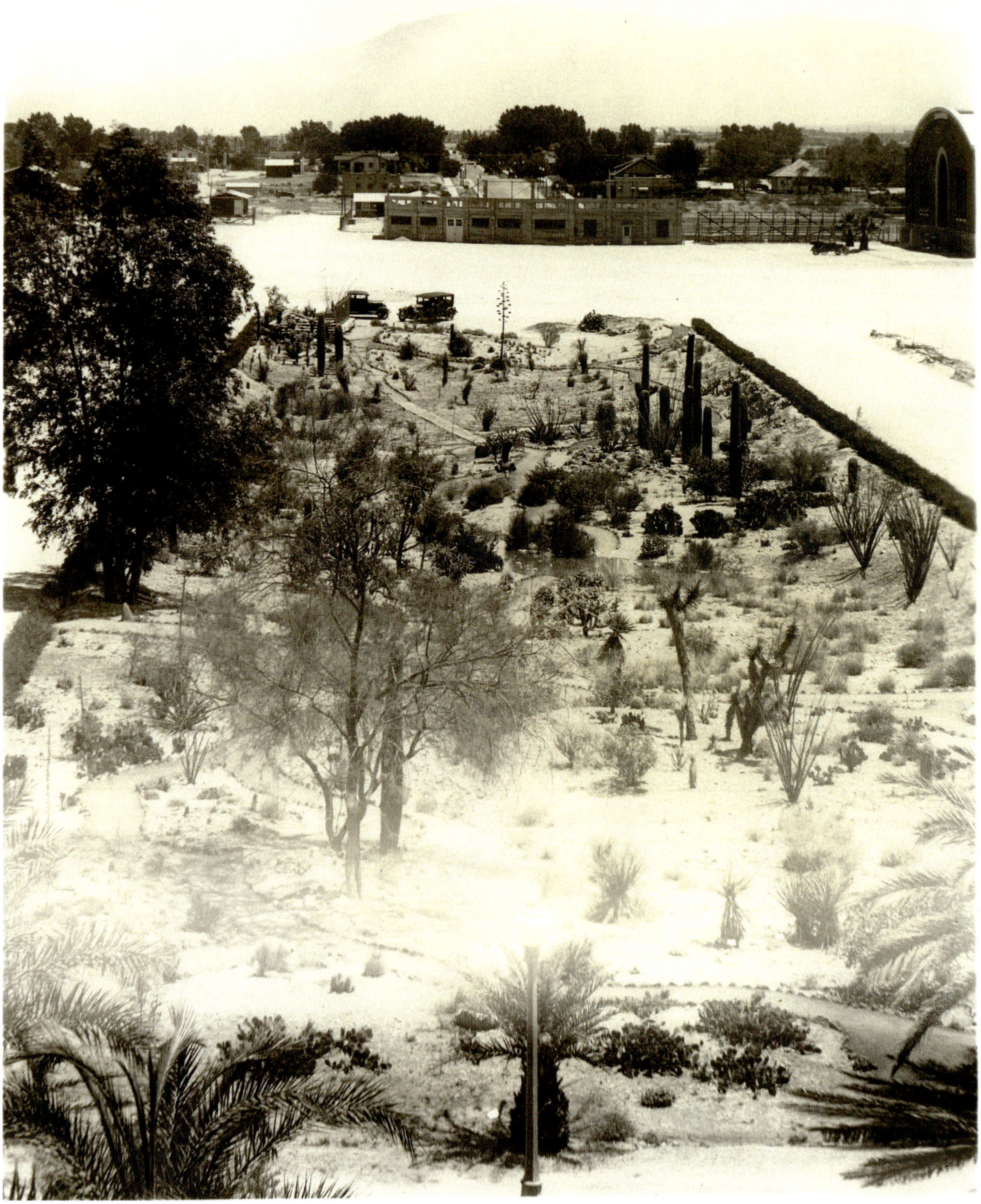

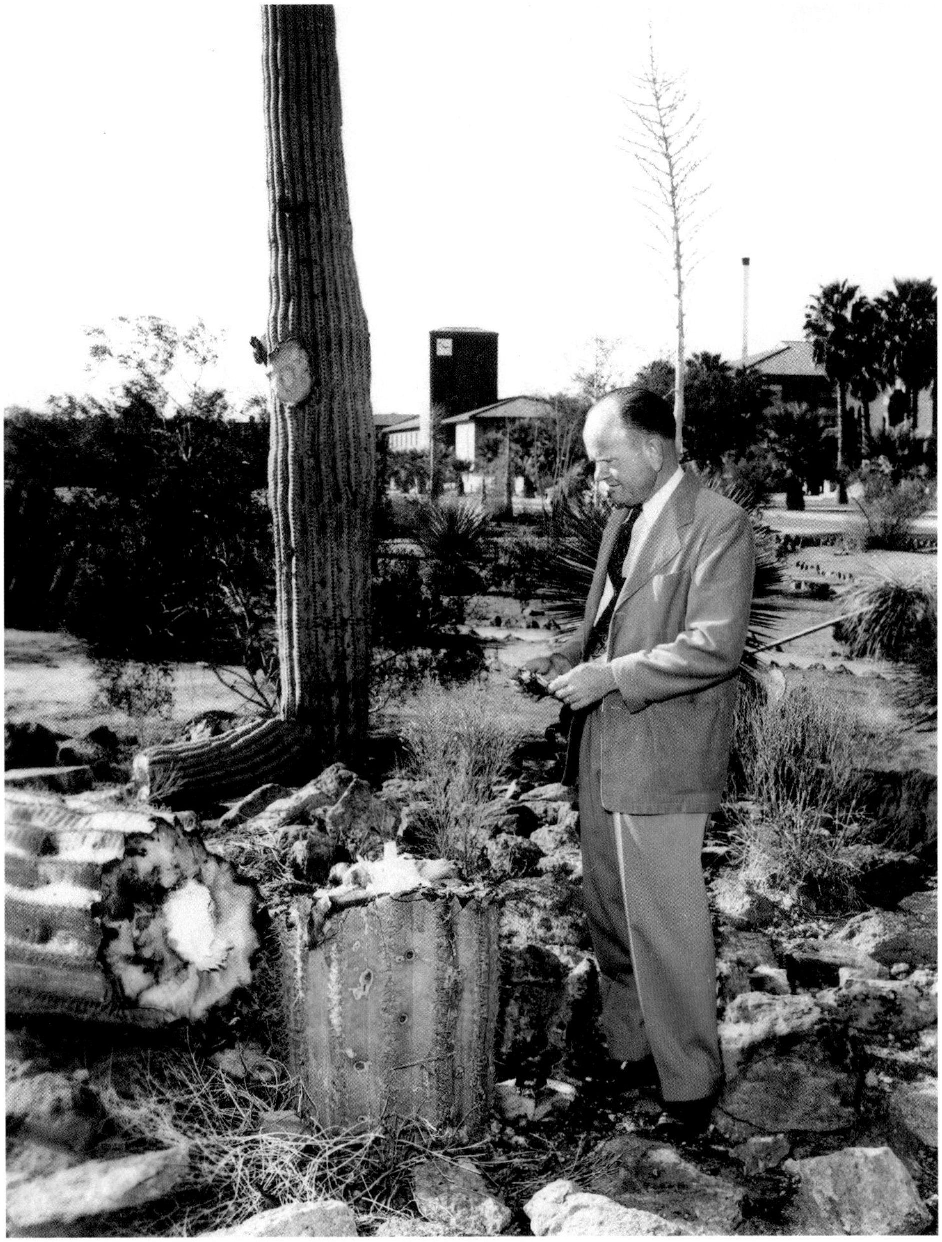

FIGURE 5.39. Copy photograph. Unidentified man inspecting vandalized saguaros, campus cactus garden, University of Arizona, Tucson, 1952.

FIGURE 5.38 (LEFT). Copy photograph, Campus cactus garden arroyo east of Main building, University of Arizona, Tucson, ca. 1937.

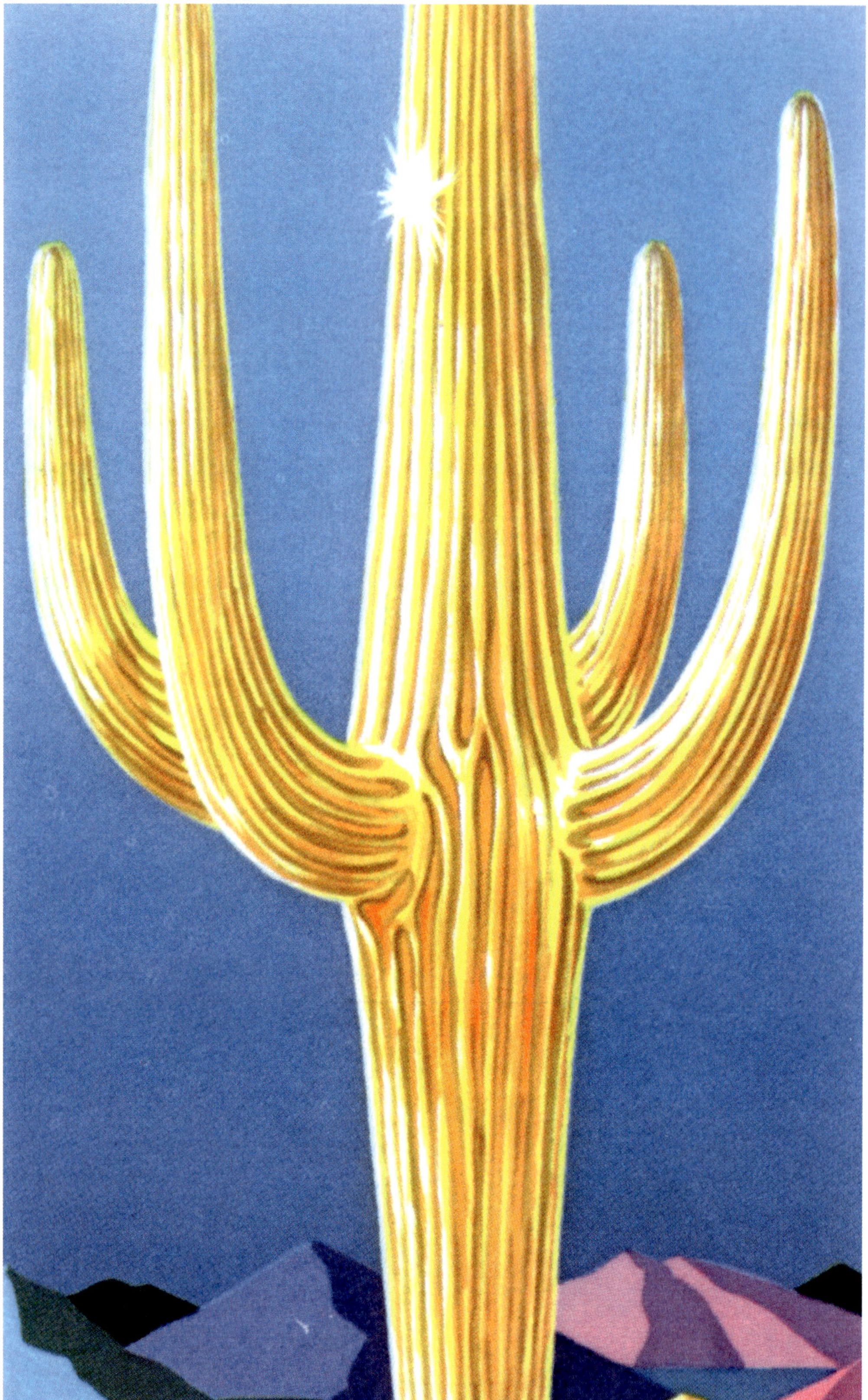

CHAPTER SIX
LIBERTIES TAKEN

From the flatlands of Phoenix to the foothills of Tucson, travel and tourism literature struggled with the saguaro's symbolic place. Hard to ignore among the attractions of a transcontinental picture portfolio, saguaros approached graphic parity with the orange groves of Phoenix's Valley of the Sun. But questions remained. Should the saguaro be on the spine of a brochure, opening to show its majestic self—or lurk behind the fold of a fruit-bearing orange tree in an irrigated field? Less a problem for photography than the plastic arts of commercial illustration, the saguaro temporized slightly off-center on the covers' edge (Figures 6.1 and 6.2).

Railroad passenger and guest ranch advertising popularized the idea that the desert was not just something to ride through, but a worthy destination to ride to. Each had an interest in deploying saguaros with which members of the traveling public could become comfortable, wrap up in, and even sleep with—ideas reinforced with graphic business stationery and room décor (Figures 6.3 and 6.4).

Competing for east-west passenger traffic, the Santa Fe and the Southern Pacific railroads opted for native botany in resort settings. Published figures for railroad publicity became available for analysis in the advertising trade press of the early 1920s. In terms of graphic excellence the advertising output of the Santa Fe and its concessionaire Fred Harvey Company dwarfed that of the Southern Pacific. The Santa Fe enjoyed long experience in merchandising the Grand Canyon and packaged Indian Detours. For its part the Southern Pacific focused upon California destinations like the Hotel Del Monte (Hungerford 1923; Leininger 1923; Thomas 1978; d'Emilio and Campbell 1991; Howard and Pardue 1996, 2016; Zenga 2001; Orsi 2005:145, 159–160).

The liberty taken by the Santa Fe in picturing saguaros on the periphery of its operations in northern Arizona suggests how the cactus spawned a commercial range well beyond its native habitat. In one example, a Santa Fe brochure deployed a saguaro that to the knowing eye appeared as an anomaly in the mesas and purple shadows of the railroad's New Mexican palette. Beneath the saguaro's towering arms, vacationers relax in lounge chairs on a lawn. Others ride horseback in the distance (Figure 6.5; Weigle 1992; Snead 2002).

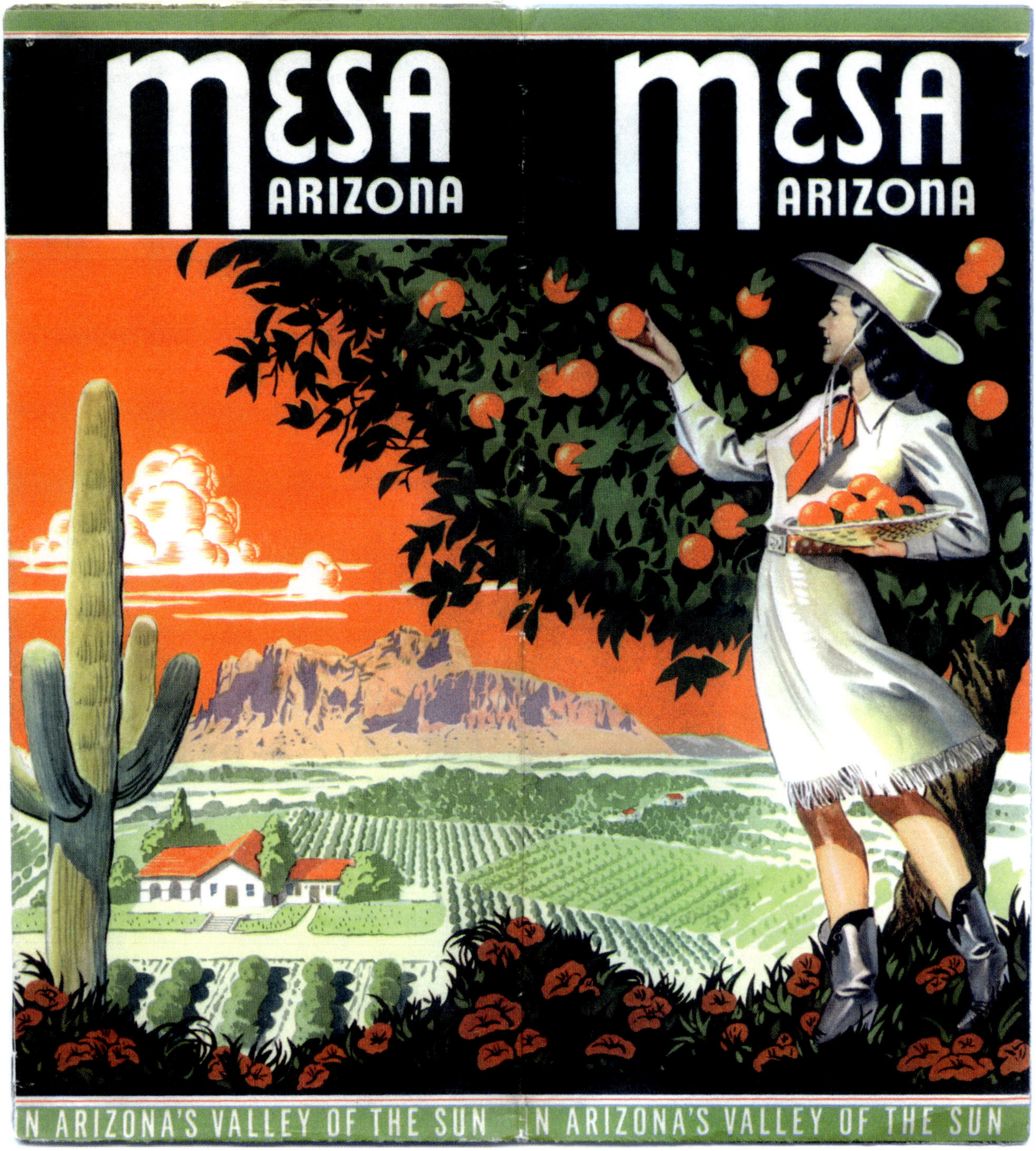

FIGURE 6.1. Brochure cover. In Arizona's Valley of the Sun. Mesa Arizona Chamber of Commerce, ca. 1940.

Unlike the Santa Fe, whose passenger line ran through Flagstaff, the Southern Pacific (SP) cut through the saguaro-laden Gila River Valley acquired by the Gadsden Purchase. Yet SP passenger advertising came late to the exploitation the saguaro, as there was little need for it. The railroad's ticket structure promoted through-train service rather than stopovers in Arizona—with the notable exception of "motor side

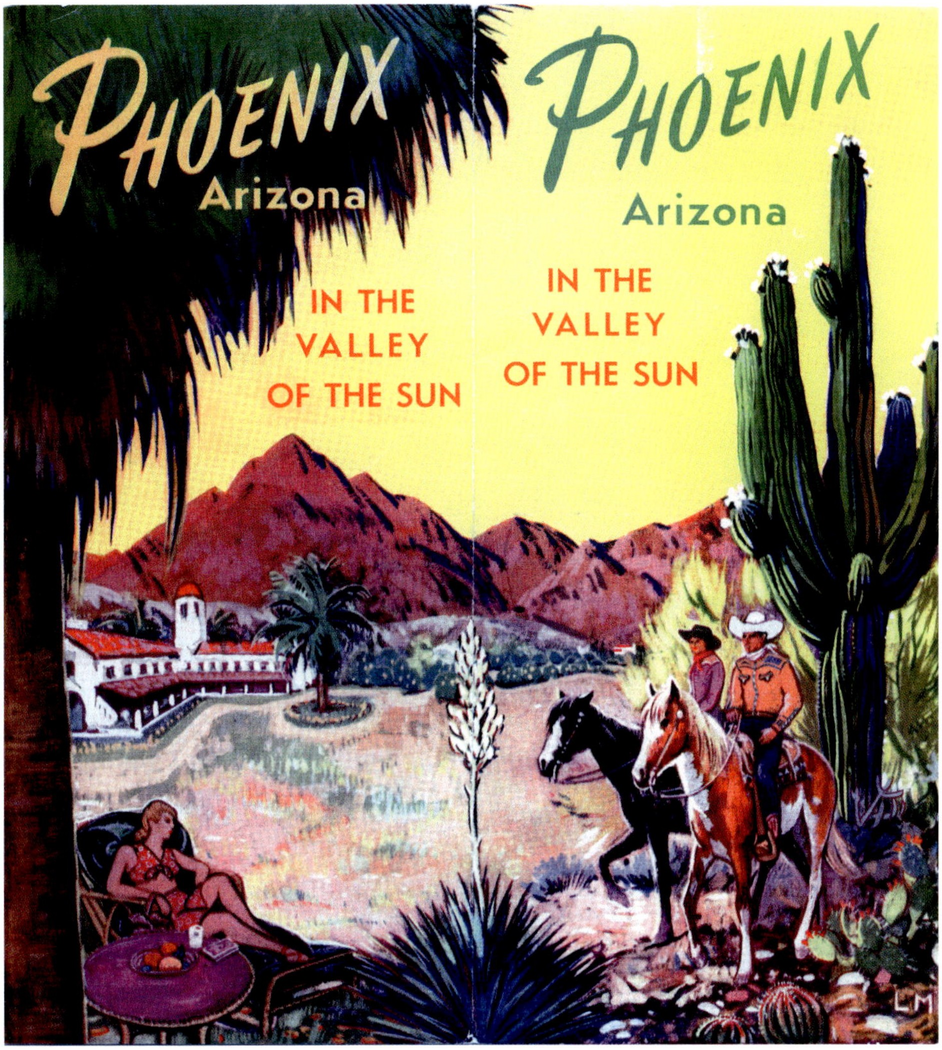

FIGURE 6.2. Brochure cover. "Phoenix Arizona in the Valley of the Sun," ca. 1954.

trips" from Phoenix to Roosevelt Dam along the unpaved construction road marketed as the Apache Trail (Figure 6.6). SP booklets did picture saguaros as one of the many attractions on its transcontinental routes operated with the Rock Island Railroad and the SP's premiere train, the Sunset Limited (Figures 6.7 and 6.8; Ingram 1927, 1929; Southern Pacific Lines 1928; Zimmerman 1985; Runte 1991). Not until the late 1920s does it appear that SP and Rock Island passenger advertising directed attention to Arizona hotels and guest ranches as a matter of policy (Figure 6.9).

In 1939, the Phoenix Junior Chamber of Commerce commissioned a prototype for city street lights cast in the accordion-fold form of a mature, fully armed saguaro (Figure 6.10). The fluted column and two arms of the steel structure, each holding a streetlamp, was the brainchild of *Arizona Republic* cartoonist Reg Manning, who sketched the design in an apparent effort to overcome his city's half-lit indifference

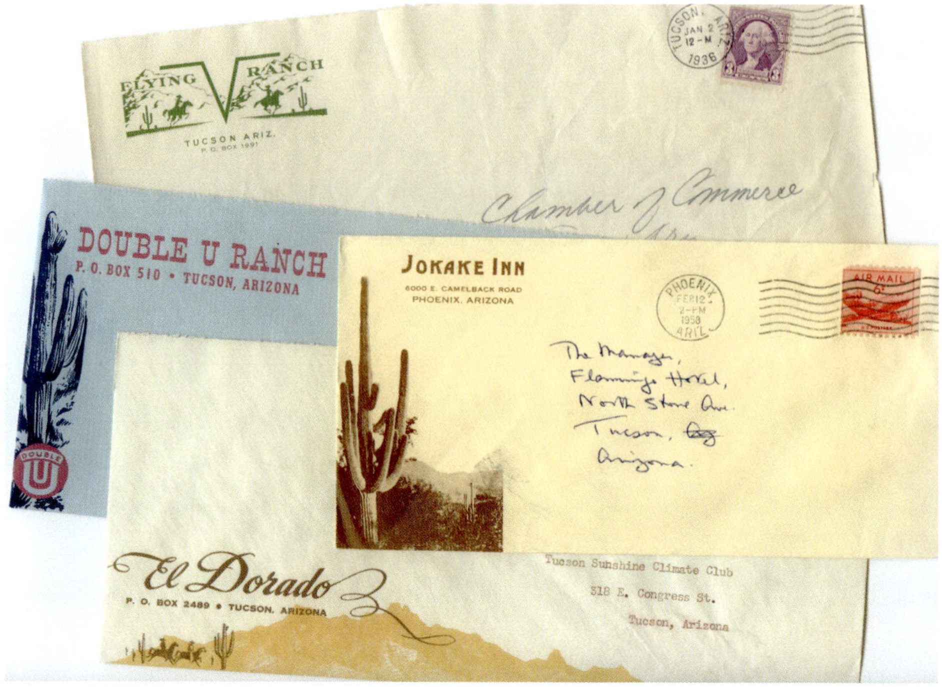

FIGURE 6.3. Envelope covers. Guest ranch cactus graphics. 1936–1958.

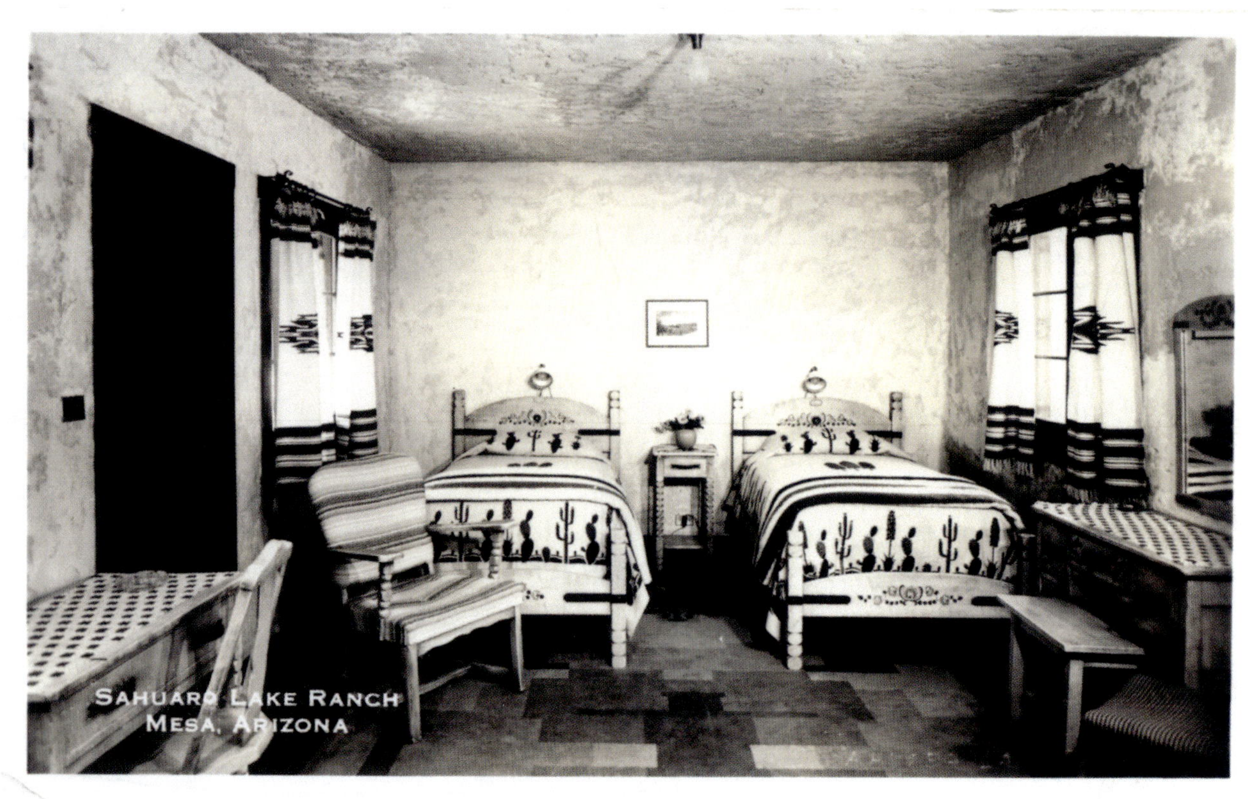

FIGURE 6.4. Realphoto postcard. Saguaro Lake Ranch, Mesa, Arizona. Postmarked 1945.

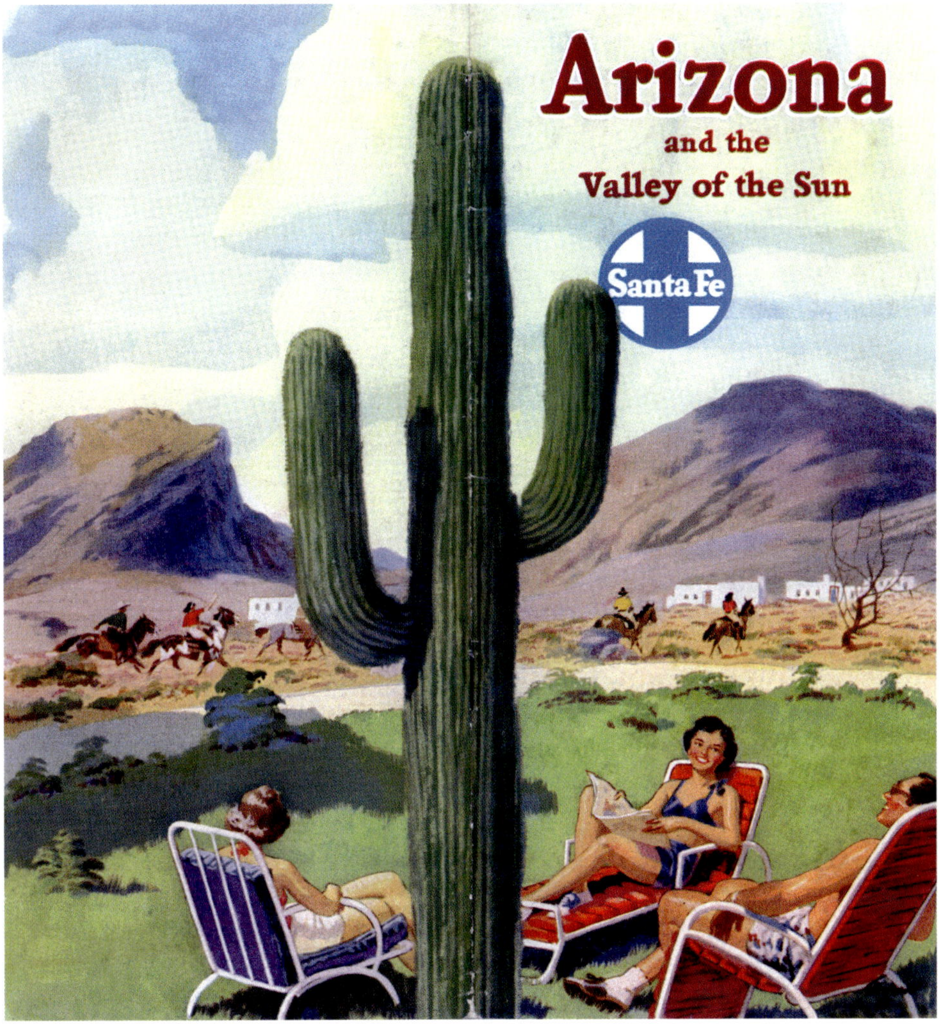

FIGURE 6.5. Brochure cover. "Arizona and the Valley of the Sun," picturing saguaro across the spine of the open brochure. Santa Fe Railroad, 1950.

to the plant. An opinion survey conducted by the Jaycees confirmed the public's desire for improved street lighting, but not Manning's saguaro design, on which opinion was evenly divided. Immortalized by Farm Security Administration photographer Russell Lee, Manning's saguaro street light developed a following if only among the Jaycees, who moved it whenever they relocated their office about the city. Never giving up on the idea of a cactus-themed streetscape, Manning suggested that barrel cactus replicas be outfitted with lamps to illuminate pedestrian crossings on busy street corners and at the airport landing strip (Figure 6.10; Manning 1939; Arizona Republic 1940, 1948; Fuller 1960; Unger 1964).

Manning's lighthearted proposals suggest how staging and manipulation of the desert habitat could be accomplished with clever art, illustration, and photomontage. One of the period's more noteworthy examples of the photomontage technique included a booklet that described the construction of the Colorado River

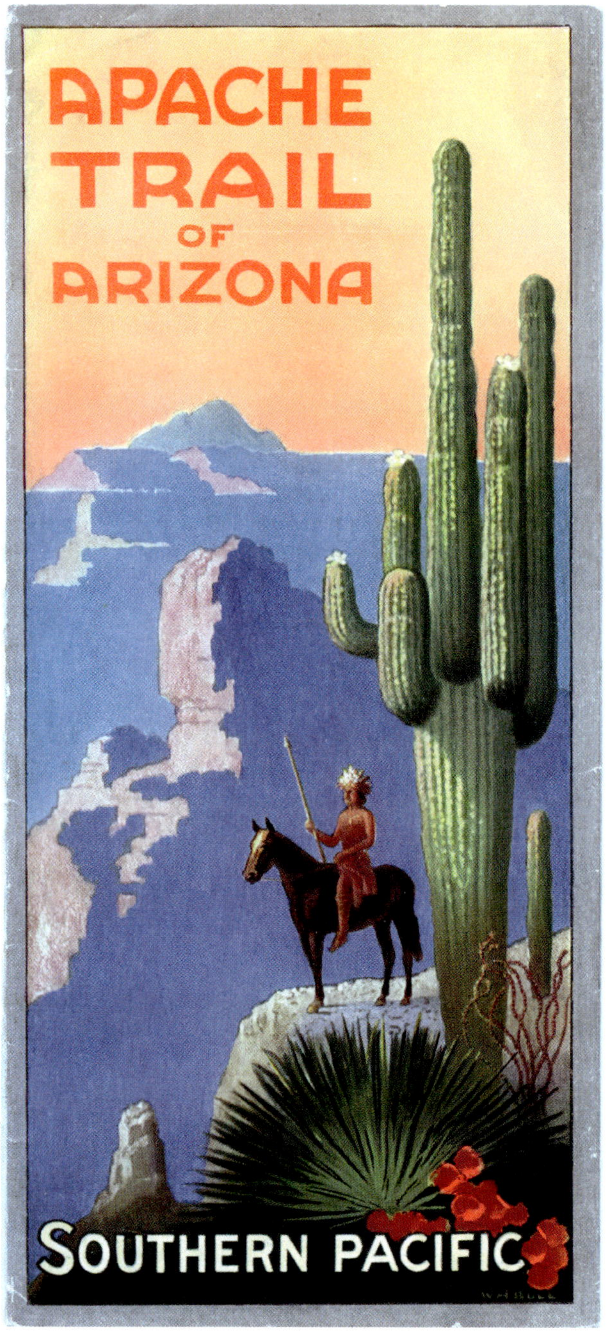

FIGURE 6.6. Pamphlet cover., by artist William Howard Bull. "Apache Trail of Arizona." Southern Pacific Railroad, 1924.

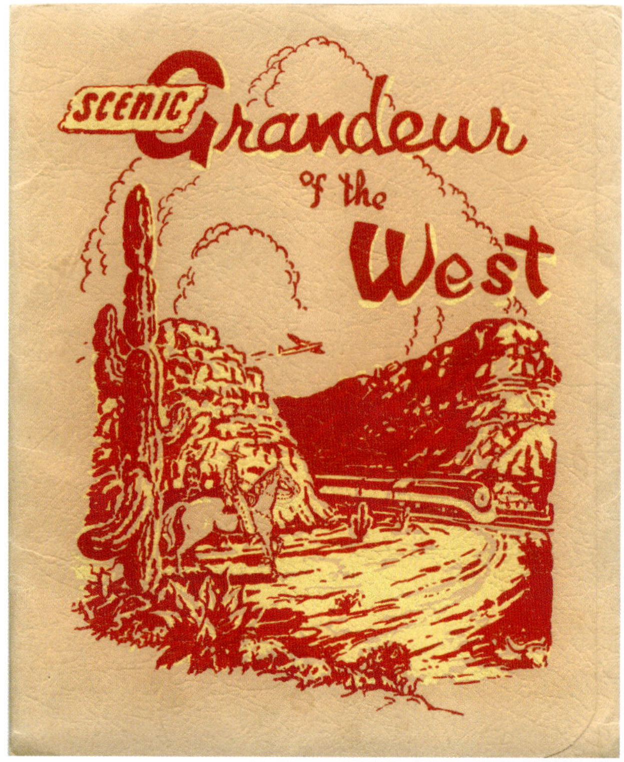

FIGURE 6.7. Portfolio cover. "Scenic Grandeur of the West." Southern Pacific Railroad, 1943.

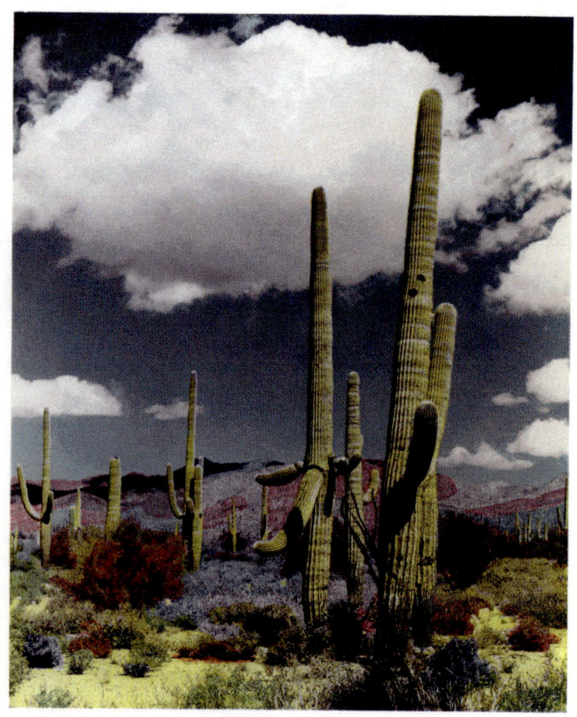

FIGURE 6.8. Photoengraving. Giant Cactus, Arizona, from portfolio "Scenic Grandeur of the West." Southern Pacific Railroad, 1943.

Aqueduct (Figure 6.11). Published by the Metropolitan Water District of Southern California, the booklet was part of a regional public education campaign that deployed newspapers, magazines, films, and exhibits to win and maintain political support for the aqueduct (Los Angeles Times 1931; San Bernadino County Sun 1931). Entitled, "Water for Thirteen Cities," the booklet linked saguaros and downtown Los Angeles in a photomontage. The water district prepared the booklet for distribution at the 1935 San Diego Panama Pacific International Exposition. There the water district's exhibit featured an operating diorama of the aqueduct. On cue, the diorama flooded with water, dramatizing the pumping of the Colorado River from an intake dam then under construction at Parker, Arizona. The water raced through open conduits and closed tunnels to southern California cities concluding at coastal Los Angeles (Los Angeles Times 1935). The photograph of the two stately saguaros on the booklet's title page received equal graphic treatment in the airbrushed montage that seamlessly connected the desert with a bird's-eye view of downtown Los Angeles. The photomontage of saguaros, elevated passenger trains, tall buildings, cars, trucks, and pedestrians might have seemed odd at first. For one, it did not present the saguaros as found objects in an alfalfa field or on the margins of an orange grove. Rather, as stately symbols of southern California's salvation and future water supply. Though no one complained about the saguaro's symbolic appropriation across state lines, the water was another matter (Starr 1993).

The tipping-in and pasting-up of saguaro montages became something of an art in the brochures and booklets issued by the travel and tourism boosters of southern Arizona. The Phoenix Chamber of Com-

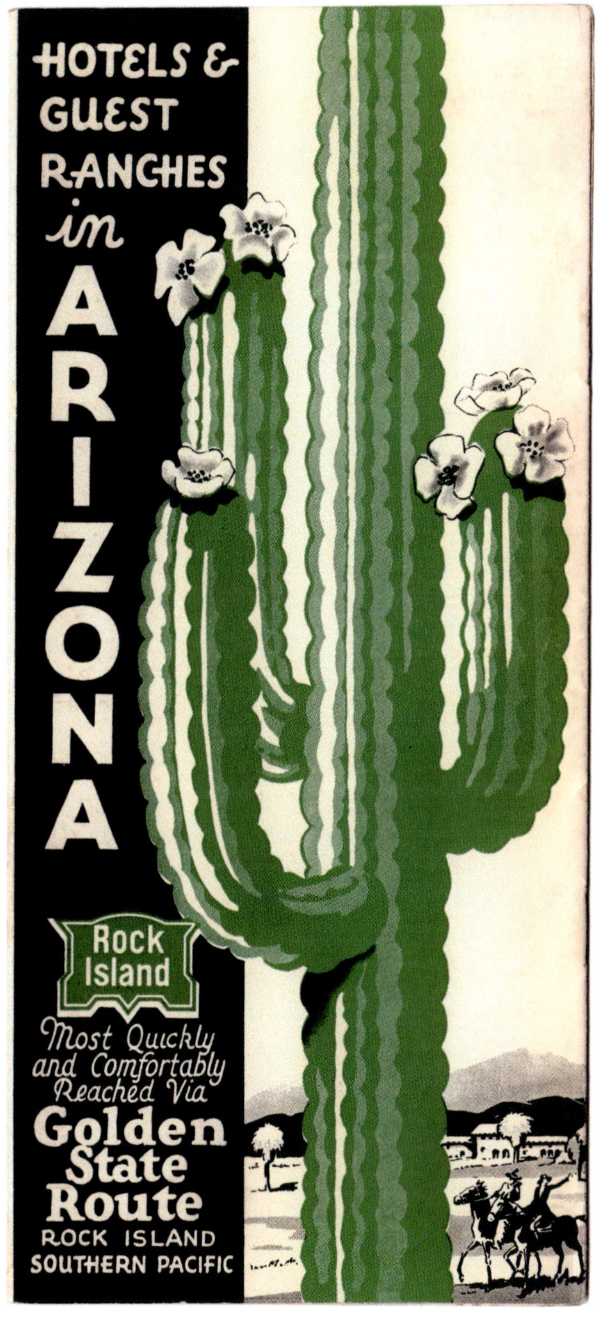

FIGURE 6.9. Brochure cover. "Hotels and Guest Ranches in Arizona." Rock Island Railroad, 1933. A guide to hotels and guest ranches featuring an abstract flowering saguaro. The Rock Island Railroad ran passenger trains from Chicago through Tucson and on to California with the Southern Pacific Railroad.

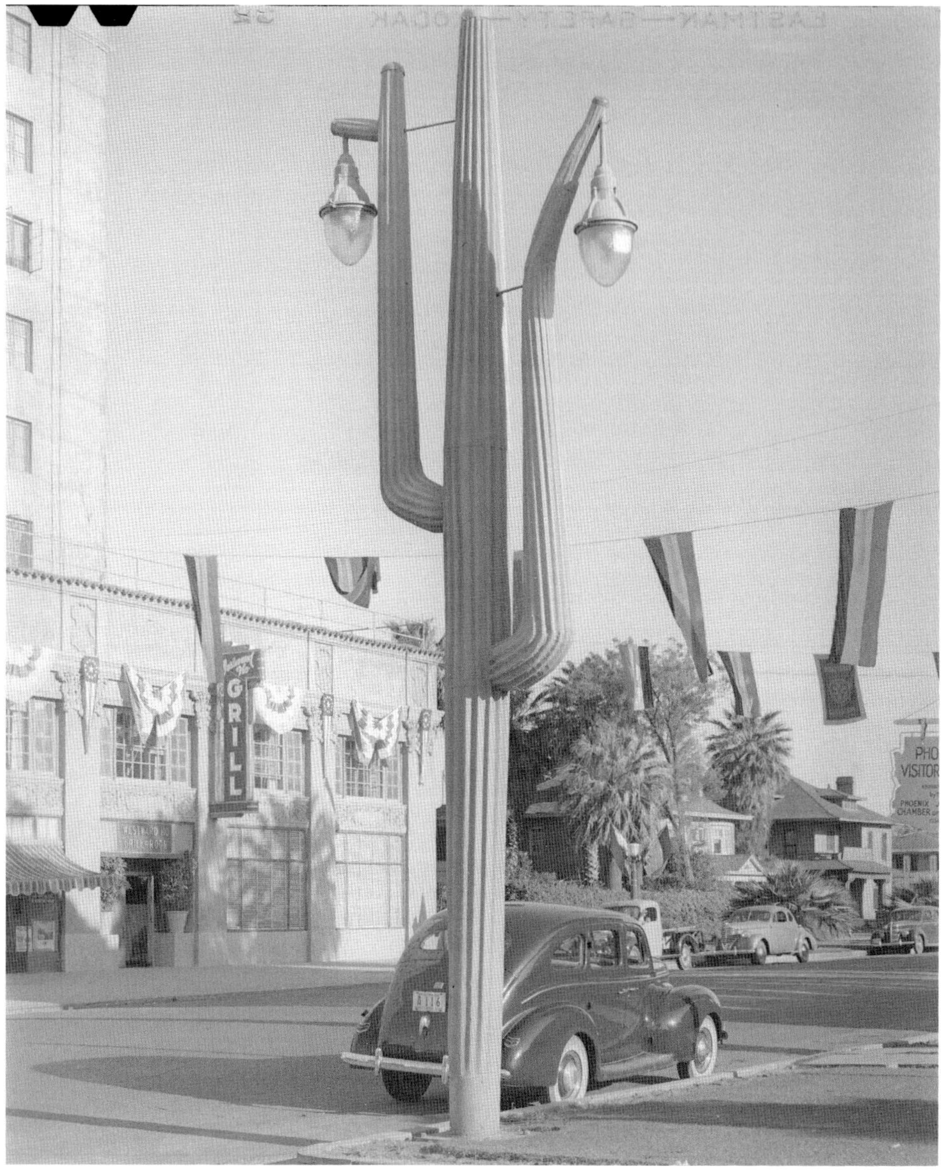

FIGURE 6.10. Photograph by Russell Lee. Cactus light standard in front of hotel in Phoenix, Arizona, 1940.

merce, for example, took liberties with a saguaro photographically rendered as a signboard for the cover of its 1947 accommodations guide. A model seated on a horse burdened with sporting goods posed with a saguaro pasted up with wayfinding arrows for golf, tennis, riding, sun bathing, swimming, and loafing (Figure 6.12). An alternate version moved the arrows around on the plant with a different model and sporting goods astride the same horse (Figure 6.13). Apparently no saguaros were harmed in posting the arrows, given the apparent skill of the art director that may have led some to believe that the cactus had been posed in such a way.

The wholesale photo manipulation of a Tucson golf course suggested that the saguaro's commercial range was becoming as expansive as its accordion folds. In

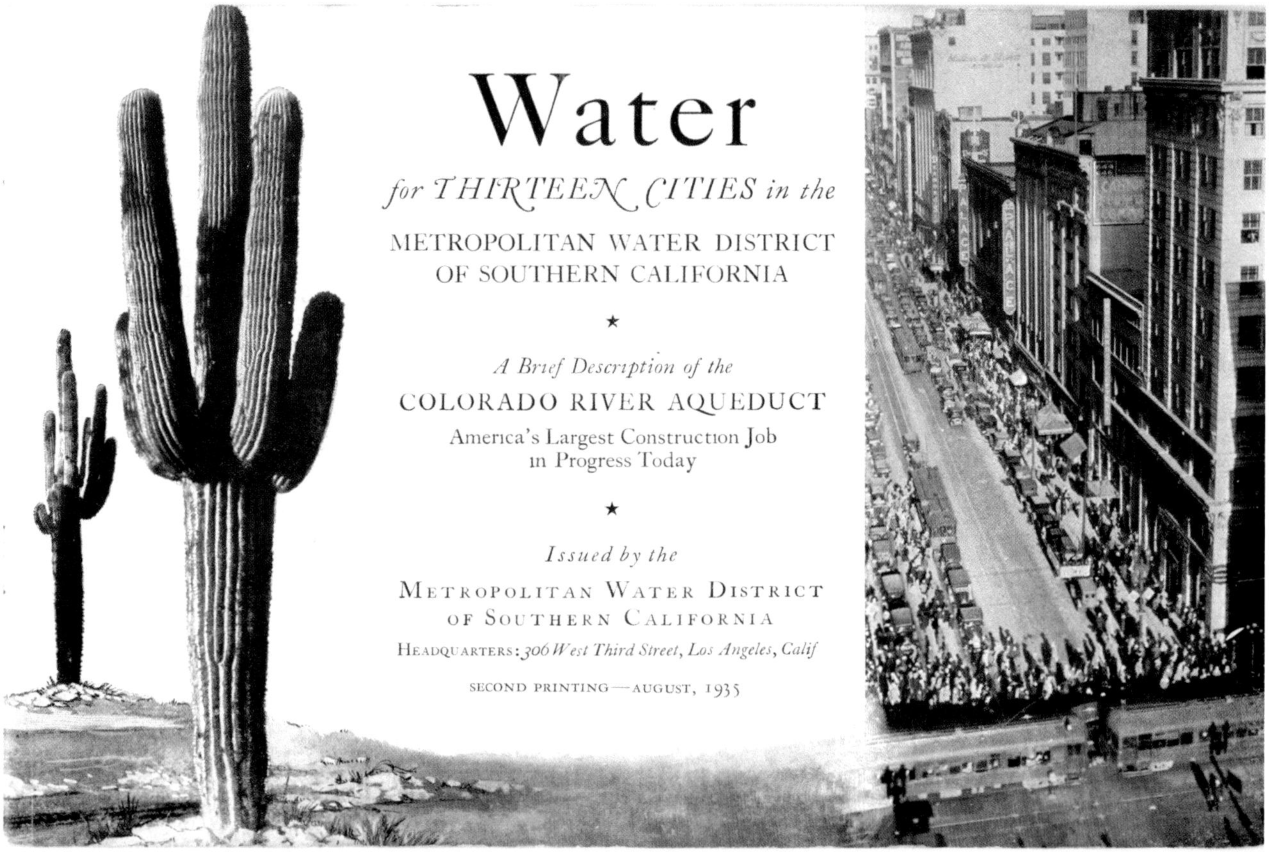

FIGURE 6.11. Booklet photomontage title page. "Water for Thirteen Cities." Exhibit Souvenir, California Pacific International Exhibition, San Diego, California. Metropolitan Water District of Southern California, 1935.

1950, the Sunshine Climate Club returned to national advertising, dusting off its keyed coupon campaign with McCann-Erickson. Building its pitch around winter golf, the SCC's first ad pictured the greens of Tucson Country Club. Though the greens were real, a photomontage of tipped-in saguaros erased the golf course's oak and mesquite trees—a paste-up trick explained in a newspaper report for the benefit of club members who may have wondered what had happened (Arizona Daily Star 1950). The tipped-in saguaros modeled a desert golf course that was understood to be everywhere and nowhere at the same time (Figure 6.14).

Sunshine Climate Club booklet covers no longer pictured the homes of Tucson's residential district, nor did they discuss the therapeutic aspects of "man-building" *per se*—rather, leisure and relaxation in ranchy saguaroland that softened appearances and elevated expectations. A new crop of complimentary booklets for the SCC's keyed coupon campaign credited photographer Esther Henderson and others. Subsequent editions credited Henderson and her husband, Chuck Abbott, the photographer whom the SCC's Roy Drachman retained for guest ranch publicity.

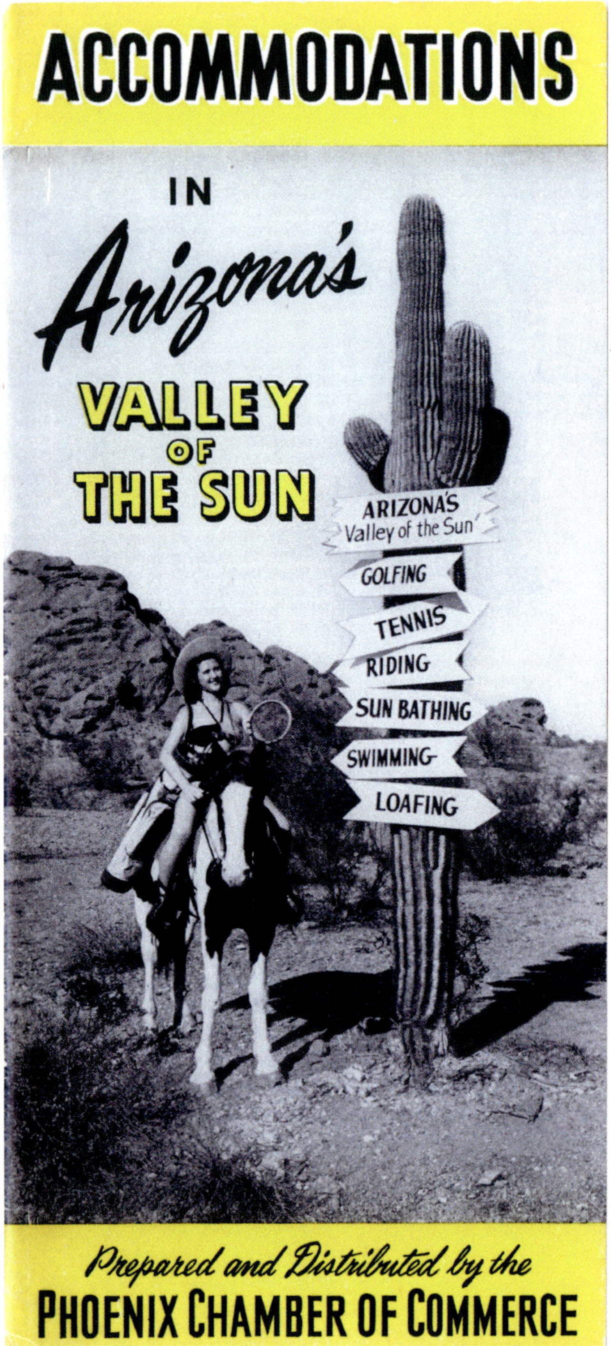

TOP: FIGURE 6.12. Brochure cover. "Accommodations in Arizona's Valley of the Sun." Phoenix Chamber of Commerce, 1952.

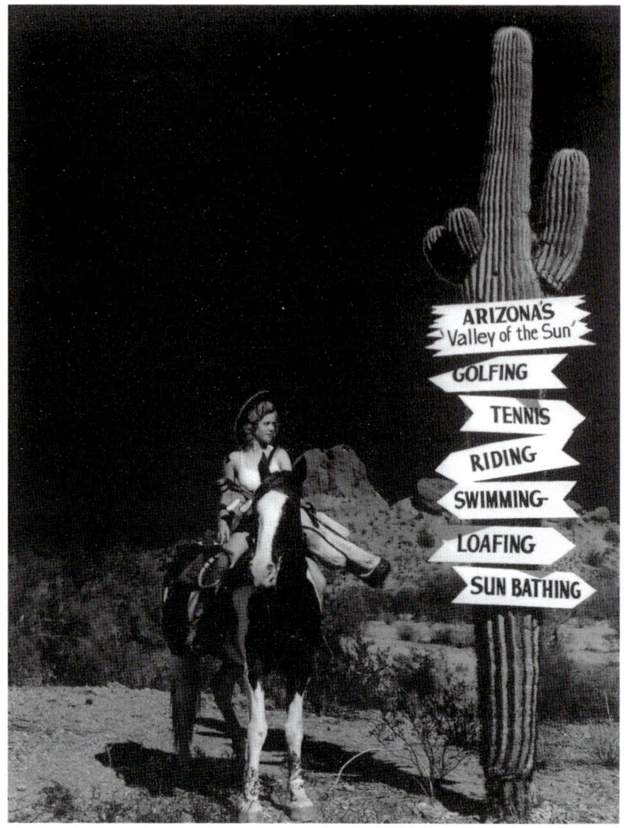

RIGHT: FIGURE 6.13. Photoengraving. Model on horse with photographically posted saguaro, ca. 1952.

The Herberts' Western Ways Feature Service became an important incubator of commercial photography. Ray Manley, a young photographer from Cottonwood, Arizona, came to Charles Herbert's attention in the pages of *Arizona Highways*. Invited to Tucson, Manley joined Western Ways, living in the Herberts' guest house and taking over Abbott's guest ranch photo chores (Henry 1993). Western Ways maintained files on some fifty guest ranches, which kept Manley busy posing the guests (Figure 6.15; Western Ways 1949).

It was Manley who perfected the classic ranch color composition for widely published postcards and commercial brochures. With allowances for an occasional horseback ride through the saguaro-dotted hills, there were but two scenic variations: by day, the swimming pool; by night, the campfire floodlit behind conveniently placed rocks (Figures 6.16 and 6.17).

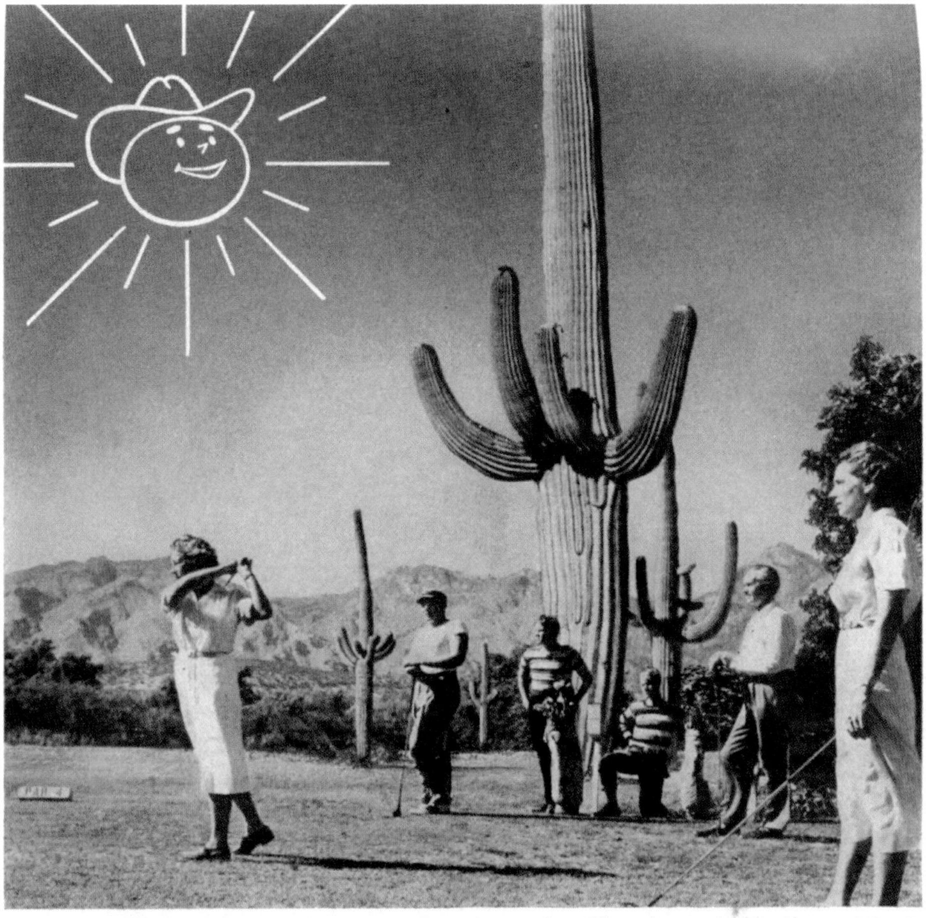

FIGURE 6.14. Advertisement (detail, photographically tipped-in saguaros). Sunshine Climate Club. 1950. "More sunshine in Tucson than any other resort city."

It was through the postcard, the brochure, newspaper, and weekly illustrated magazines that the liberties taken in staging saguaros captured the imagination as a western symbol. By the early 1960s, there was little question that the saguaro symbolized a free, open, and welcoming land, and, if one were lucky enough, the leisure time in which to enjoy it (Stern and Stern 1993). The frenetic action of Chuck Jones' Road Runner cartoons come to mind, as do the pulp fiction novels inspired by John Ford's genre-defining western films that linked the mesa and the stagecoach (Figure 6.18). Further colonizing the Four Corners area, a popular paint-by-number desert subject pictured saguaros near the Navajo Nation's seat of government and capital at Window Rock, Arizona. (Figure 6.19).

It appears that the celebration of the fiftieth anniversary of Arizona statehood in 1962, in which the saguaro played a leading symbolic role, secured its place on the American roadside. In 1961, the Arizona State Legislature authorized the creation

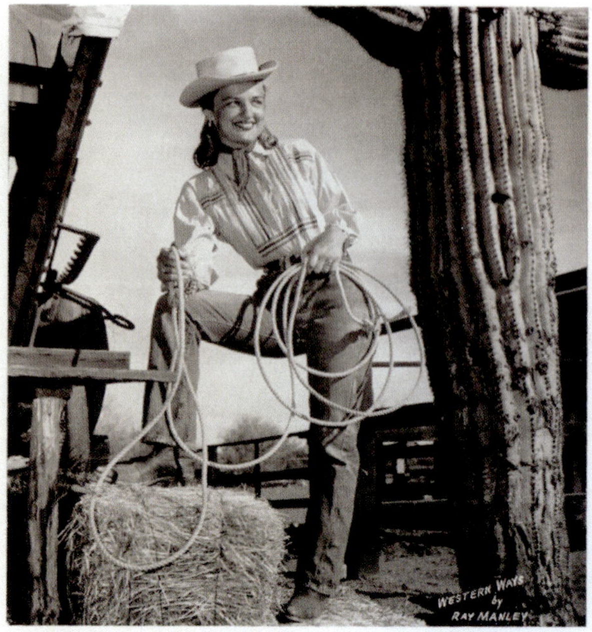

Western Ways

GUEST RANCH PROMOTION SHOTS

by *Ray Manley*

1. PLEASE YOUR GUESTS
2. SATISFY NATIONAL EDITORS
3. PUBLICIZE YOUR INDUSTRY

RIGHT, FIGURE 6.15. Advertisement (detail). "Guest Ranch promotion shots by Ray Manley." Western Ways, 1951.

BELOW, FIGURE 6.16. Postcard. Double U Ranch. Kodachrome original by Ray Manley. Published and distributed by Bob Petley, Phoenix, Arizona. Postmarked 1953.

OPOSITE, FIGURE 6.17. Postcard. Saddle and Surrey Guest Ranch, Tucson. Published by Ray Manley Commercial Photography, Tucson, Arizona, ca. 1958.

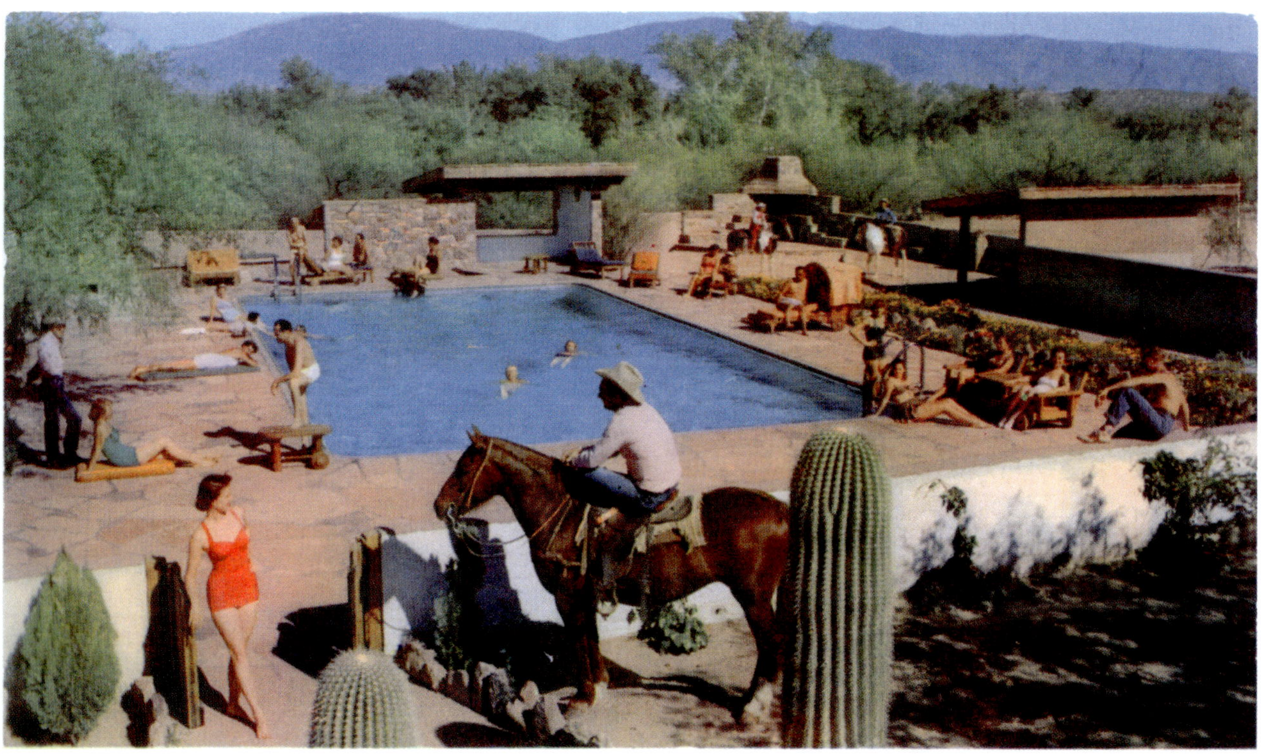

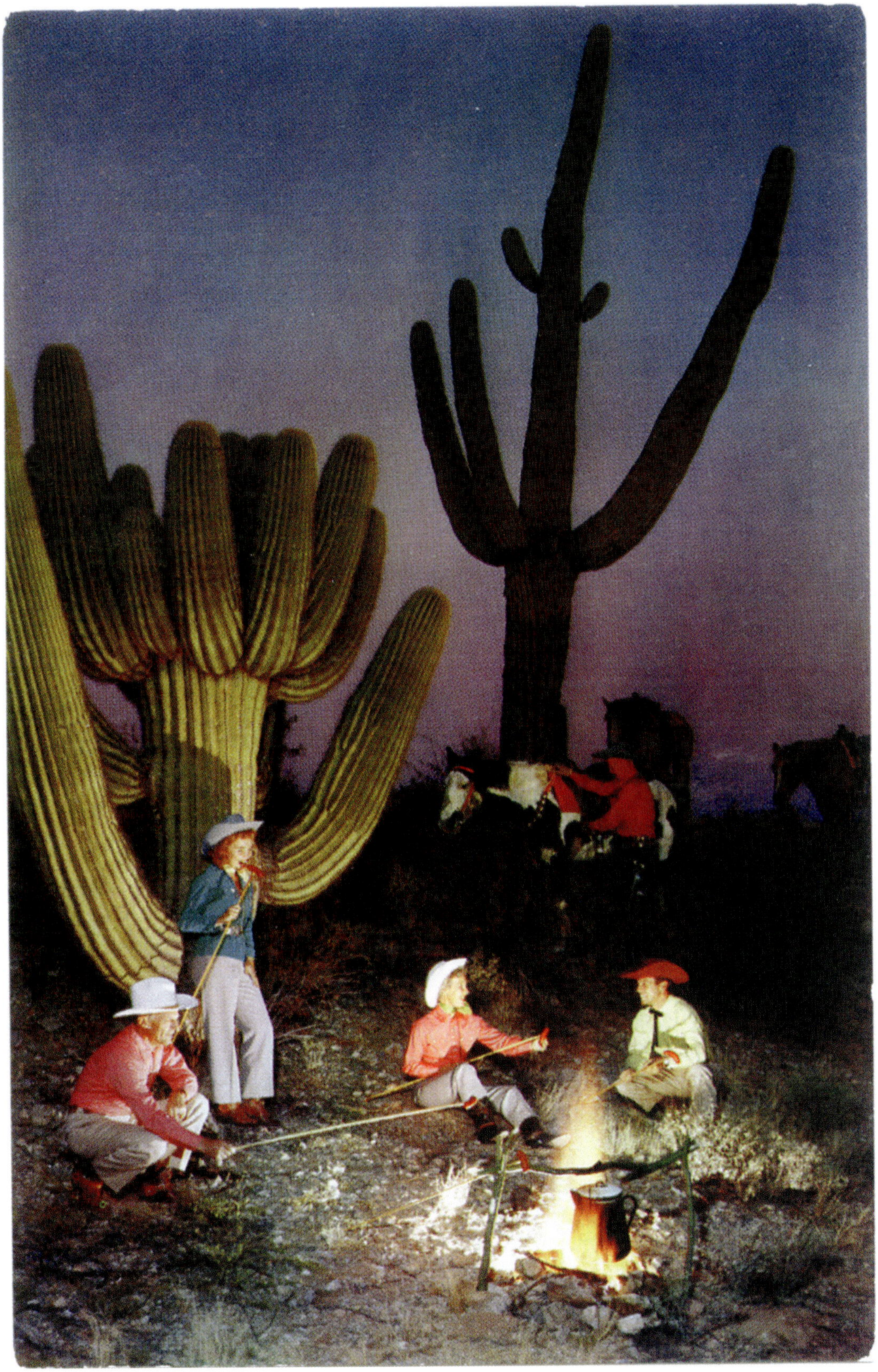

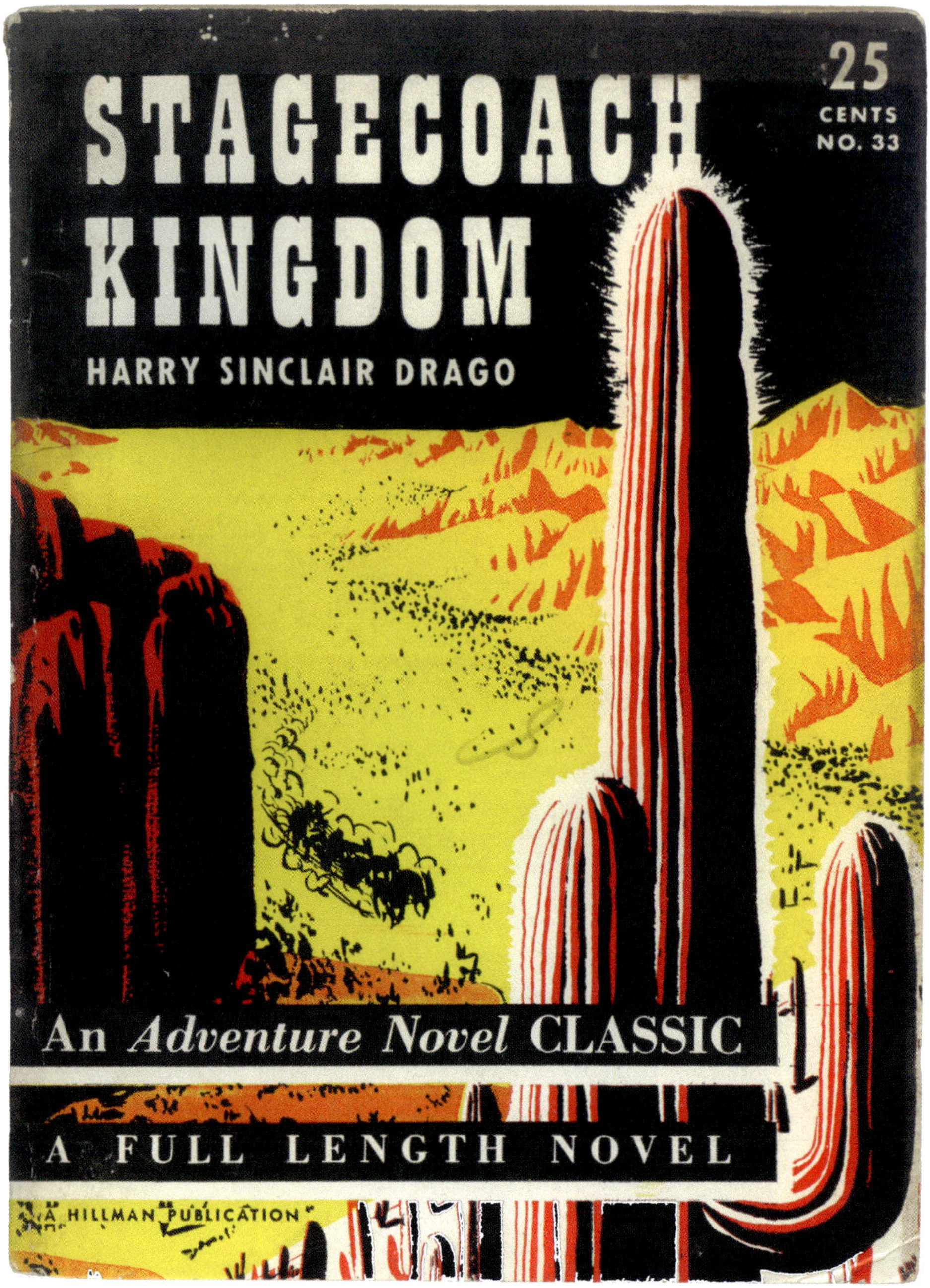

FIGURE 6.18. Novel cover, *Stagecoach Kingdom*, by Harry Sinclair Drago. Novel Selections, New York, 1943.

of a commission to celebrate the state's semicentennial golden anniversary in 1962, and its territorial centennial in 1963. Failing to establish an advertising budget, the creation of an official commemorative seal was left to Audrie Morris, an artist in the telephone directory department at Mountain States Telephone and Telegraph Company. Morris' seal artwork graced the annual telephone book (Arizona Republic 1961; Arizona Daily Star 1962; Arizona Daily Sun 1962; Avery 1962). The design passed over mining and agriculture iconography for the visual vocabulary of the out-

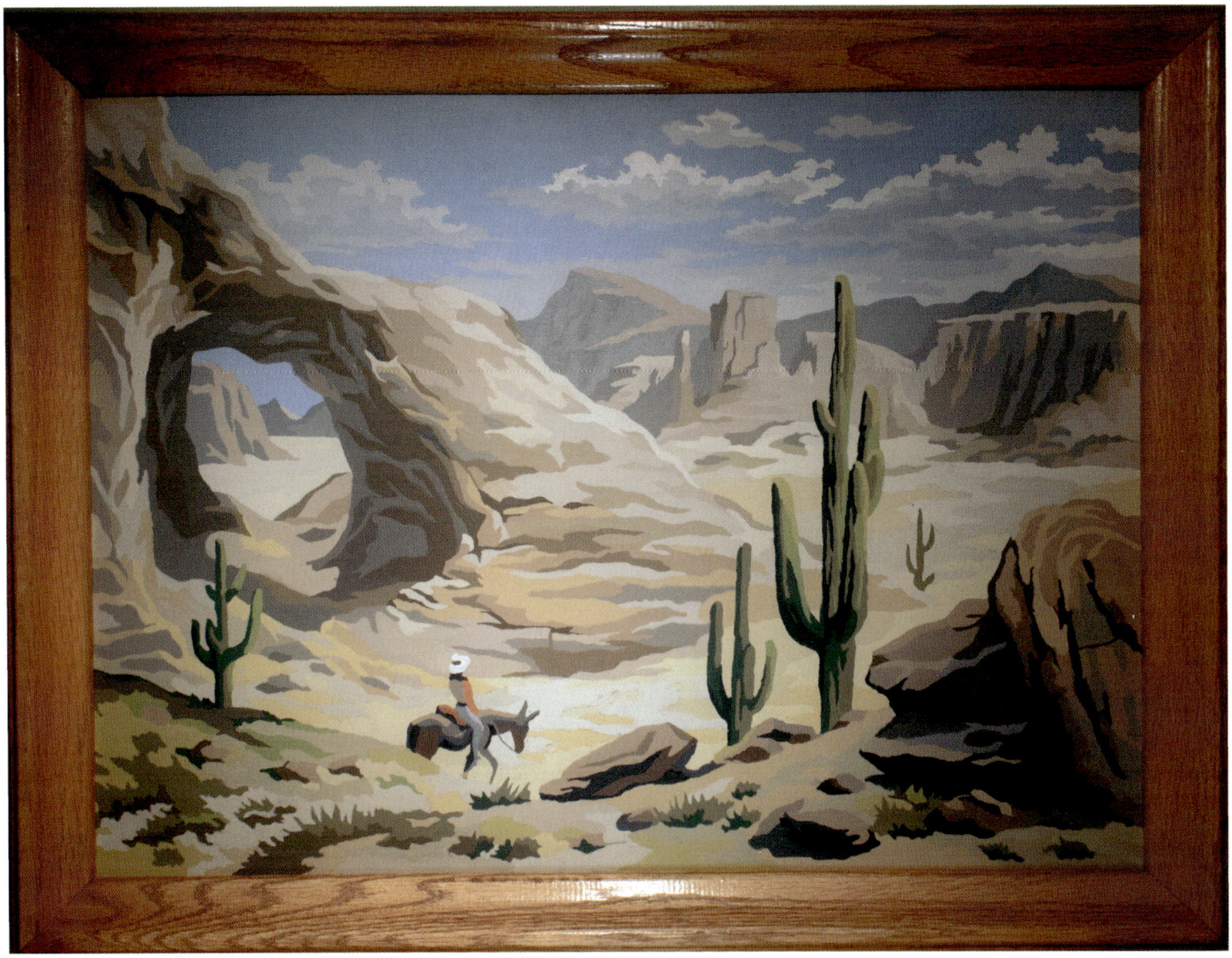

FIGURE 6.19. Paint-by-number painting, Painted Desert. Craft Master New Artist Series, 1964.

of-doors represented by the Grand Canyon, the cactus forest, and a saguaro flower (Figure 6.20). Similarly left to the talents of its artists and illustrators, the Arizona Highway Department published a special double issue of *Arizona Highways* and placed a gold saguaro on the cover of the state's official road map. Accompanied by figurative examples of the state's sparkling mineral wealth gleaming all around it on the ground, the gold saguaro appeared as if it too had just sprung from the ground fully formed and ready for exploitation (Figure 6.21).

A case can be made for the proliferation of roadside saguaro imagery in the United States inspired by Arizona's state semicentennial and territorial centennial celebration. Though the evidence is anecdotal, in 1963, for example, the debut of a monumental neon saguaro sign in the northern suburbs of Boston, Massachusetts, anchored a western-themed steak house and helped it maintain a loyal following (Figure 6.22; Boston Globe 1964; Miller 1988; Seelye 2013). The saguaro signs closest to the plant's native habitat took the least liberties with it. The neon sign for

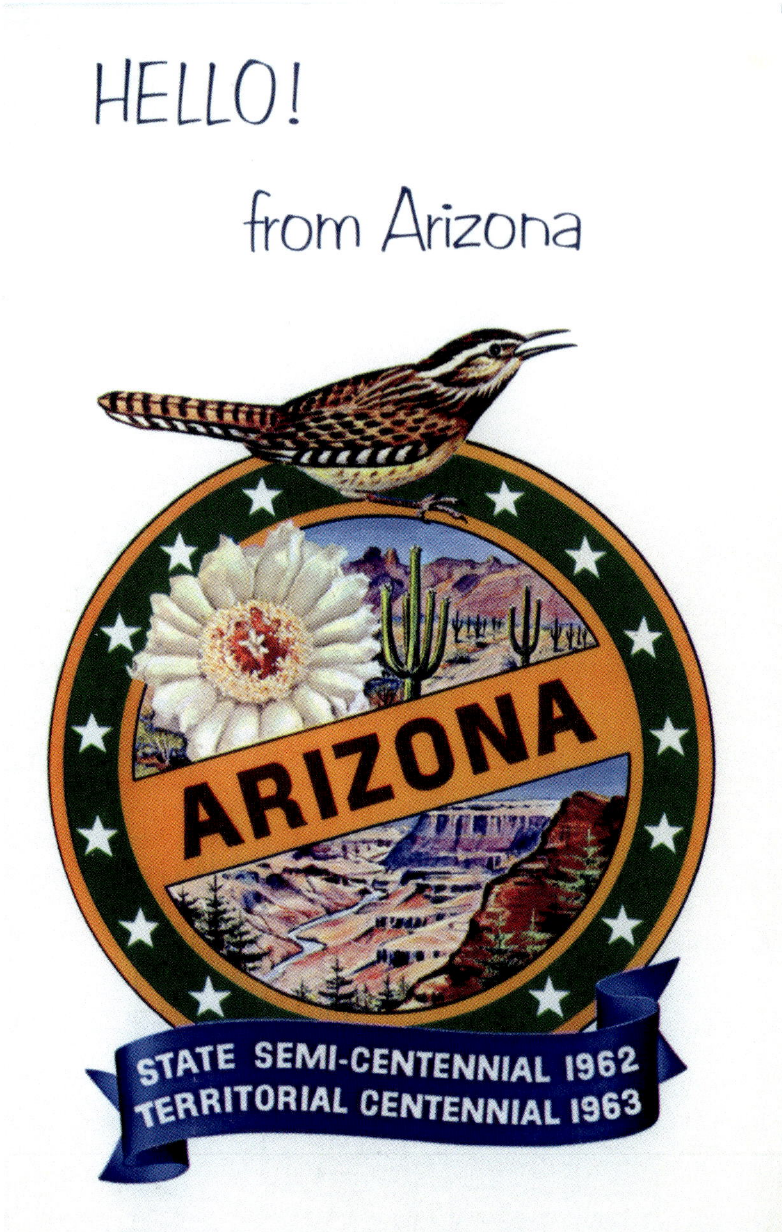

FIGURE 6.20. Postcard. Hello! From Arizona. Arizona Development Board Phoenix, Arizona, 1962. The art direction of a special seal to commemorate the state's semi-centennial and territorial centennial jettisoned the extractive imagery of the state seal for renderings of the state bird (a cactus wren), flower (a saguaro), and the Grand Canyon.

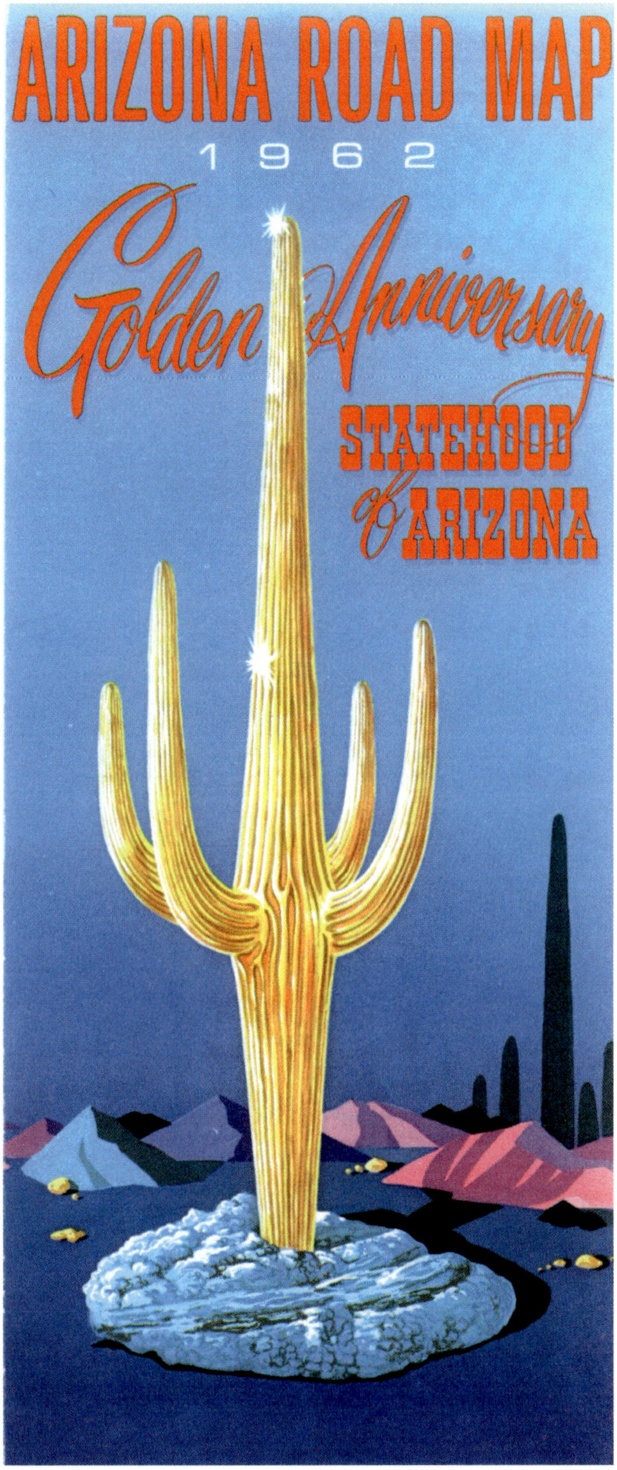

FIGURE 6.21. Road map. "Golden Anniversary Statehood of Arizona." Arizona Highway Department, 1962.

the Traveler's Motel in Yuma, Arizona, for example, made an accurate showing, as did the Cactus Motel on Colfax Avenue's motel row in Denver, Colorado (Figures 6.23 and 6.24). The anthropomorphically playful neon saguaro of the El Rancho Motel in High Springs, Florida, waved a Mexican hat (Figure 6.25). A similar example of the habitat of the absurd could be seen in the roadside mashup of a saguaro, a sleeping Mexican, and an Indian arrow directing attention to the Las Vegas Motel in Eatontown, New Jersey (Figure 6.26).

One of the more meaningful and long-lived examples of the saguaro's semicentennial–centennial era sign art may be seen today at the Westland Shopping Center in Richmond, Virginia (Figure 6.27). Essentially a sheet metal box on a riser at the edge of a parking lot, the saguaro-shaped neon sign dates to the construction of the shopping center in 1963. The land was then far west of Richmond, hence the shopping center's name. As one writer explained, "The Westland Shopping Center has a sign in the shape of a cactus because at the time, they were way out west" (Brissman 2019). Rendering the familiar plant's arms and folds in detail, the saguaro beckons to shoppers and motorists passing in the moment.

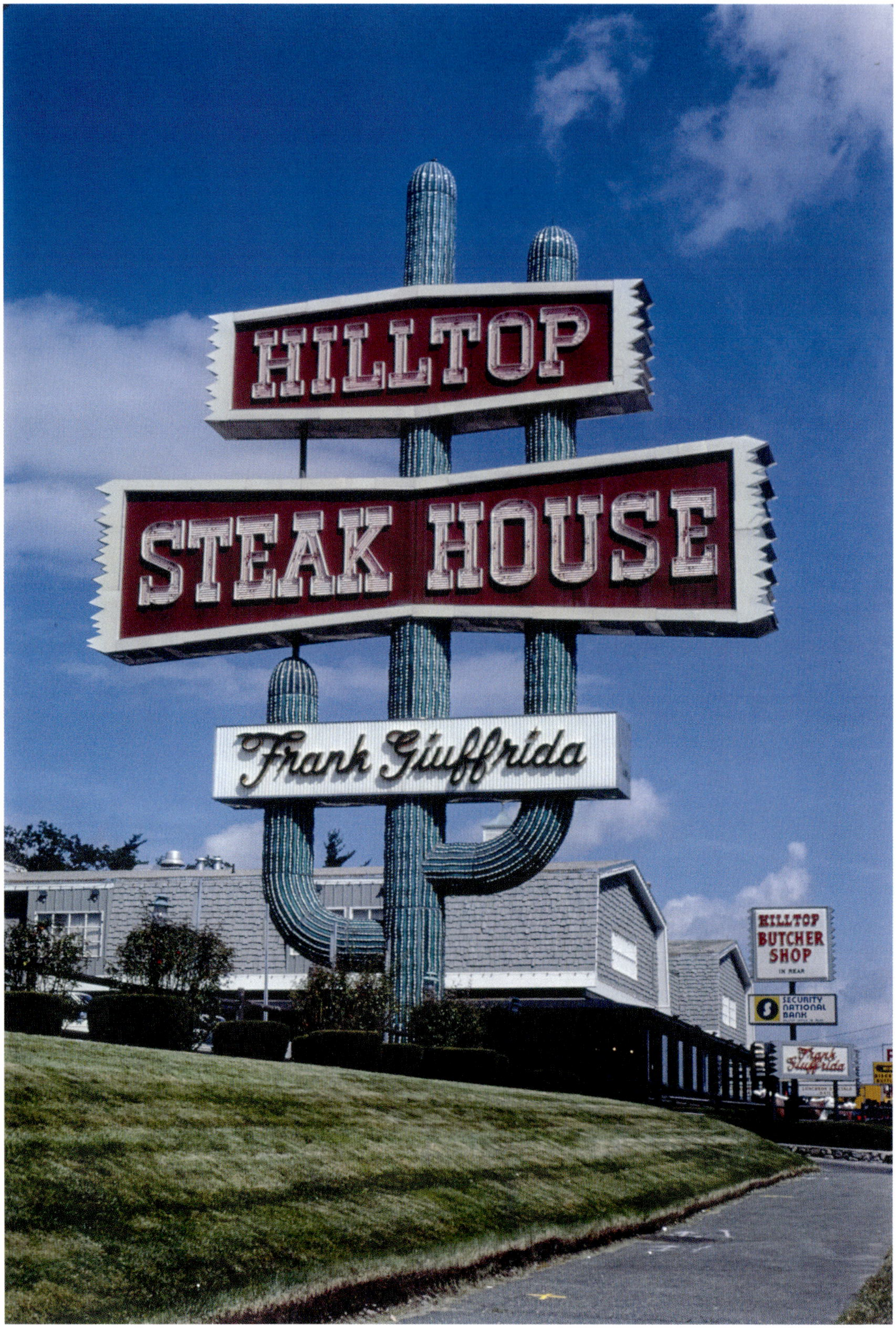

FIGURE 6.22. Hilltop Steak House, Saugus, Massachusetts. Photograph by John Margolies.

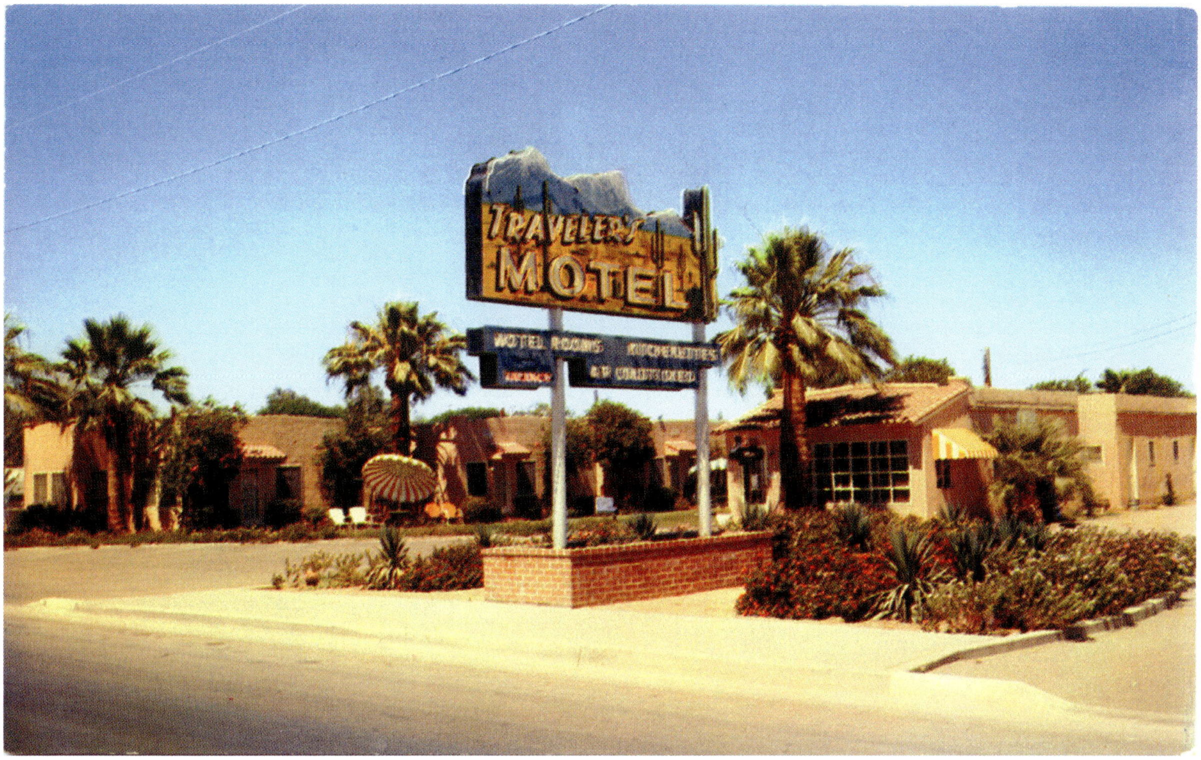

FIGURE 6.23. Postcard, Traveler's Motel, Yuma, Arizona, ca. 1960. Photo by Bob Von Luchene, published by Petley Studios, Phoenix, Arizona.

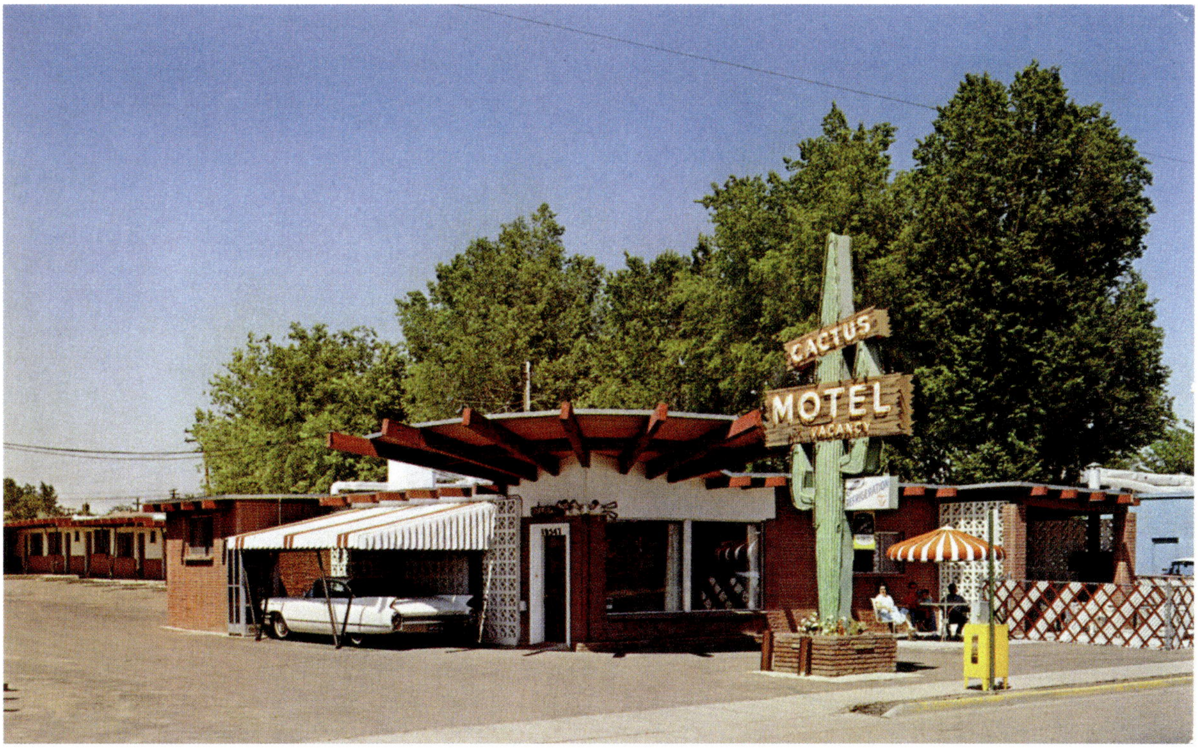

FIGURE 6.24. Postcard, Cactus Motel, Denver, Colorado, ca. 1962. Distributed by Clark A. Brandenburg, Denver, Colorado.

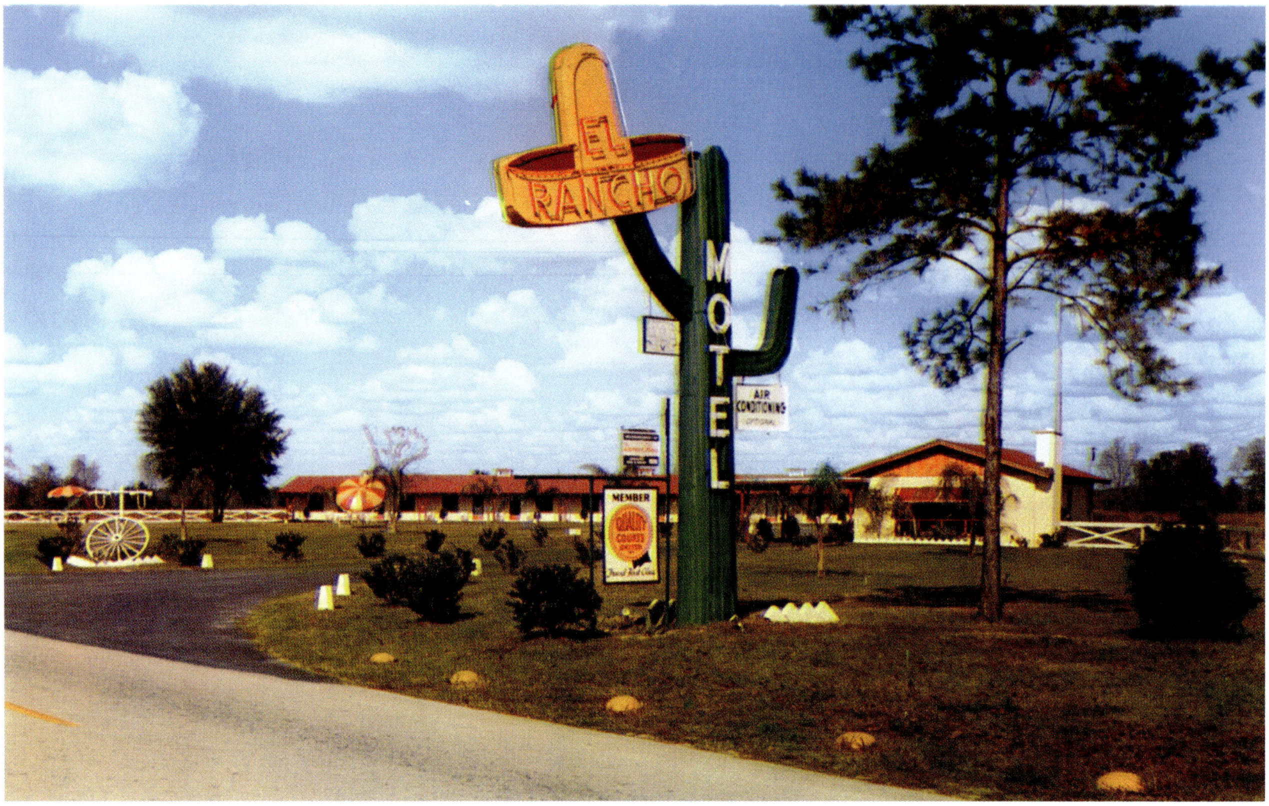

FIGURE 6.25. Postcard, El Rancho Motel, High Springs, Florida, ca. 1958. Dexter Press, West Nyack, New York. Published by Jack Taylor.

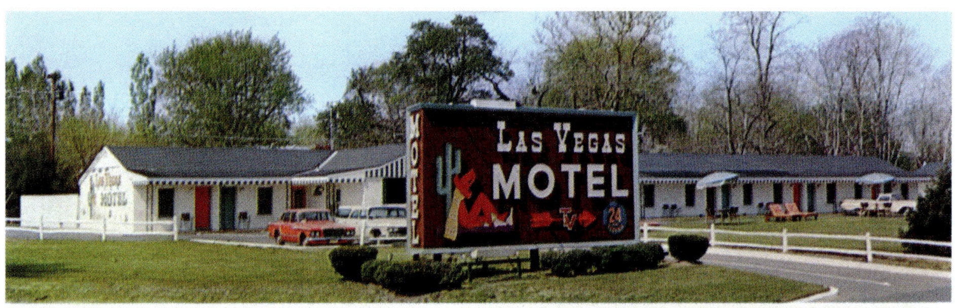

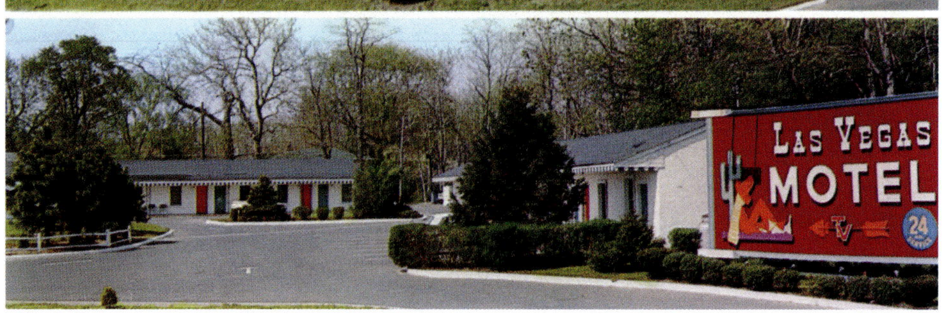

FIGURE 6.26 Postcard, Las Vegas Motel, Eatontown, New Jersey, ca. 1962. Published by N.P.C. Studios, Avon-by-the-Sea, New Jersey.

OPPOSITE PAGE: FIGURE 6.27. Westland Shopping Center saguaro sign, Richmond, Virginia. 2010. Photograph by Dean Jeffrey.

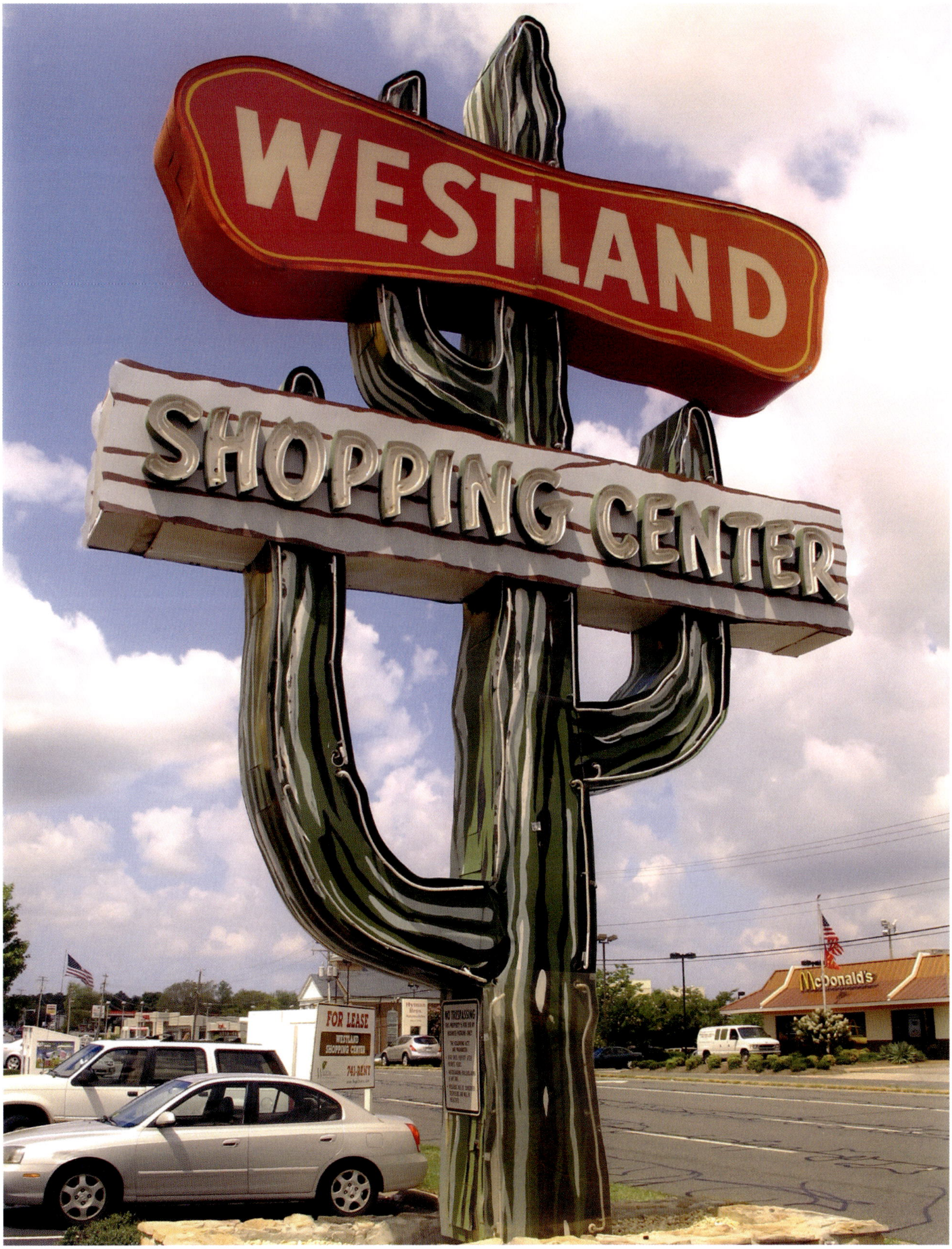
WESTLAND
SHOPPING CENTER
FOR LEASE
WESTLAND SHOPPING CENTER
NO TRESPASSING
McDonald's

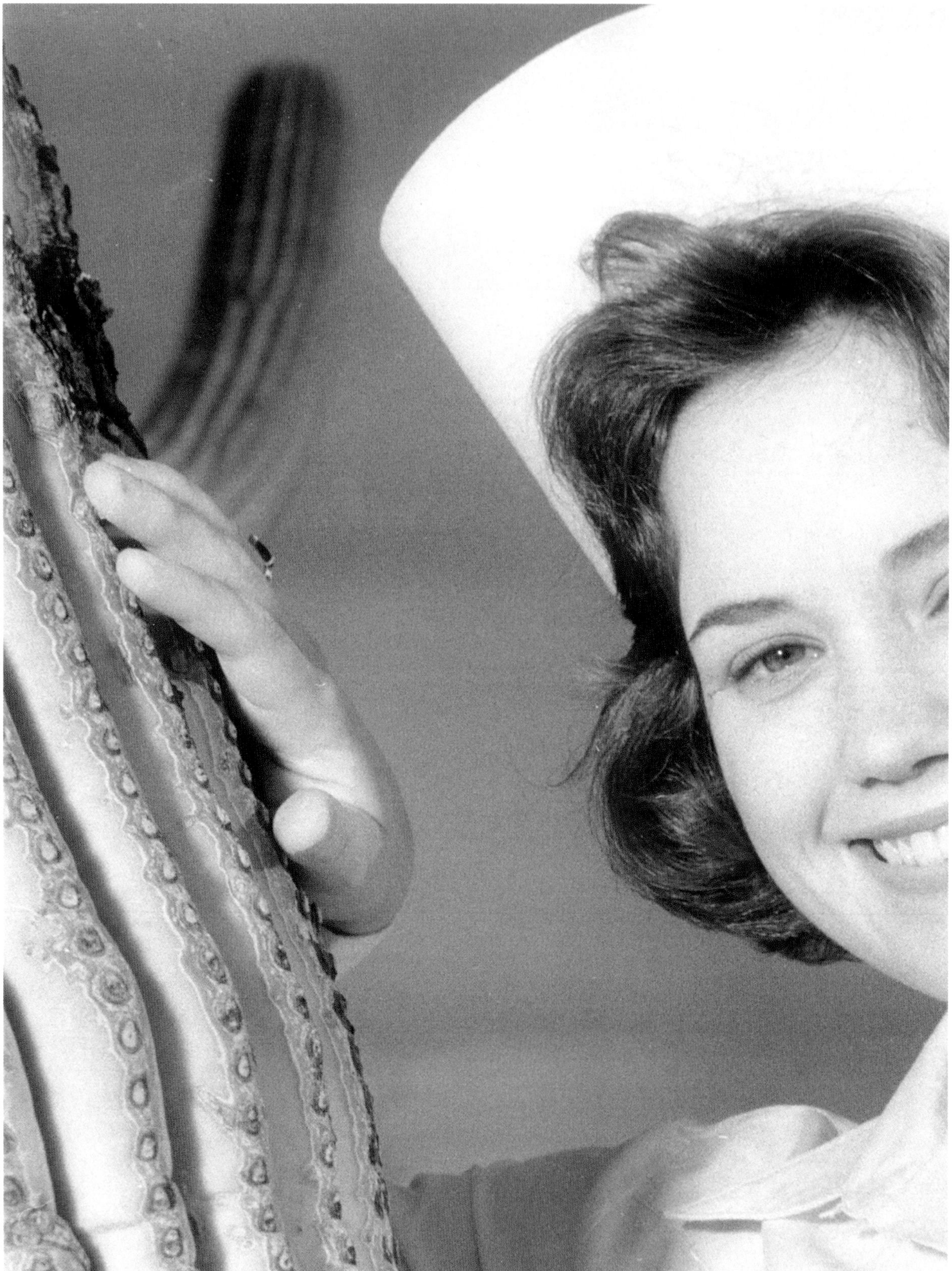

CHAPTER SEVEN
THE VIRAL RANGE

In the world of advertising and popular promotion defined by social media, the saguaro is having a moment. From the fashion runways of Milan to the California music scene, the saguaro projects strength, durability, and fun—infused with the ethos of inviolable conservation. Individuals who share saguaro memes and the plant's right-sized props are captured. Whether in a Fifth Avenue holiday window display or an event expressly staged for taking selfies, the saguaro is having its moment.

Long before the advent of social media, the saguaro's ability to arrest attention was exploited by feature photography as an alternative to paid advertising. In the late 1930s, the most successful of viral photographic images pictured cactus "sun-suits" cut from the growth tips of unfortunate saguaros. Devised to catch the reader off guard—to stop and stare, not unlike the earliest desert travelers who first encountered the plant—became the fulltime function of the cactus gag photograph. The cactus gag erased the distance between the individual and the plant. The first gag was wearing parts of the saguaro. The second was posing on the arm of a saguaro. The most curious plants sought for poses counted among the oldest specimens in and about Saguaro National Monument. Becoming a photographic genre peculiar to southwestern travel and tourism, the cactus gag played out against the cactus scare of the early 1940s caused by the death of large, droop-armed specimens that called into question the saguaro forest's capacity to reproduce itself.

The saguaro's emergence as a social meme dates to the ascendant cactus gags of the early 1940s. *Holiday* (1947b) magazine, for example, stated that the saguaro had "been as important and effective to Arizona as the orange grove to Florida" in creating a "synthetic west." *Holiday's* editors praised the photographers who posed models standing, sitting, and leaning on saguaros. *Holiday* concluded that the pose had become so much a part of popular culture that "the pioneer publicity man who devised the idea of posing a bathing-suited model perched on a spiny saguaro deserves a statue hewn in cheesecake."

Though no statues were forthcoming, the leading candidates for commemoration included Tucson's Charles Herbert and Roy Drachman. As the leader of the

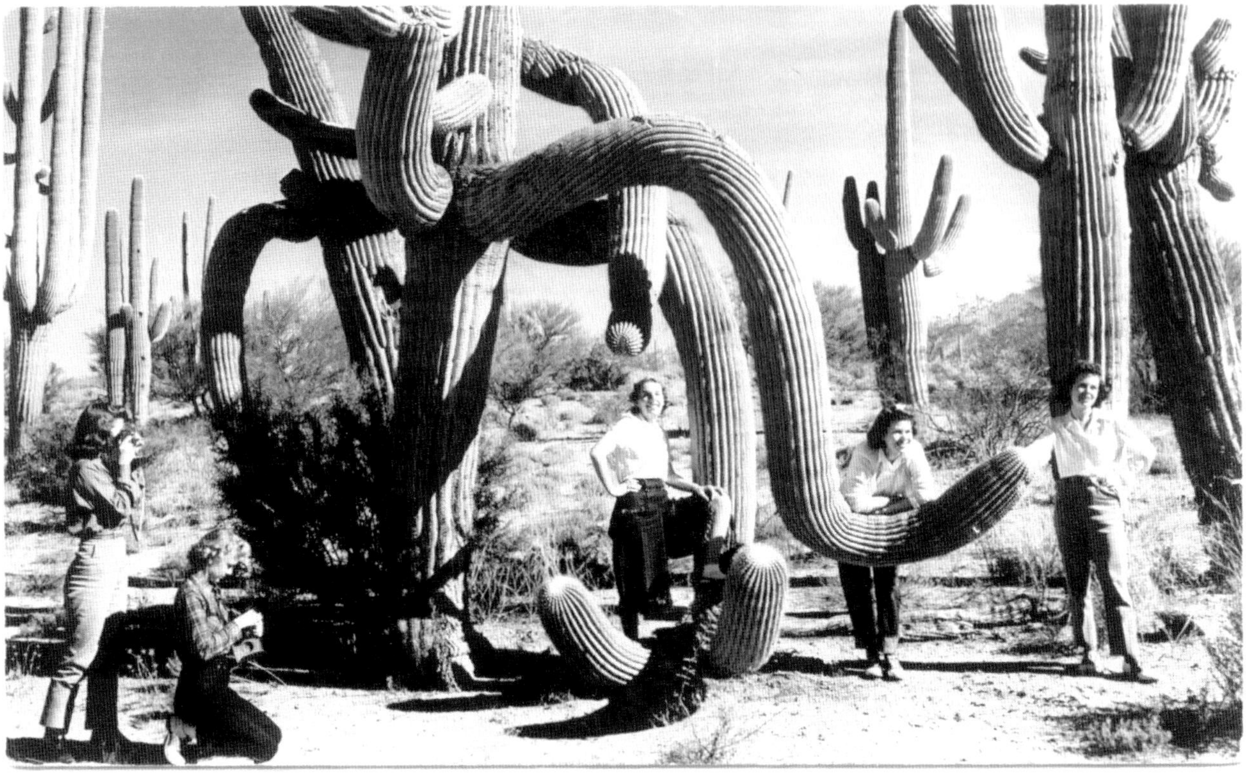

FIGURE 7.1. Postcard. Saguaro National Forest. Published by Western Ways Features, Tucson, Arizona. Postmarked 1947.

Sunshine Climate Club, Drachman knew well that its members reluctantly spent money on national advertising, despite the success of the campaign for which the club had been organized in the early 1920s (Arizona Daily Star 1929b). The situation had hardly improved by 1947 (Tucson Daily Citizen 1947b). Over the years, the club's Los Angeles-based advertising agency H.K. McCann (later McCann-Erickson) had gone to great lengths to prepare cost analyses to build a case for unending investment in advertising. Instead, Drachman and Herbert looked to viral photographs—though they did not use the word *viral*—to win the attention of newspaper and magazine editors without the further expense of an advertising agency.

To keep Tucson in the news during the national defense emergency of the early 1940s—and Herbert in pictures, Drachman conceived the first in a series of publicity stunts staged for the cameras. Drachman built a nighttime bonfire at San Xavier del Bac Mission and invited Native Americans to toss artifacts bearing purported Nazi swastikas into the blaze. Never mind that the Indian symbol of friendship was not a swastika, or that Drachman had arranged with an artist to dummy-up fabrics and cardboard boxes as artworks for the stunt. Newspaper reports of the Indians' denunciation of Hitler appeared across the country with a Tucson byline. Photographs of the bonfire, however, did not. Before the ink had dried on the press clippings, a controversy started among tribal leaders about faux Nazi symbolism (Atlanta Constitution 1940; Boston Globe 1940; Hartford Courant 1940; Los Angeles Times 1940; Washington Post 1940; Drachman 1980:59–60).

Drachman and Herbert quickly moved on to the next stunt, a woman's "sun-suit" made of cactus. Inspired by Herbert's experience as a publicity photographer in Florida, the so-called "cactus sun-suit" involved cutting the growth tips of two saguaro arms hollowed into the bowl-size cups of a woman's bathing suit. A skirt of prickly pear pads suspended from a belt-like wire completed the ensemble. Unlike the bonfire, photographs of the cactus sun-suit lit up the press wires.

University of Arizona coed Merri Ciocetti modeled the outfit in the desert. Between March 1940 and April 1941, photographs of Ciocetti's "sun suit of Arizona cactus" appeared in some 300 newspapers across the nation, most often with the caption, "Touch Me Not." The photo's notoriety culminated in a full-page *Life* magazine feature with new photographs and the original Ciocetti picture credited to photographer Robert Burns, a Tucson newspaperman whom Drachman had retained for Climate Club publicity (Figure 7.2). A strip of three photographs including a second cactus sun-suit modeled by Mary Thorbecke was credited to Herbert (Chicago Daily Tribune 1940; San Bernardino Sun 1940; Life 1941).

The staging of Ms. Thorbecke's sun-suit featured the narrative props that Herbert used to illustrate a story: a camera on a tripod and a Stetson-wearing "cowboy photographer" later identified as Chuck Abbott, the Climate Club's newest guest ranch photographer (Western Ways n.d. [1939?]; Flick 1998). Drachman modeled a shiny rodeo parade shirt while attending to the thorny details of Ms. Thorbecke's cups (Figures 7.2 and 7.3).

Life (1941) played the story for laughs, captioning Drachman's adjustment of the cups "artfully shaped by nature for this purpose . . . made safe for Mary" by Roy. But it was the third photo at the bottom of the page, an outtake from the previous year's shoot with Merri Ciocetti, that anticipated the spineless future of the cactus gag. *Life* (1941) explained: "The top of the barrel cactus . . . has been sawed off and fitted with a smooth board so that Merri will not be riddled with prickles" (Figure 7.4).

As a viral confection the cactus sun-suit was admittedly novel. Yet even Drachman might have admitted that as a proposition it simply didn't work; in fact, it was

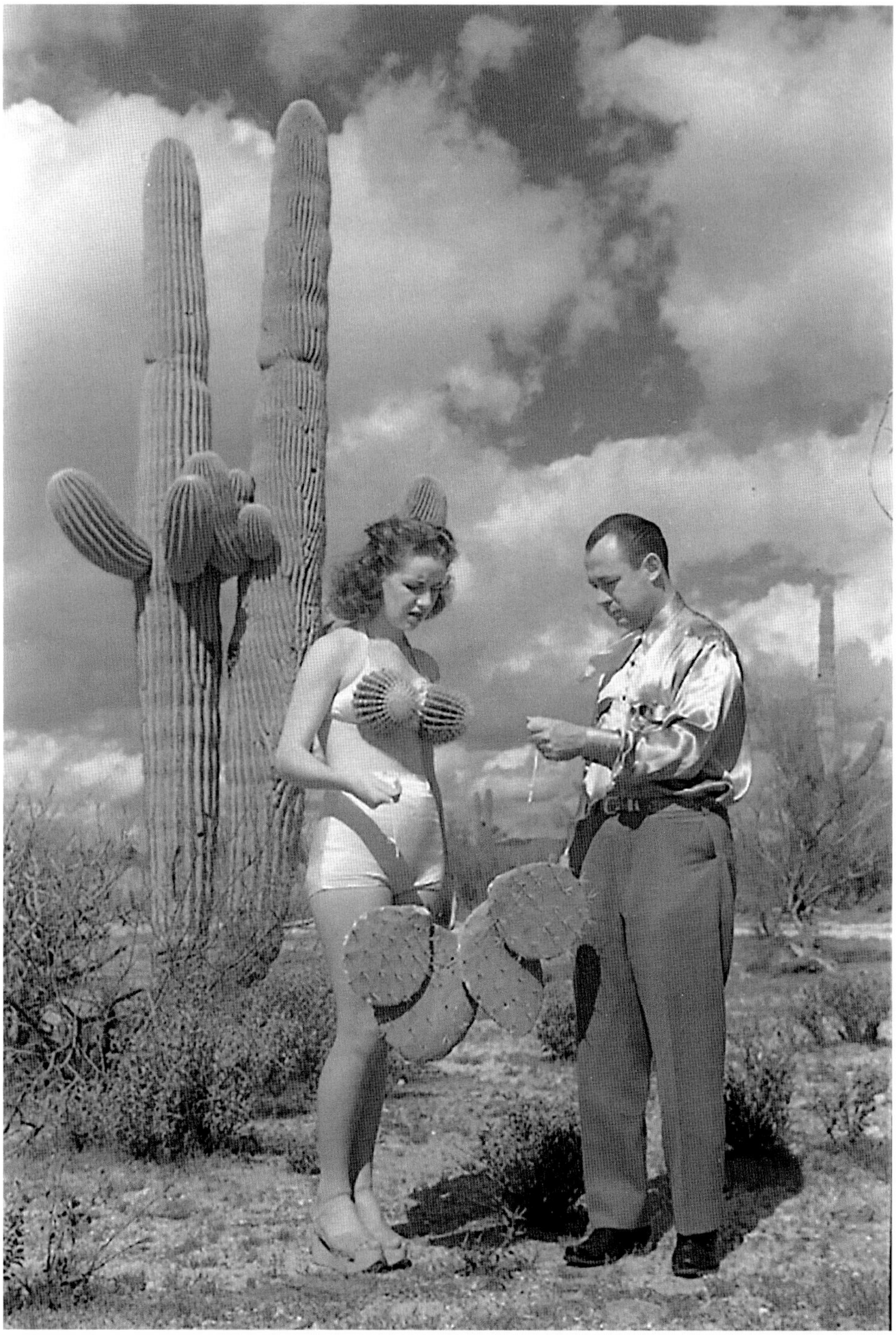

FIGURE 7.2. Mary Thorbecke and Roy Drachman prepare the "cactus sun-suit" for its *Life* magazine debut, 1941. Photograph by Charles Hebert.

off-putting. The *Chicago Tribune,* one of the first newspapers to run the photo in the spring of 1940, appears to have contributed the headline, "Touch Me Not." Appearing in 300 newspapers didn't make it any better. As *Life* (1941) explained, "the publicity hasn't brought many tourists to Tucson, but it makes the people of Tucson feel fine." Whether that was true or not, what the Ciocetti photograph did do was

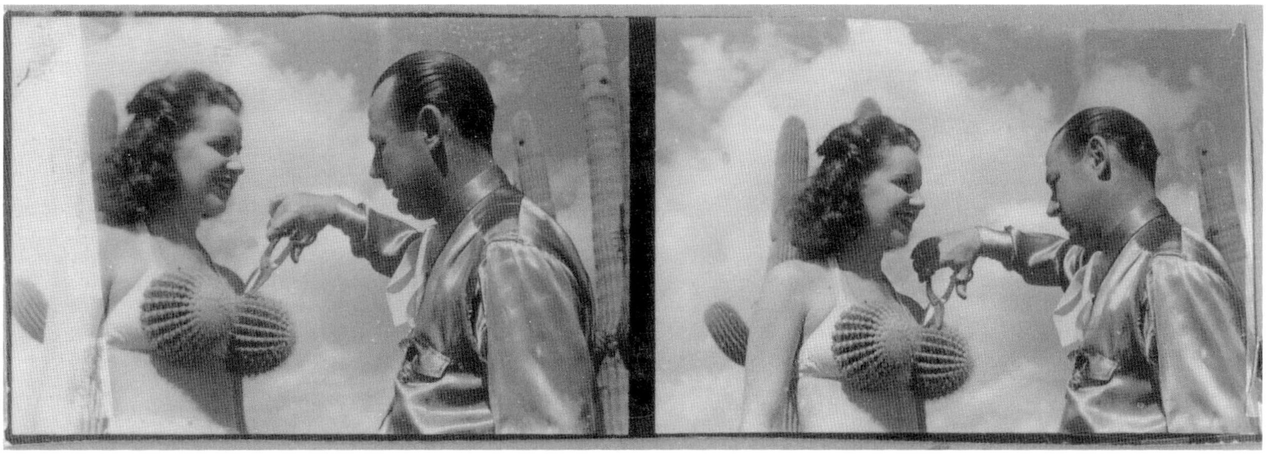

FIGURE 7.3. Mary Thorbecke and Roy Drachman. 1941. Photograph by Charles Herbert.

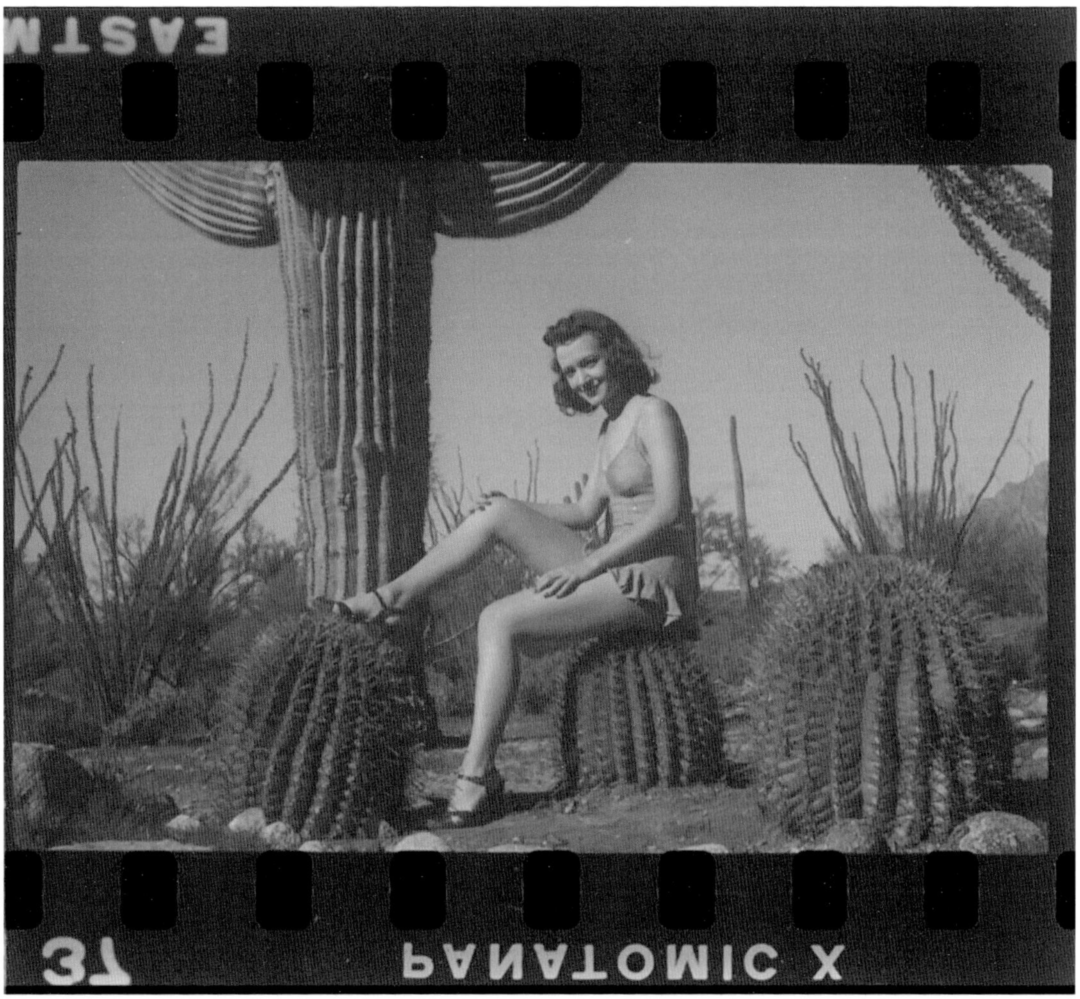

FIGURE 7.4. Merri Ciocetti sitting on barrel cactus. 1940. Photograph by Charles Herbert

confirm the power of self-generated photo publicity for a generation of Sunshine Climate Club leaders. Photos, mostly cactus gags, could run concurrently with paid advertising if and when the club chose to return to it.

The Second World War forestalled the club's admixture of photo publicity with paid advertising. McCann's proposed copy themes soothed jangled wartime nerves, while the persistent image of Merri Ciocetti circulated in and out of the pockets of soldiers, including a German prisoner of war. The caption of the captured clipping cited the cactus sun-suit as an example of Americans' "utter lack of taste." The writer misidentified the setting as Florida, a mistake sure to have raised the ire of Corporal Roy Drachman had he found out about it on his stateside army base (Arizona Daily Star 1945).

Drachman's emphasis upon photo publicity might have been a boon to Tucson's photographers. In Chuck Abbott, however, they saw a competitor. At Drachman's behest, Abbott had relocated from Palm Springs just in time, as it turned out, to appear in *Life*. The Stetson-wearing Abbott had built a following at Palm Spring's Desert Inn as a trailside cook and photographer, conducting horseback rides and barbeques, photographing the resort's guests at leisure. Their pictures appeared by wire service in their hometown newspapers in the dead of winter with a byline from sunny Palm Springs. Expected to do the same for Tucson's guest ranches, Abbott's residence as a photographer proceeded under Drachman's directive of "less but better use of advertising" (Desert Sun 1936; Arizona Daily Star 1940a, 1940b).

Tucson's photographers were not impressed. They deputized photographer Esther Henderson to talk the situation over with Drachman. Henderson had taken up photography in New York City in the early 1930s after a successful career as professional dancer. Settling in Tucson with her father in 1936, she opened a portrait studio. Henderson noted a decline in new business that she attributed to Abbott's guest ranch portraits. Confronting Drachman, she exclaimed, "What the hell is he? A cook or a photographer?" (Richards and Bell 2005:13). She later came to terms with Abbott. They went on a Sunday picnic with their cameras. Within a year they married. Henderson continued her portrait work and joined Abbott in landscape photography. Her work was among the first published in the newly revamped *Arizona Highways* under editor Raymond Carlson (Arizona Daily Star 1941; Pyle 1942; Cooper 1973; Drachman 1980:58–59; Henderson

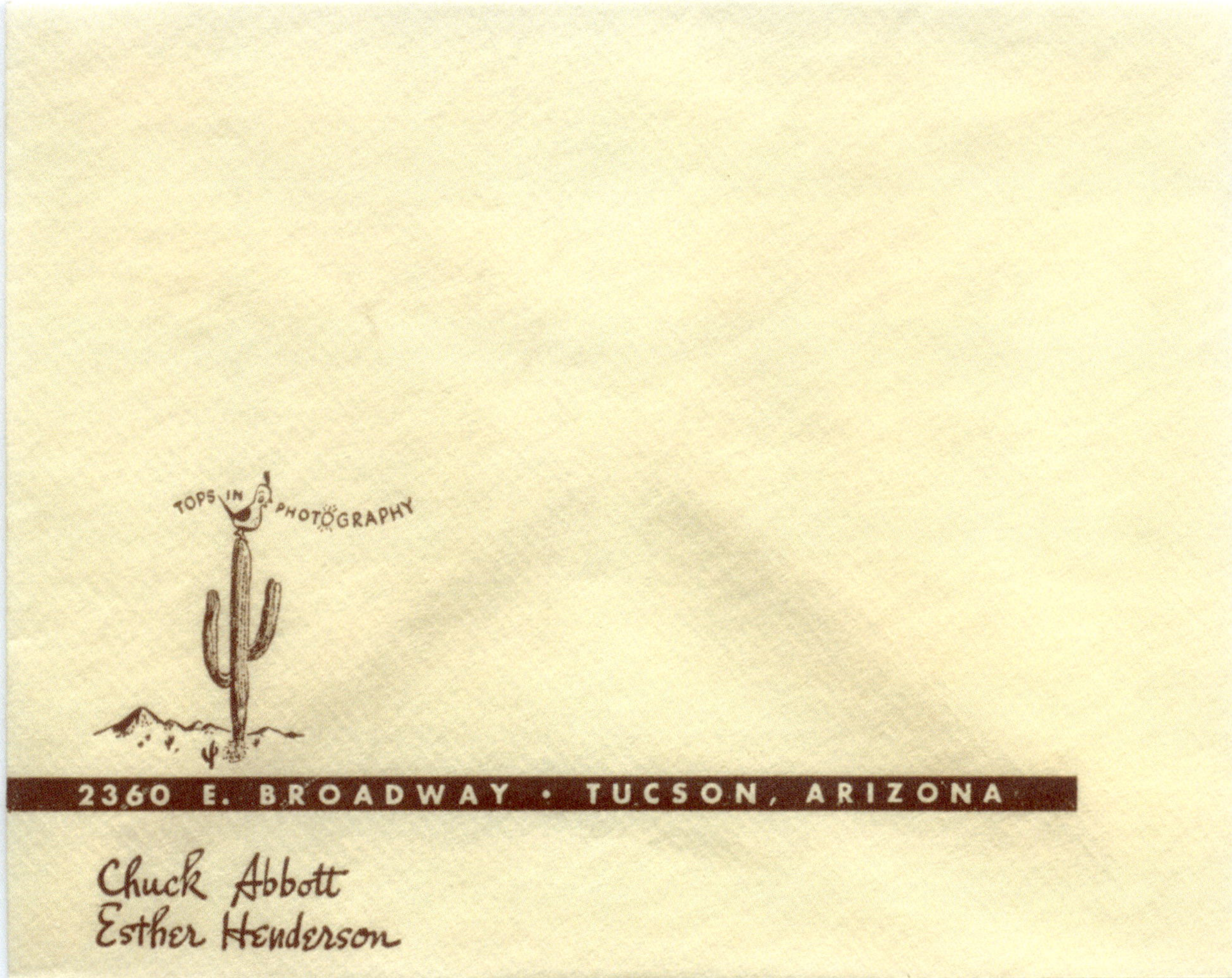

FIGURE 7.5. Envelope. "Tops in photography." Photocenter, Tucson, Arizona, 1947.

1987; Richards and Bell 2005). After the war, the Henderson-Abbotts opened the "Photocenter," a full-service retail camera shop. The store merged the couple's interests in a 20 × 60 foot outdoor studio with "western atmosphere." In a nod to Abbott's former life as a cook, the store's prop patio featured a chuck wagon and several species of cacti for "character background" portraits. The props including the saguaro were real (Tucson Daily Citizen 1947c). The Cabat-Gill advertising agency designed the store's logo, a bird on a saguaro singing, "Tops in photography" (Figure 7.5).

Like the Herberts and the Henderson-Abbotts, many others contributed to the region's postwar tourist economy. Between 1950 and 1960, tourism and manufacturing overtook mining and agriculture as Arizona's leading industries (Casaday 1950; Shirer 1954; Arrington 1969; Klein 1993). By now the saguaro's role as a symbol

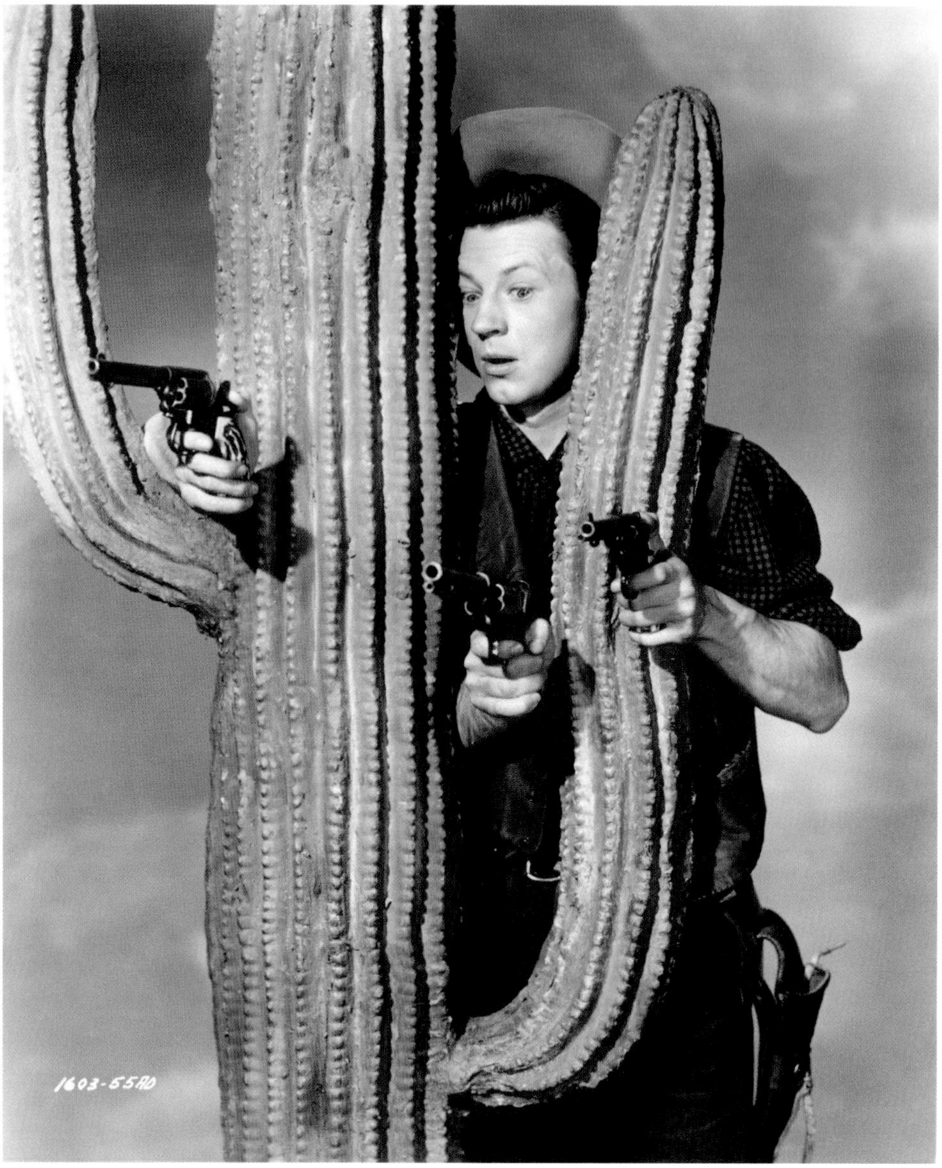

FIGURE 7.6. Copy photograph. Donald O'Connor, *Curtain Call at Cactus Creek.* Universal Pictures Company, 1949.

OPPOSITE PAGE, FIGURE 7.7. Copy photograph. All Year 'Round Club of Southern California. 1957. Model seated among golden barrel cacti at the Huntington Museum and Library Desert Garden, San Marino, California.

was well-understood—and it is not far off the mark to suggest that its iconography occupied a manufacturing category of its own, led by cactus gags of all kinds from Hollywood to San Marino (Figures 7.6 and 7.7; Year 'Round Club of Southern California 1957).

But all was not well in saguaroland. The apparent decline of the cactus forest in and around Saguaro National Monument could be measured in photographs taken over time (Hastings and Turner 1965; Cohn 2003; Turner et al. 2003:168–191; Webb et al. 2007; Ahnmark and Swann 2008). Not until the 1960s did botanists realize that the saguaro forest admired by Wright, Shantz, and others was in fact a remarkable nineteenth-century bumper crop that matured in the late 1930s—and

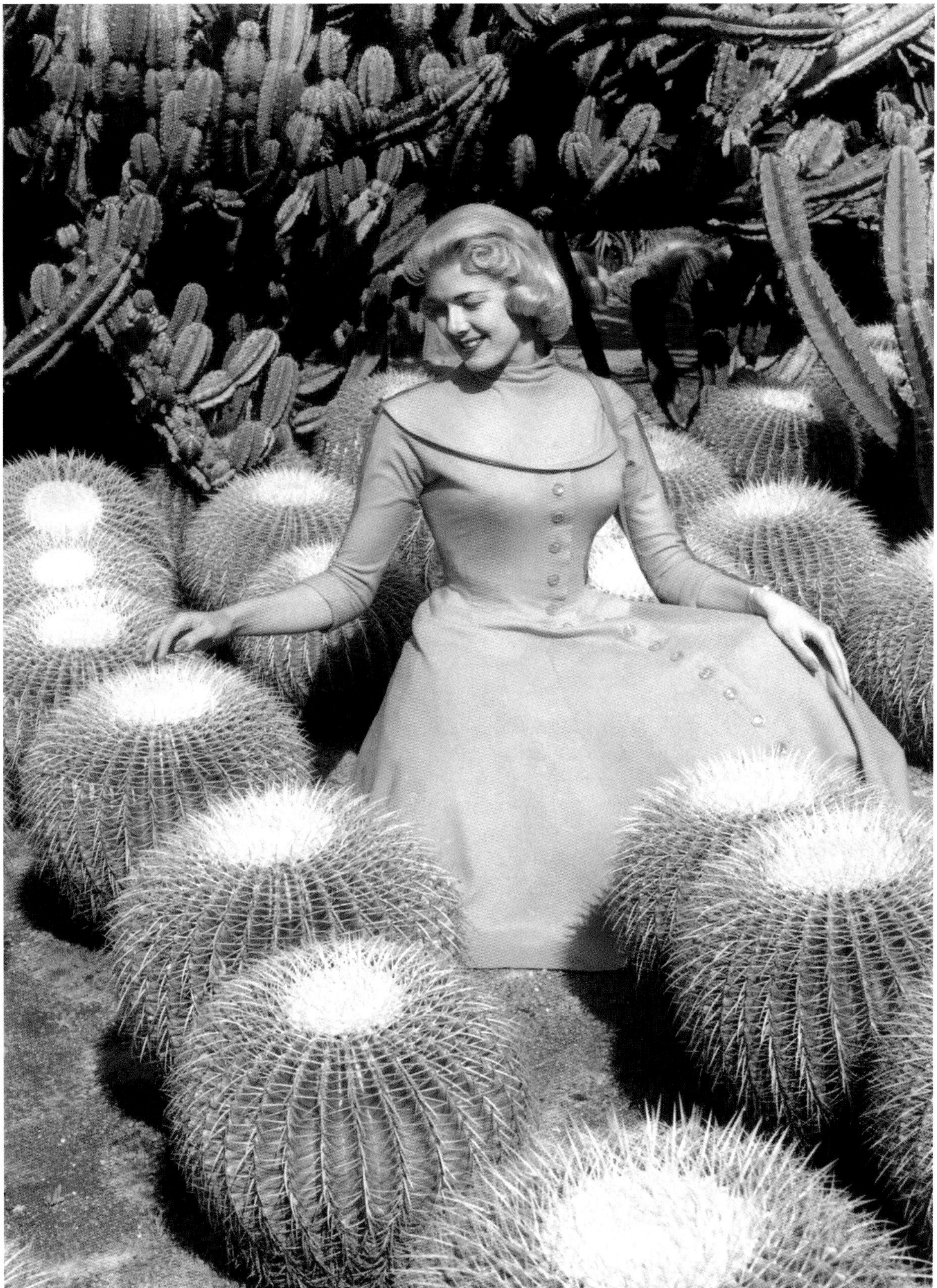

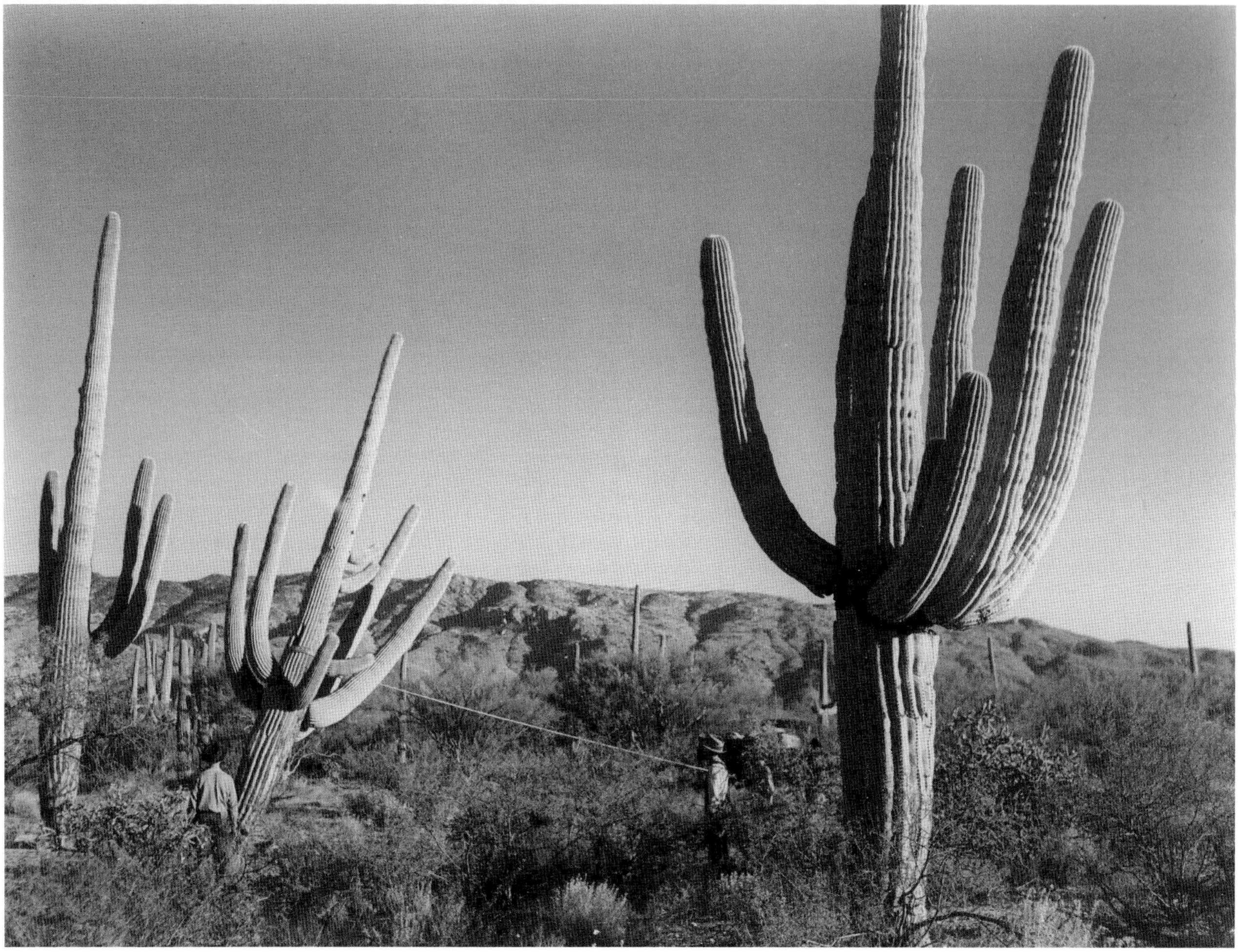

FIGURE 7.8. Cactus disease control. Pulling down a saguaro with a wire attached to a tractor, Saguaro National Monument, 1941.

was slowly returning to normal by dying off. The forest's largest and oldest saguaros began to be disfigured by black spots—bacterial necrosis—the natural conclusion of the plants' life cycle that botanists misread as a disease (McAuliffe 1996; Burtner 2011:133–139).

The saguaro forest's apparent decline led to hand-wringing among the publicists who had burnished their creative reputations in furthering the plant's iconography. "Their trademark is tumbling down," one feature writer explained. Even the saguaros "used as background for a beautiful smiling girl, are already rotted inside, the only evidence being the holes which appear in the arms and body of the cactus" (Western Ways n.d. [1941?]). For many, the loss was personal. The Herberts, for example, wondered if the saguaro around which they had framed the view-with-a-room would die before its time. The Herberts helped sound the alarm, dramatizing cactus disease with a steady stream of illustrated news features about the botanists who sought to find a cure.

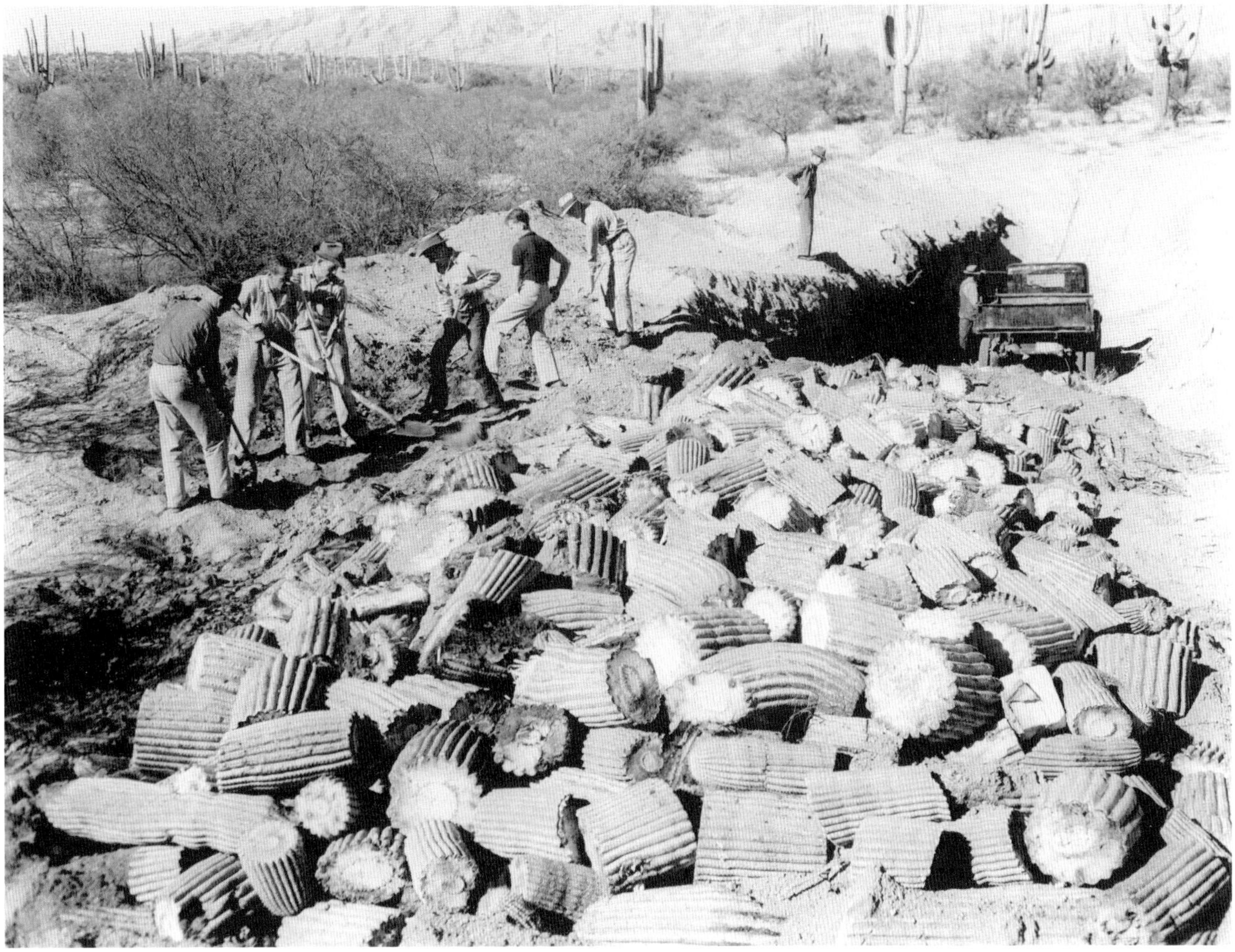

FIGURE 7.9. Cactus disease control. Workers preparing to bury cut sections of saguaros, Saguaro National Monument, 1941.

The diagnosis of black spot as a disease and the ensuing panic about it reflected the saguaro's iconic status—and upon whose remediation reputations awaited to be made. The errant diagnosis led to a moth believed to be responsible for the tissue-rot that weakened the plants. The cure would destroy spotted plants for the protection of those not yet afflicted. Botanists roped off a test section at the edge of Saguaro National Monument. Park rangers pulled down spotted saguaros with a cable and tractor, cut their trunks and arms into pieces, doused them with insecticide and buried them. Park officials judged the destruction of the plants to be too expensive (Figures 7.8 and 7.9; Western Ways n.d. [1942?]). Experiments with topical remediation ensued. The visualization of end-of-life cactus care pictured university botanists who had published theses on the so-called disease tending to the stumps of amputated arms from ladders and daubing bleach solution while seated in the arms of the plant for feature news photographs (Figure 7.10; Lightle 1947; Boyle 1948; Western Ways 1956; Herbert 1960).

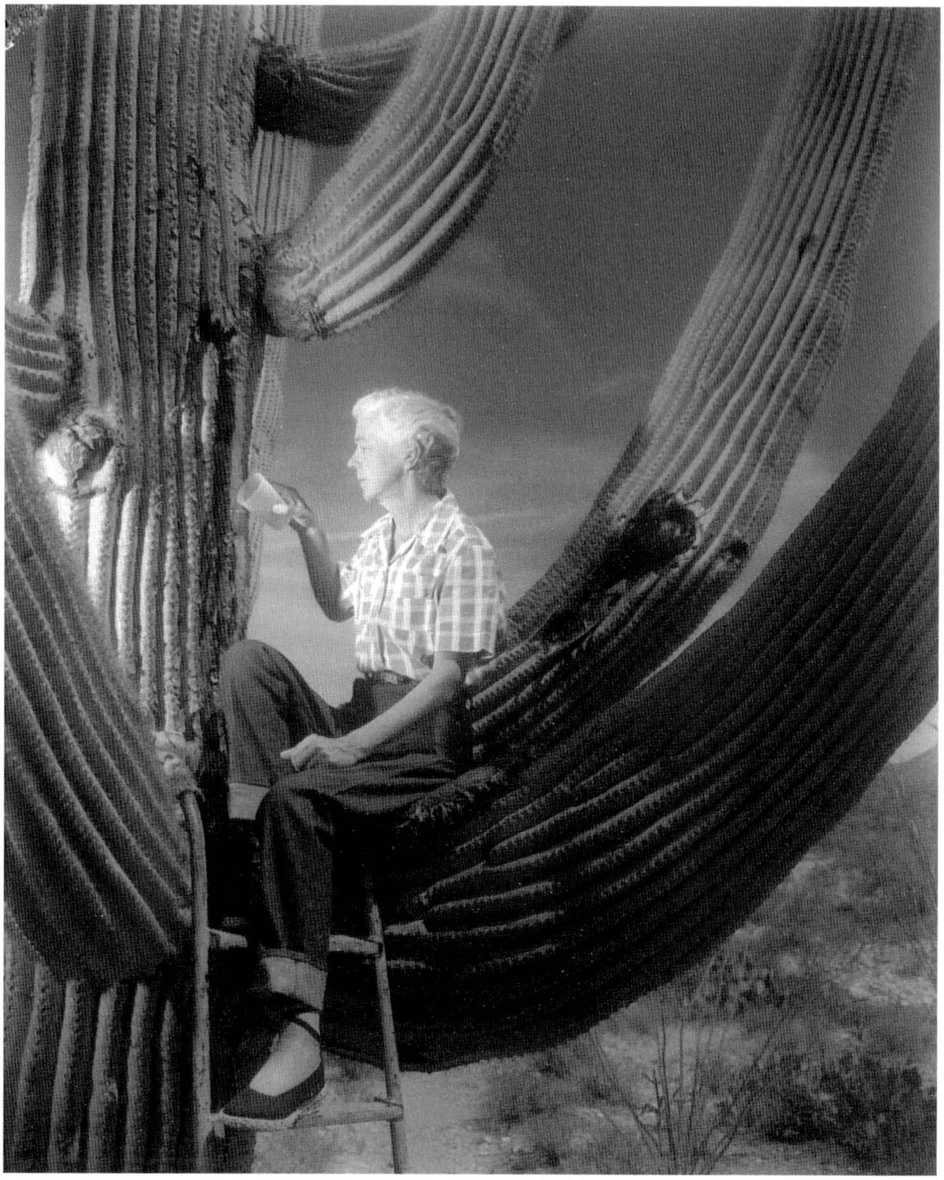

FIGURE 7.10. Dr. Alice Boyle treating a saguaro with bleach solution. 1956.

Today's botanists chalk up the lapse in propagation that so alarmed the botanists of the early 1940s to the saguaro forest's natural life cycle. The saguaro was indeed reproducing itself—not drifting toward extinction hastened by bacterial necrosis. Spotty seasonal rainfall and winter freezes had tempered growing conditions since the bumper crop of the nineteenth century (Young 1968; Cohn 2003; Turner et al. 2003:276). This was particularly true for the necrosis that visited the aging plants after a severe winter freeze in 1937. Black-spot was not a disease and certainly not spread by moths. Rather, the spots indicated the destruction of plant tissue hastened by a freeze of several days' duration that caused the plants' arms to droop, sometimes touching ground (Steenbergh 1969; Steenbergh and Lowe 1977).

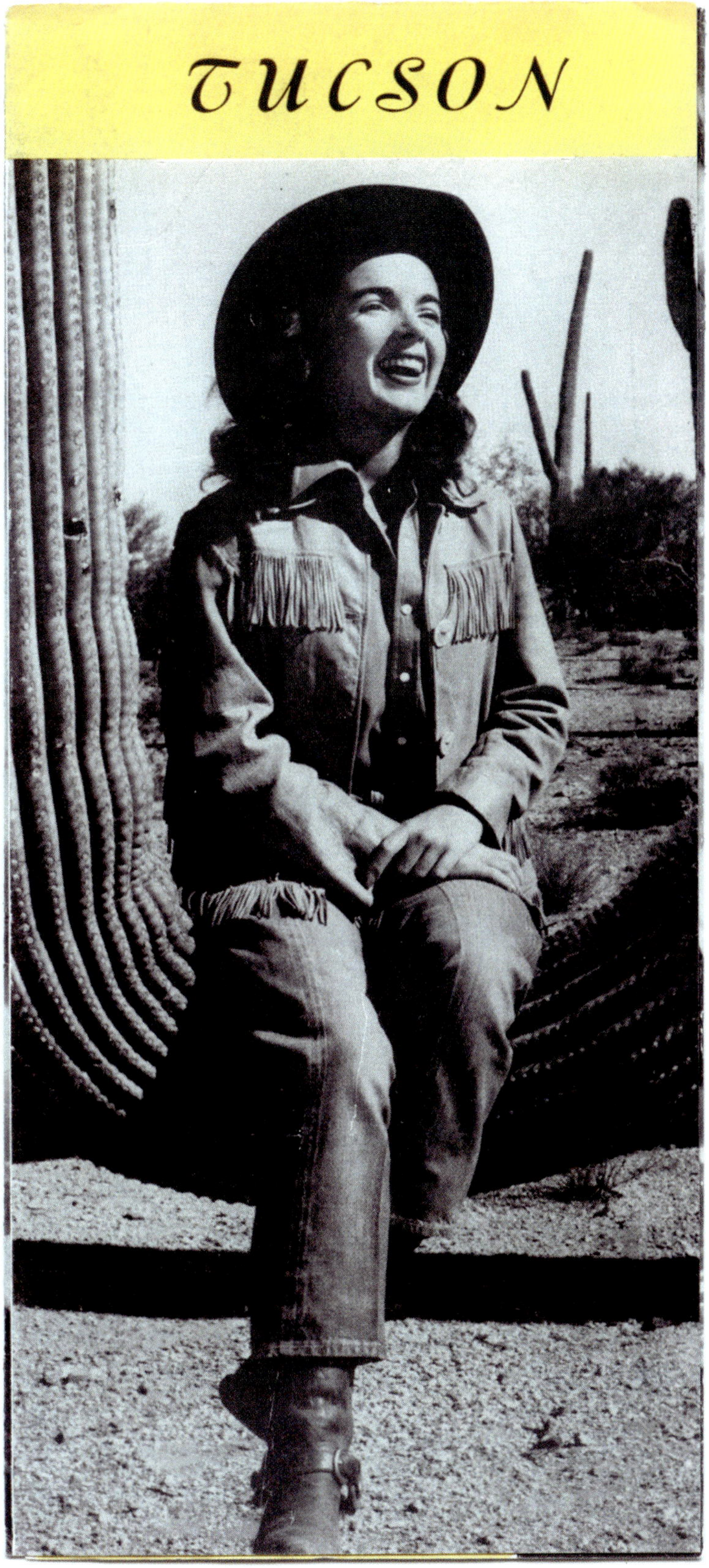

FIGURE 7.11. Brochure cover. Tucson Arizona Sunshine Climate Club, ca. 1952.

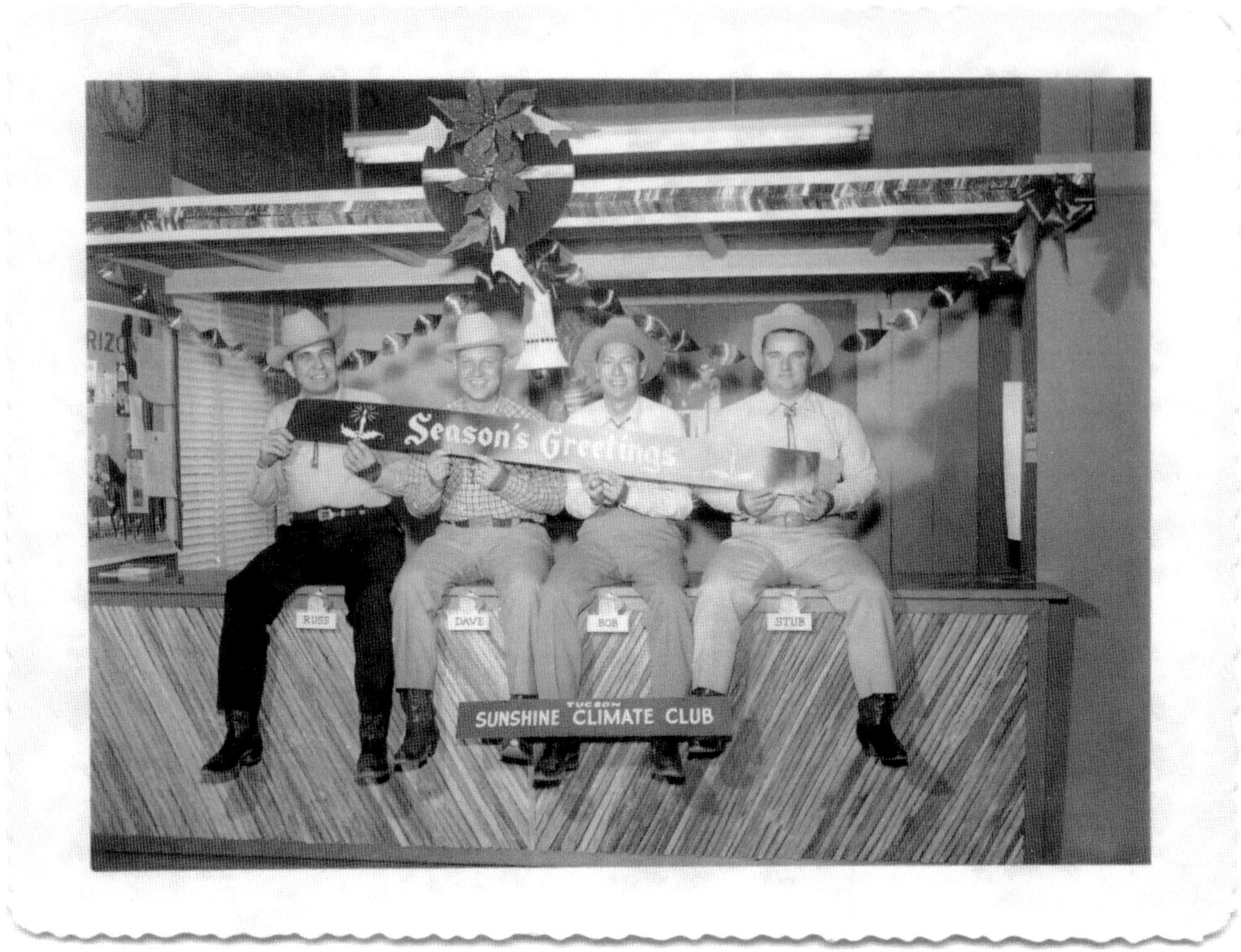

FIGURE 7.12. Photograph. Seasons Greetings. Sunshine Climate Club, 1955. The officers of the Tucson Sunshine Climate Club pose on their saguaro rib office counter for their annual holiday card. Left to right: Russell L. Soden, business manager; Dave Steger; Robert M. Ridell, Jr., publicity director; and C.L. "Stub" Ashcraft, Jr., general manager.

Such was the legacy of the nineteenth-century bumper crop upon whose arms the photo publicity of the mid-twentieth century literally sat (Figure 7.11). Sunshine Climate Club photos pictured bathing suit models sitting, standing and leaning on the arms of saguaros, instilling the club's fun-in-the-sun sales message with a welcoming proposition that broached no distance between visitor and plant.

At the Climate Club's office on Tucson's Miracle Mile, a new generation of boosters took up the campaign. A caption sheet for a now lost photograph explained, "The walls are papered with newspaper and magazine clippings of gag photos of Arizona cacti" (Figure 7.12; Western Ways n.d. [1950?]).

The promulgation of club leaders' desert fantasies relied as much upon the ubiquity of droop-armed saguaros as the cultivation of high school fashion models that appealed to the nation's photo editors—along with photos of the city's annual rodeo, horseback rides, and the desert's flowering beauty spots. The saguaro pose led them all (Figure 7.13).

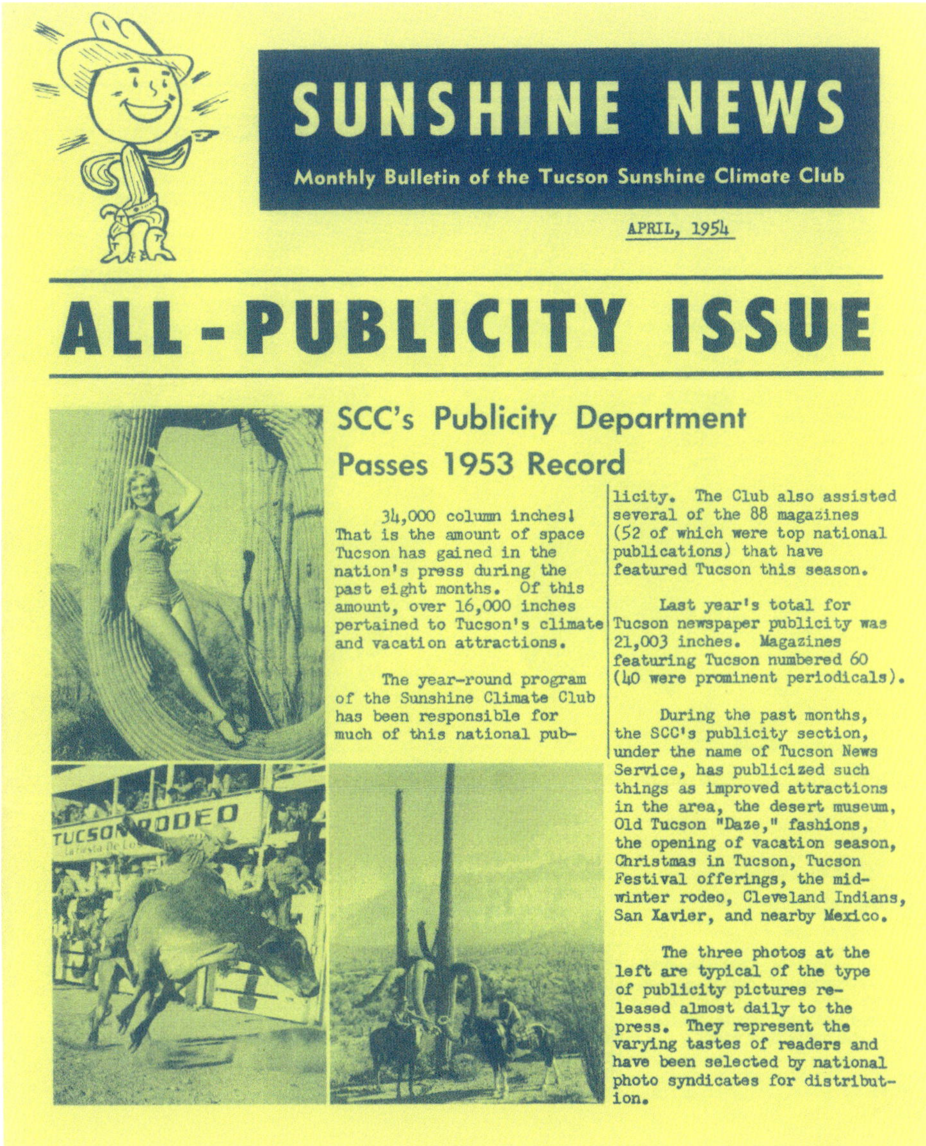

SUNSHINE NEWS

Monthly Bulletin of the Tucson Sunshine Climate Club

APRIL, 1954

ALL-PUBLICITY ISSUE

SCC's Publicity Department Passes 1953 Record

34,000 column inches! That is the amount of space Tucson has gained in the nation's press during the past eight months. Of this amount, over 16,000 inches pertained to Tucson's climate and vacation attractions.

The year-round program of the Sunshine Climate Club has been responsible for much of this national publicity. The Club also assisted several of the 88 magazines (52 of which were top national publications) that have featured Tucson this season.

Last year's total for Tucson newspaper publicity was 21,003 inches. Magazines featuring Tucson numbered 60 (40 were prominent periodicals).

During the past months, the SCC's publicity section, under the name of Tucson News Service, has publicized such things as improved attractions in the area, the desert museum, Old Tucson "Daze," fashions, the opening of vacation season, Christmas in Tucson, Tucson Festival offerings, the mid-winter rodeo, Cleveland Indians, San Xavier, and nearby Mexico.

The three photos at the left are typical of the type of publicity pictures released almost daily to the press. They represent the varying tastes of readers and have been selected by national photo syndicates for distribution.

FIGURE 7.13. Newsletter cover. *Sunshine News* All-Publicity Issue. Sunshine Climate Club, April 1954.

The organization of the "Sunshiners Modeling Club" in 1948 formalized recruitment and training. Retitled the "Sunshine Modeling Club" in 1949, a core group of 30 to 35 young women from Tucson High School made themselves available for publicity, news features, and advertising programs. The qualifications for club membership included "poise, personality, courtesy, willingness to learn, work, and cooperate with others" (Chambers n.d. [1955?]). The Sunshine Climate Club cleared all assignments through the school counselor. Photographers paid models at commercial rates, with all monies turned over to the school for fashion and beauty supplies used in the program. All models working on assignments were chaperoned. The Sunshine Model Club became an active part of the school's vocational

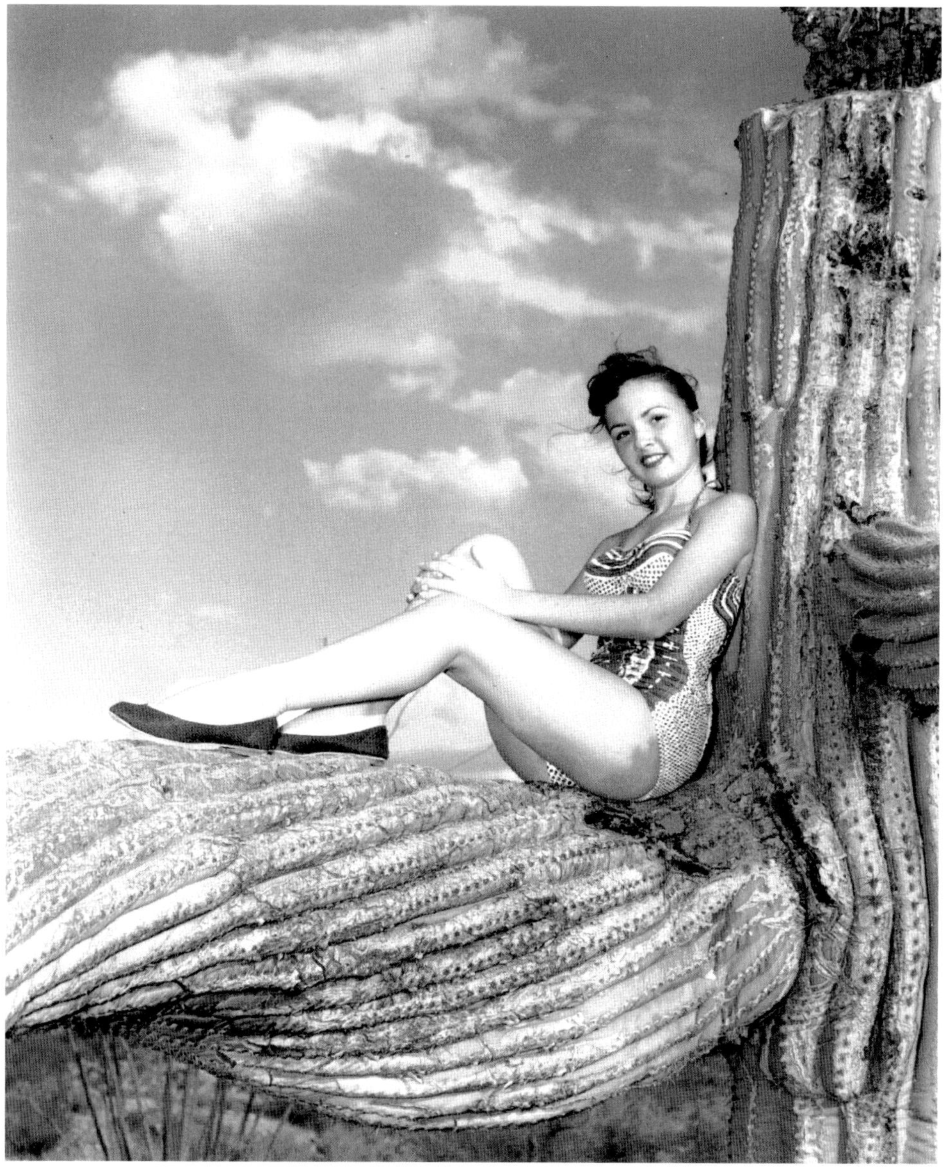

FIGURE 7.14. Copy photograph. Swimsuit model. Sunshine Climate Club, ca. 1958. Some helpful data on how to become a good Sunshine Club model.

training program, recruiting members with an annual competition that combined the elements of a high school cheerleading program with that of a beauty pageant. In 1954, for example, some 200 young women tried out for one of the 15 places in the ongoing group of 30 models. A panel of judges including the faculty advisor, placed each contestant in the desired size range, from 10 to 14. The daily newspaper listed each winning contestant's name and size. The seven-member panel of judges, made up mostly of the photographers who would be taking the pictures, included Western Ways' Peter Balestrero and Ray Manley (Figures 7.14–7.16; Western Ways 1954). Miss Margaret Booher of the school's Home Economics Department, who preferred "Home Making Department," became the club's first faculty advisor, fol-

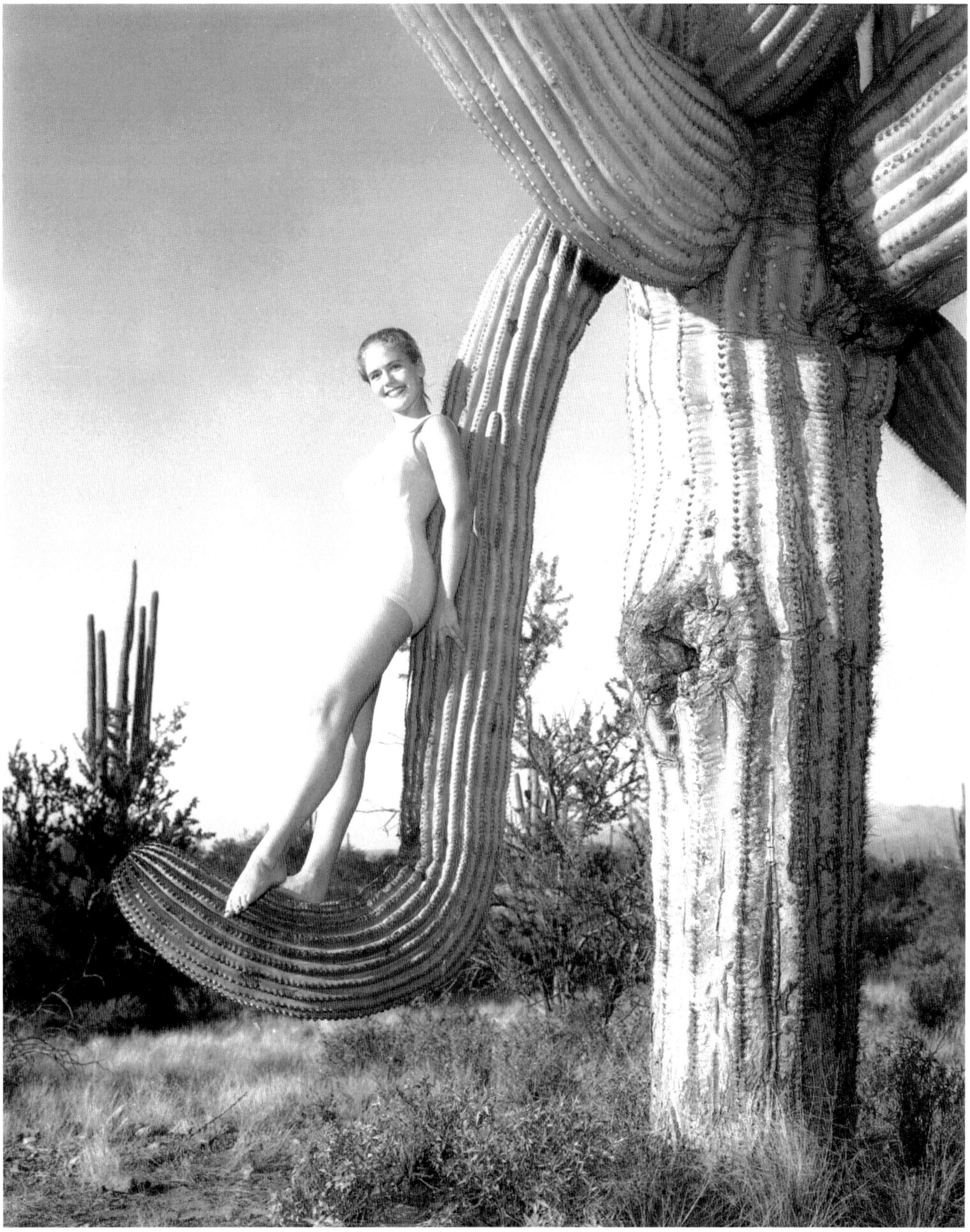

FIGURE 7.15. Copy photograph. Swimsuit model. Sunshine Climate Club, ca. 1958.

lowed in 1955 by Miss Mildred Bready, the school's typing instructor (Chambers n.d. [1955?]).

Tucson's "model high school" received its own press coverage in national magazines. Reporters took note of the "girl pool" of "especially selected, carefully trained high school youngsters who perform a novel civic duty…usually in bathing suits." Photographers made choices for assignments by scrutinizing the 8 × 10 bathing suit photographs kept on file in the Sunshine Climate Club office or by visiting a model club meeting (Sturgis n.d. [1954?]; Riddell n.d. [1960?]).

To view pictures of individuals sitting and standing on the drooping arms of the legacy plants was to ask for the suspension of disbelief. The unwitting may have overlooked the absence of spines and still found the photographs remarkable. The witting may have turned away knowing the loads of attention that hastened the plants' demise. Yet the photographs are still doing their job—causing viewers to stop to stare at their cheesy composition and jaw-dropping aplomb—breathtaking, for all the wrong reasons.

Whether clipped, padded, or tipped-in, the saguaro became the lubricant of a distinctly modern Southwest into which anyone might climb. Ruth Egermayer, the wife of the park superintendent of Saguaro National Monument, posed as a park visitor for *Arizona Highways* with a droop-armed saguaro, one of the oldest and most unusually formed specimens in the park (Figure 7.17). The photo of her leaning into the plant appeared on the cover of the park's 10-cent trail guide, inviting visitors: Come here. Come closer. See this thing (Figure 7.18).

The saguaros that sacrificed tips and spines for gag photography represent the closing of an era of environmental ambivalence at the dawn of a greater appreciation of the fragility of nature's sharpest curves. The padding of spines was a moment in the saguaro's arc from humanly relatable publicity object to the inviolable symbol of conservation and community it is today.

This then was the foundation upon which today's saguaro moment may be weighed and considered. No longer asked to suspend disbelief, we believe models wearing blouses, skirts, and shirts picturing saguaros; a wooly cable-knit mask topped with flowers; or the out-sized art at a music festival. We stop and wonder at the revival of interest in the postwar silkscreen fabrics of artist Harwood Steiger (Figure 7.19), museum exhibitions featuring the Hollywood style of Roy Rogers and Dale Evans (Figure 7.20), and the rhinestone western wear inspired by tailor Nudie Cohn (Figure 7.21; Holly 2001; Nudie 2004; DeVillemarette 2018).

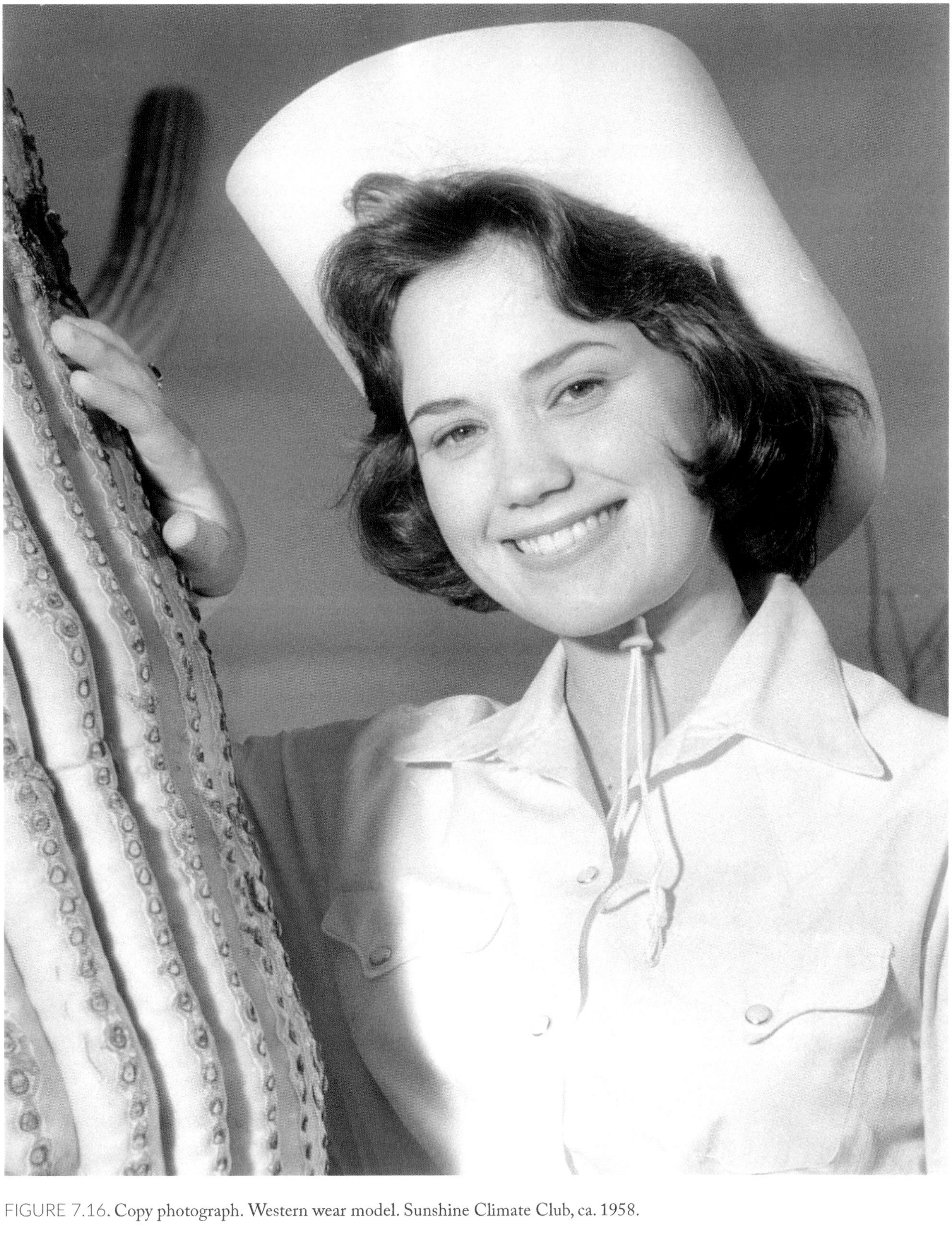

FIGURE 7.16. Copy photograph. Western wear model. Sunshine Climate Club, ca. 1958.

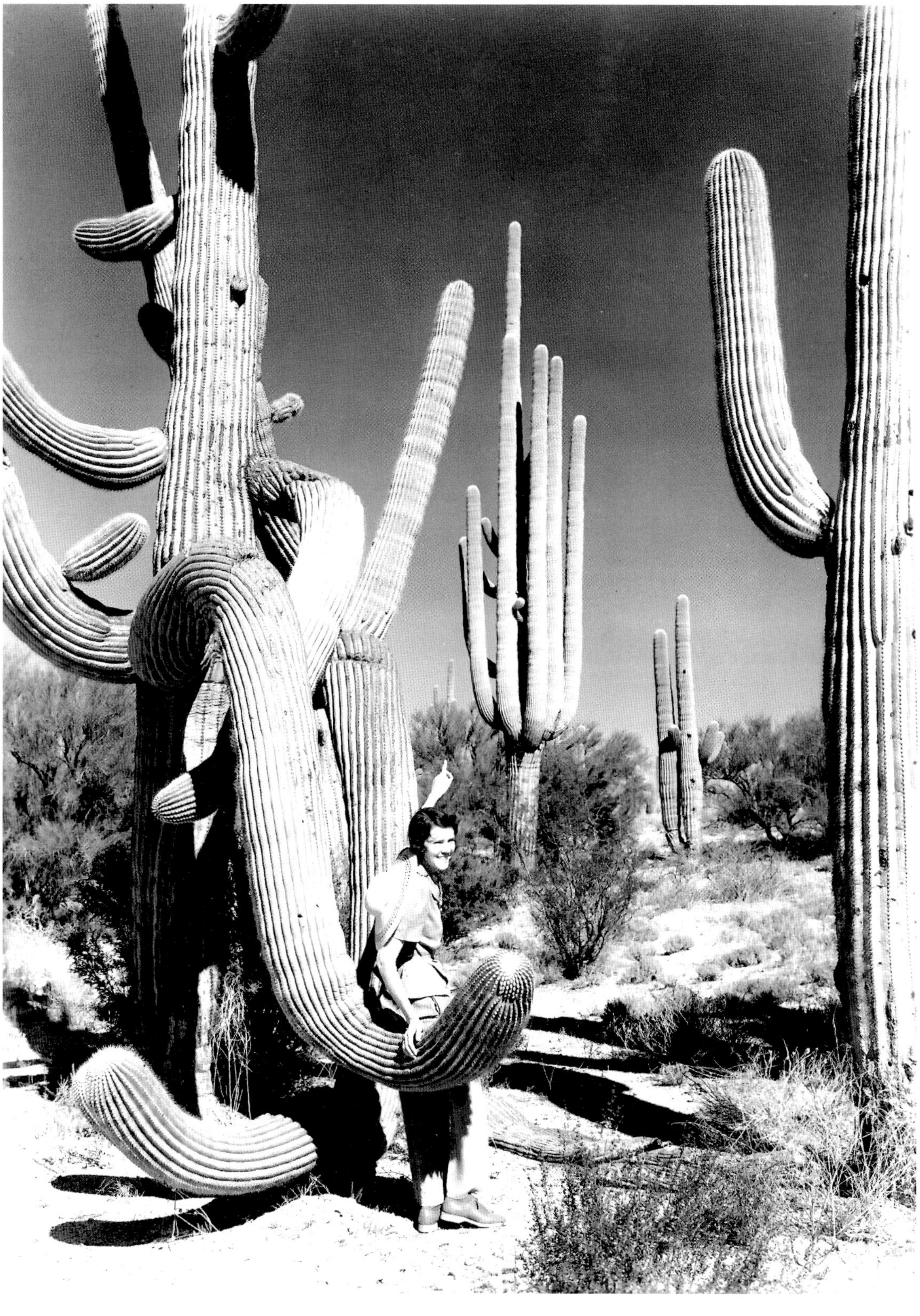

FIGURE 7.17. Copy photograph. Ruth Eggemeyer poses for *Arizona Highways*, Saguaro National Monument. 1941.

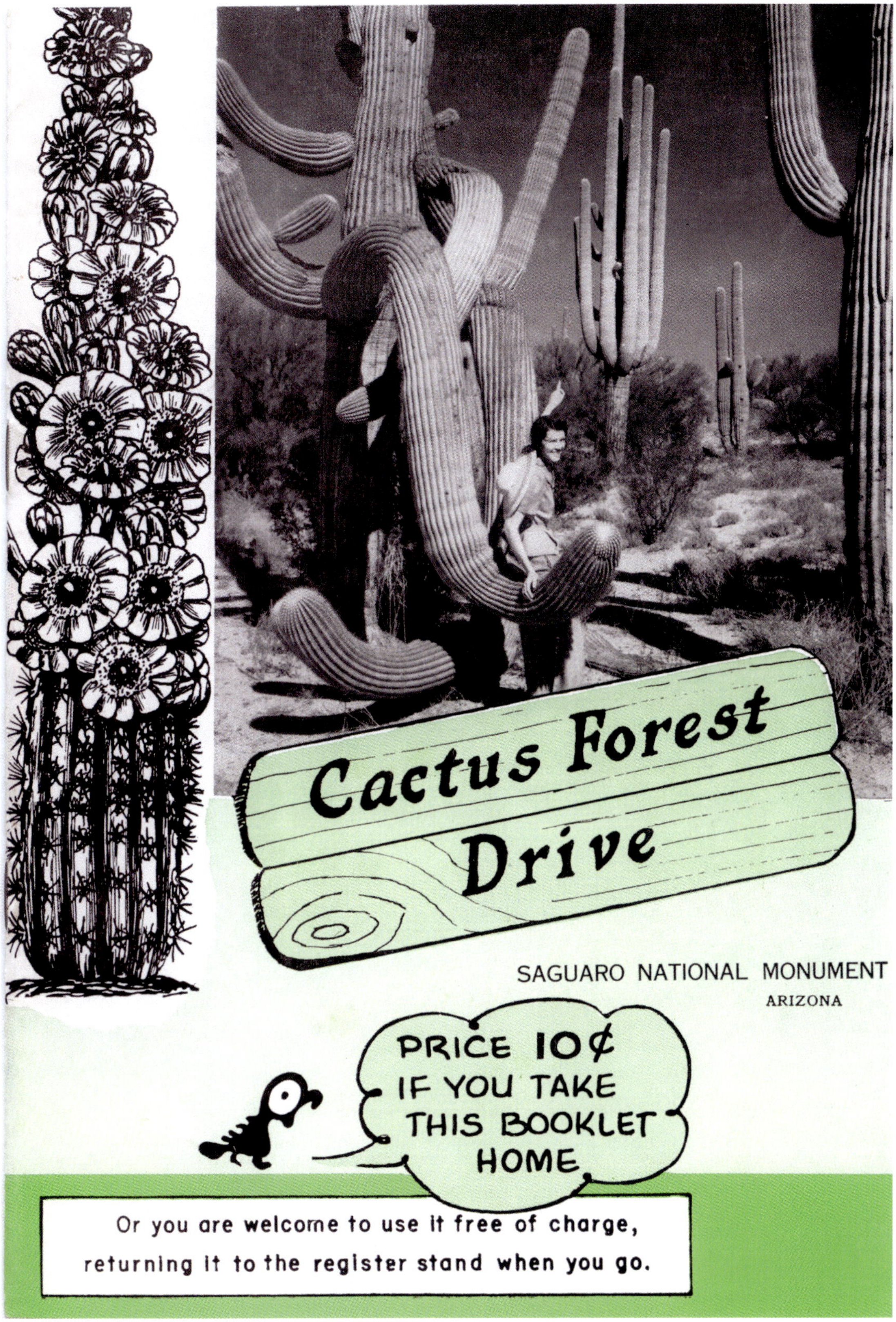

FIGURE 7.18. Tour booklet cover. "Cactus Forest Drive." Saguaro National Monument, February, 1954.

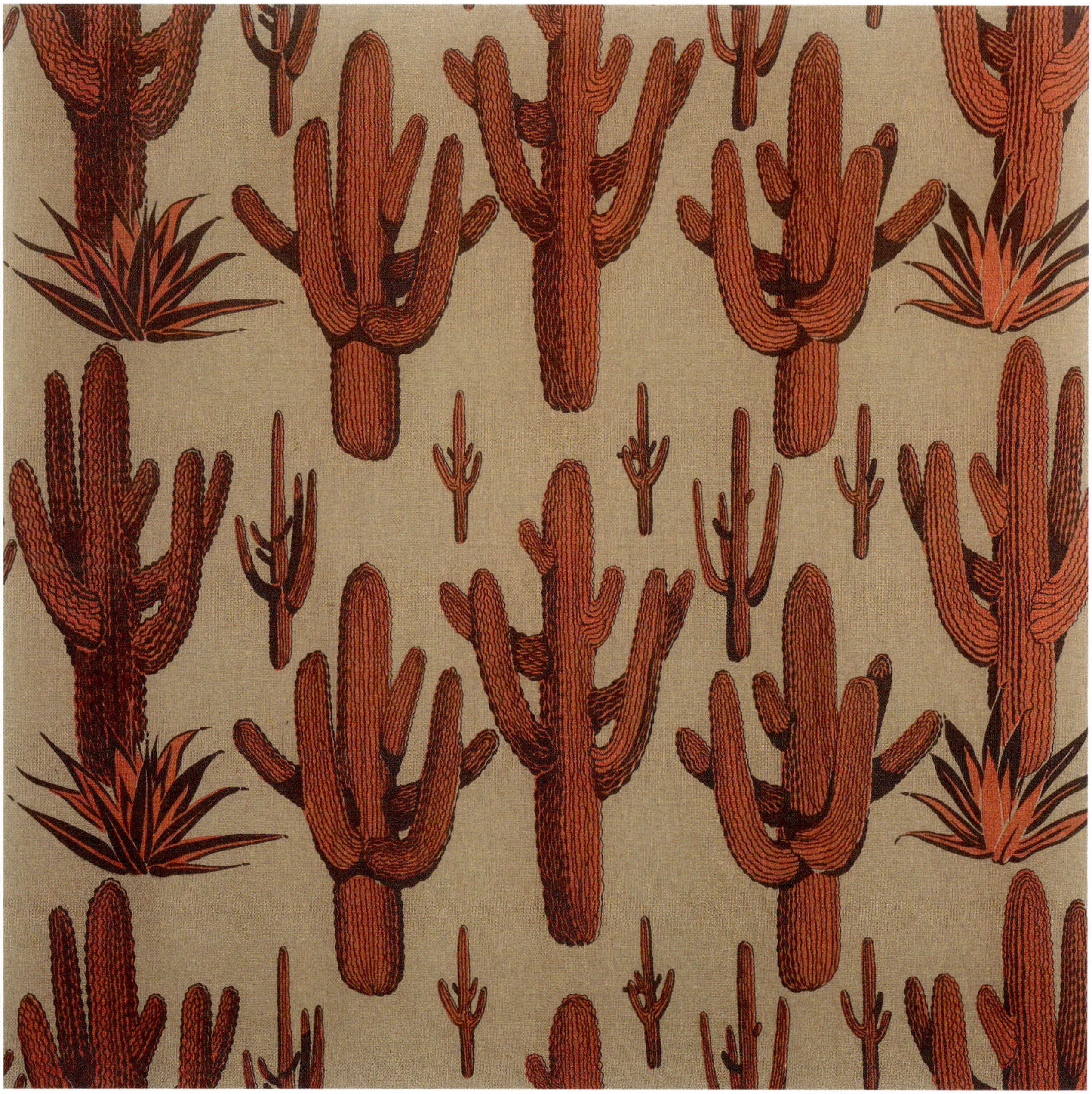

FIGURE 7.19. Fabric silk-screen print, Saguaro Land by Harwood Steiger, 1966. Photograph by Jude Ignacio and Gerardine Vargas, Tucson Historic Preservation Foundation, Tucson.

Saguaro likenesses populate clothing ranging from playful one-off designer statements to designs with mass market appeal. Representing the former, a woman models a saguaro skirt in the streets of Milan before a Prada fashion show (Figure 7.22); the latter, a London apparel maker markets a white patterned shirt whose saguaro print recalls the one that William Emory drew for George Engelmann in 1848 (Figure 7.23).

Noting the various iterations of the cactus phenomenon from earrings to T-shirts to A-line skirts, one observer of contemporary fashion notes that "the cactus

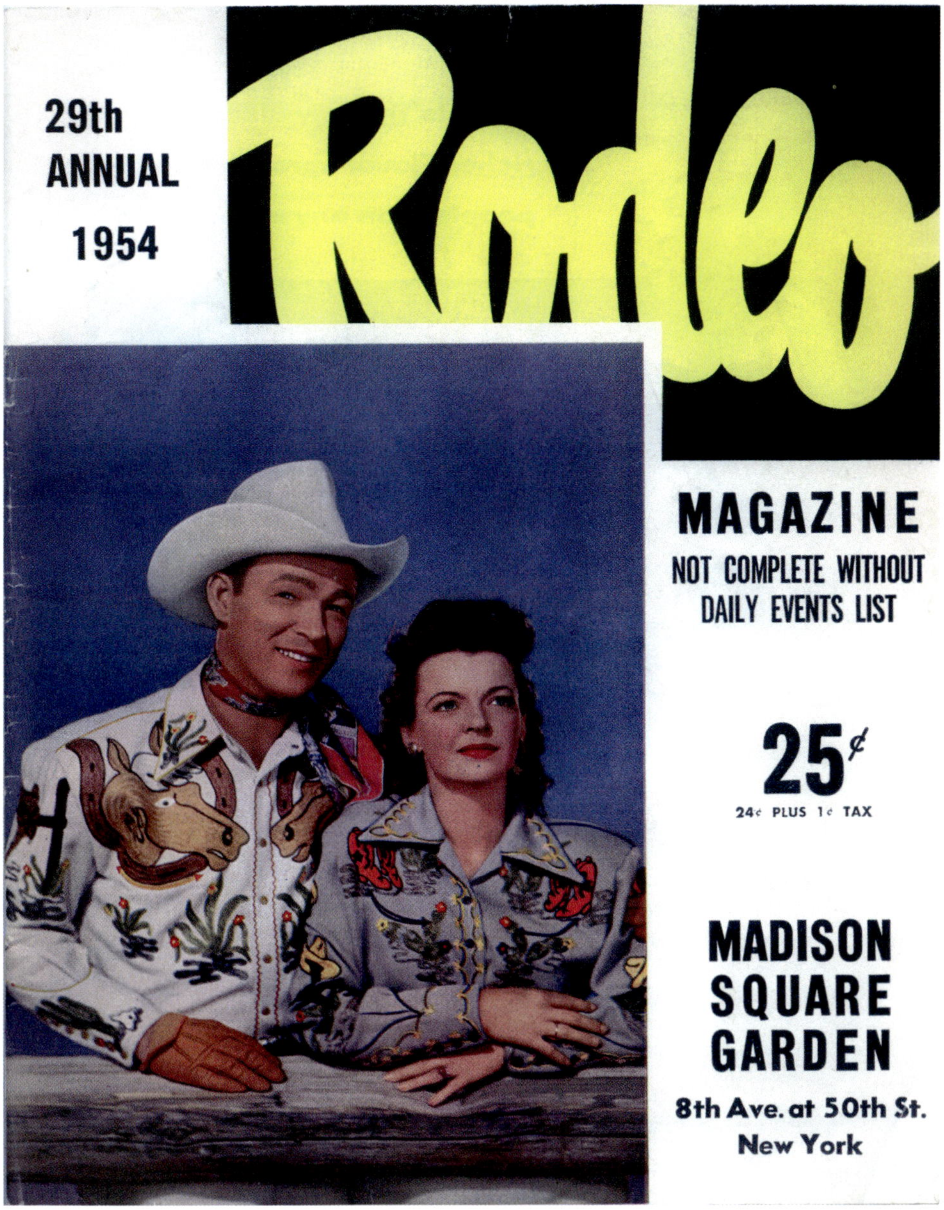

FIGURE 7.20. Magazine cover. Roy Rogers and Dale Evans. *29th Annual Rodeo*, Madison Square Garden, New York, 1954.

has officially transcended the windowsills of Kinfolk subscribers and become a statement-making look worth working into your wardrobe" (Moran 2016). While many apparel designs deploy subtle variations on the saguaro, one Chicago-based designer, Mady Berry, fully exploited the dimensional possibilities of the mature plant. Inspired by a trip to Arizona, Berry created a cable-knit saguaro sweater with bulky accordion folds, articulated arms and a columnar flower-topped headcover with eye holes (Figure 7.24). Part of Berry's saguaro-themed lineup entitled, "Sand Doesn't Have to Be Bland," her project featured a second piece reminiscent of the cactus sun-suit of the early 1940s—a "desert guide" entitled "Siren," featuring

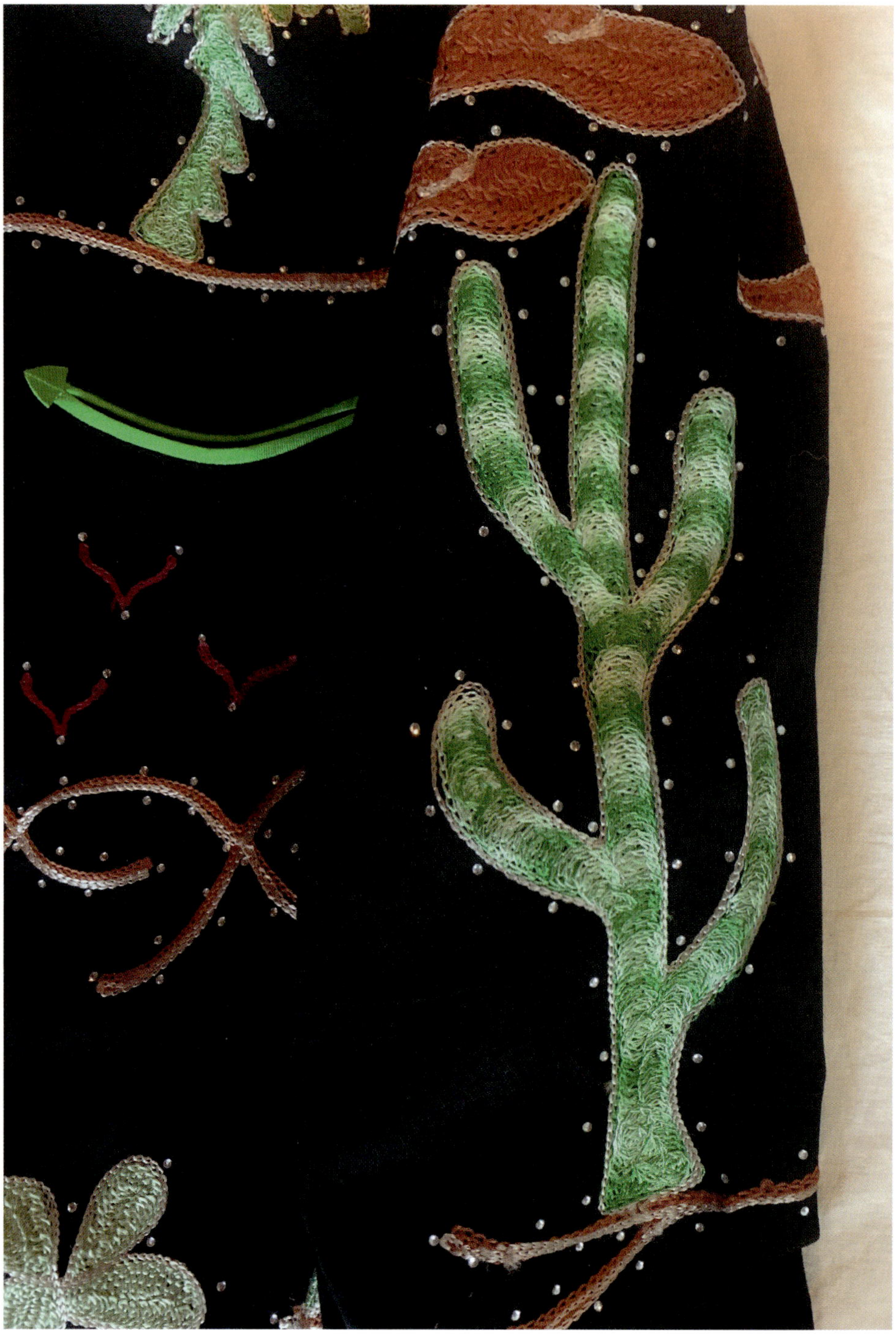

FIGURE 7.21. Photograph. Rockmount embroidered western shirt sleeve detail. 2020.

a pleated conical bra embedded with glass eyes. In an interview in *Paper*, Berry (Moran 2016) explained, "Each look [in my collection] has a set of glass eyes where they 'see' from. I thought it would be appropriate and a bit cheeky if the Siren saw from her breasts" (Figure 7.25). The immediate inspiration for the piece came from the cactus dress her mother owned in college (Moran 2016).

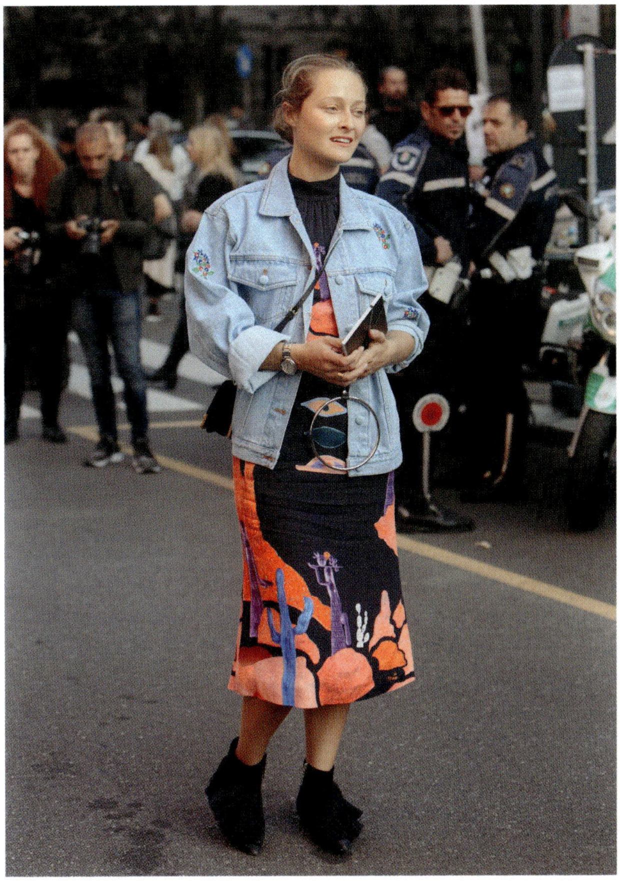

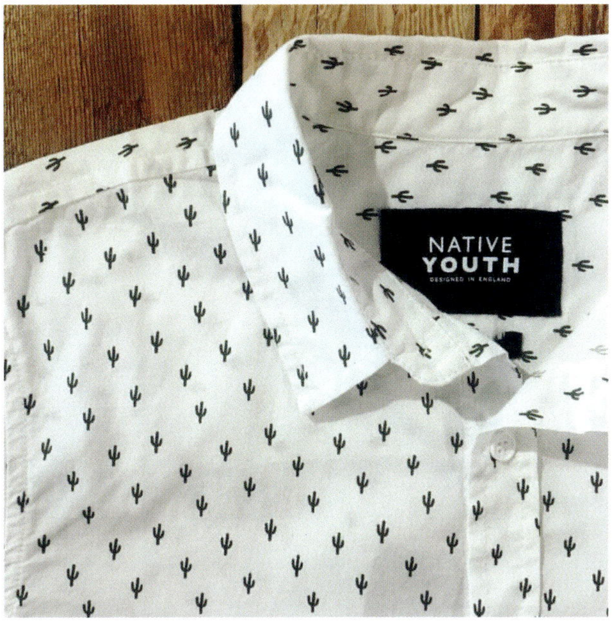

LEFT, FIGURE 7.22. Woman modeling saguaro skirt, Milan, Italy, 2015.

TOP, FIGURE 7.23. Sport shirt with saguaro print fabric. Native Youth, London, England, 2016.

Saguaro iconography accompanies plants in homes, retail spaces, and Instagrammable events. One observer declared cactus to be "the new pineapple," noting its graphic expression in homewares and paper goods products, "accessories both stylish and affordable . . . used to personalize a space" (Trendbible 2019). Another, noting the prevalence of actual cacti and images of cacti "sprouting in all corners of our culture," declared the cactus to be "the pop-up plant of our desolate age." The availability of small specimens for the home was "a welcome change from the ubiquity of the saguaro" (Cocozza 2017).

Though few contemporary observers of the phenomenon seemed aware of the history of collecting or its iconography, Patrick Sisson, writing in *Curbed* took the long view. Quoting a 1935 *New York Times* feature article by L.H. Robbins, Sisson (2018) noted that cacti had "played a big part in the human history of the romantic southwest" (Robbins 1935:SM11). Sisson (2018) concluded that the "commodification of cactuses, and their slow expansion into our homes and retail spaces, is just the latest adaptation by these rugged plants to a harsh environment. These plants aren't just Instagram fodder; they're spiny survivors."

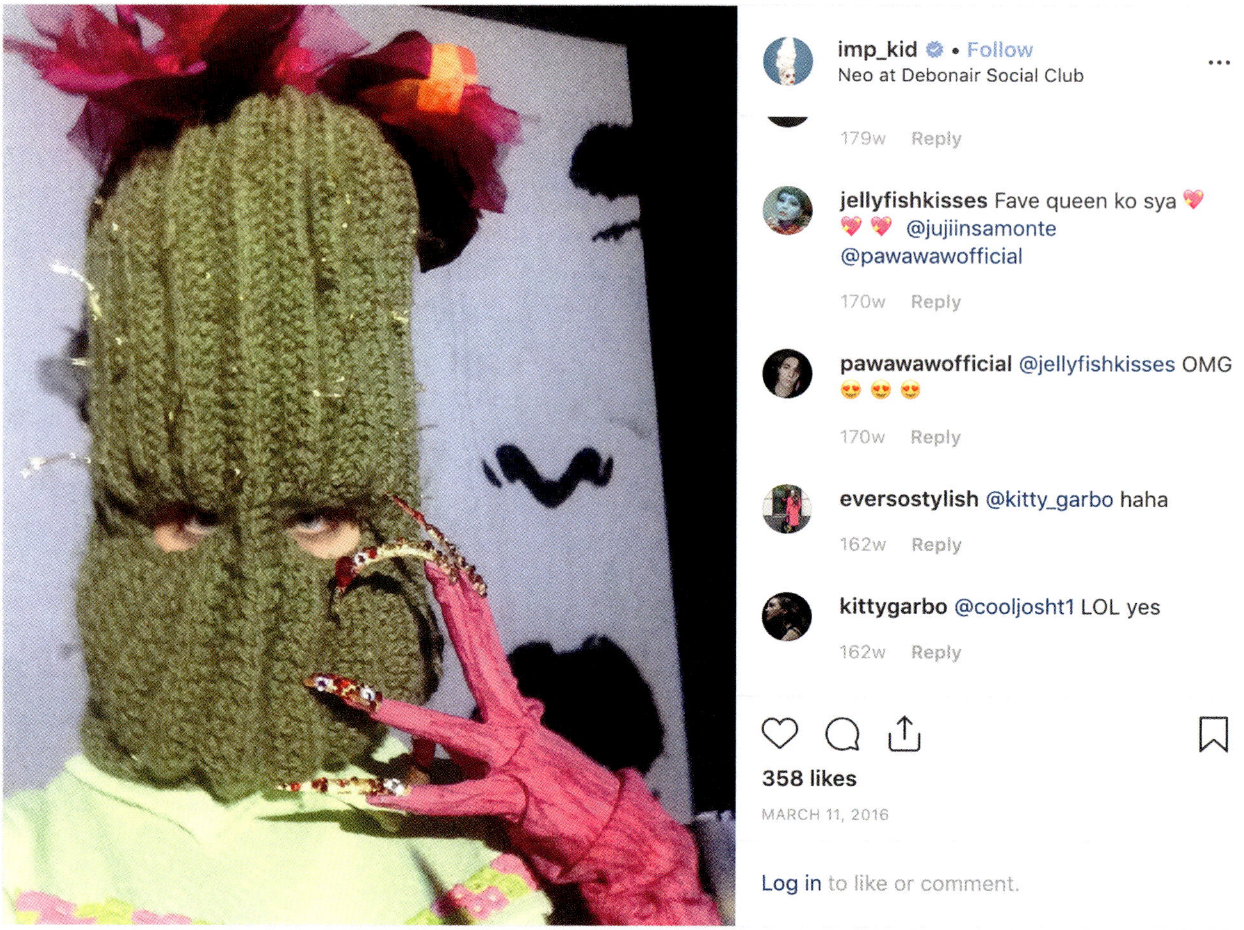

FIGURE 7.24. Instagram screen shot. 2016. Photograph by imp_kid.

OPPOSITE PAGE, FIGURE 7.25. Siren. 2016. Mady Berry. Photograph by Wyatt McSpaden.

The staging of saguaro moments inspired by the prop patios and guest ranches of yore include the boxy saguaros created for the grounds of the 2019 Coachella Music Festival (Figure 7.26). A private event during the festival staged by Instagram at the Pond Estate in Palm Springs became the social media platform's "first-ever branded house experience." The cactus-themed event offered "an Instagrammable mecca of eye-popping color" directed to the attention of "influencers, artists, and content creators" (Manifold 2019). Created by artist D'ana Nunez, the event set the stage for "photo moments." The camera-ready estate featured a pond, an island landscaped with a saguaro propped up with a brace, and other succulents and palms. The amenities brought in for the event included a curtained entrance with branded "Desert Chill" drapes that opened to the lawn, a firepit and a small bridge to the island. "Each were surrounded by real cacti and neon graphic elements intermixed to infuse a surreal element into real life scenery" (Figure 7.27; Manifold 2019; Hoffman 2019; Weinberg 2019).

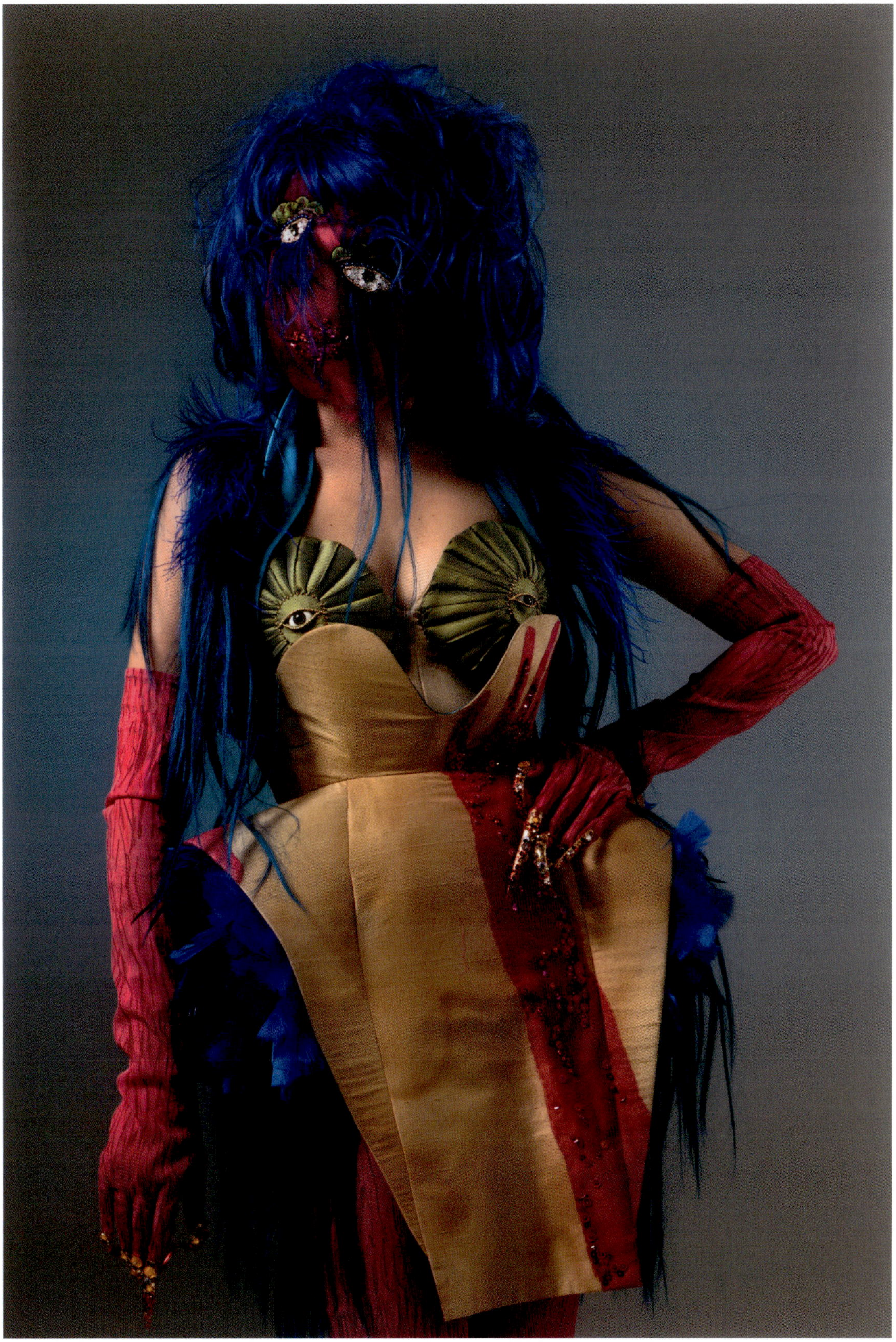

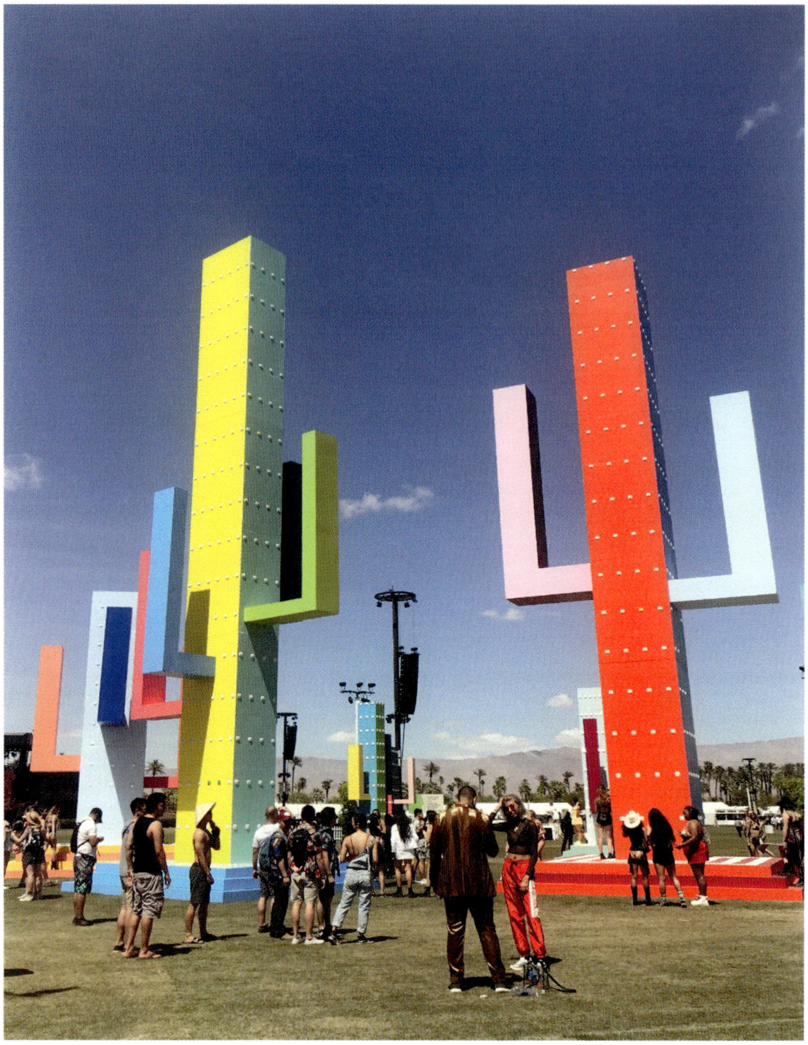

FIGURE 7.26. Coachella Valley Music and Arts Festival Saguaro installation, by Office Kovacs, 2019. Photograph by Claire Hoffman/Bizbash.

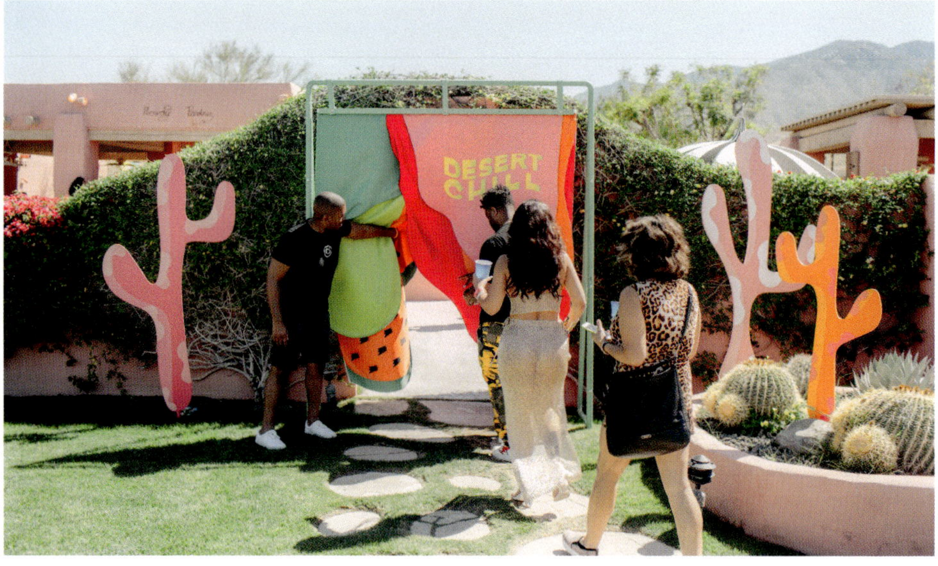

FIGURE 7.27. Pond Estate, Instagram Desert Chill party, Palm Springs, California, 2019. Photograph by Manifold.

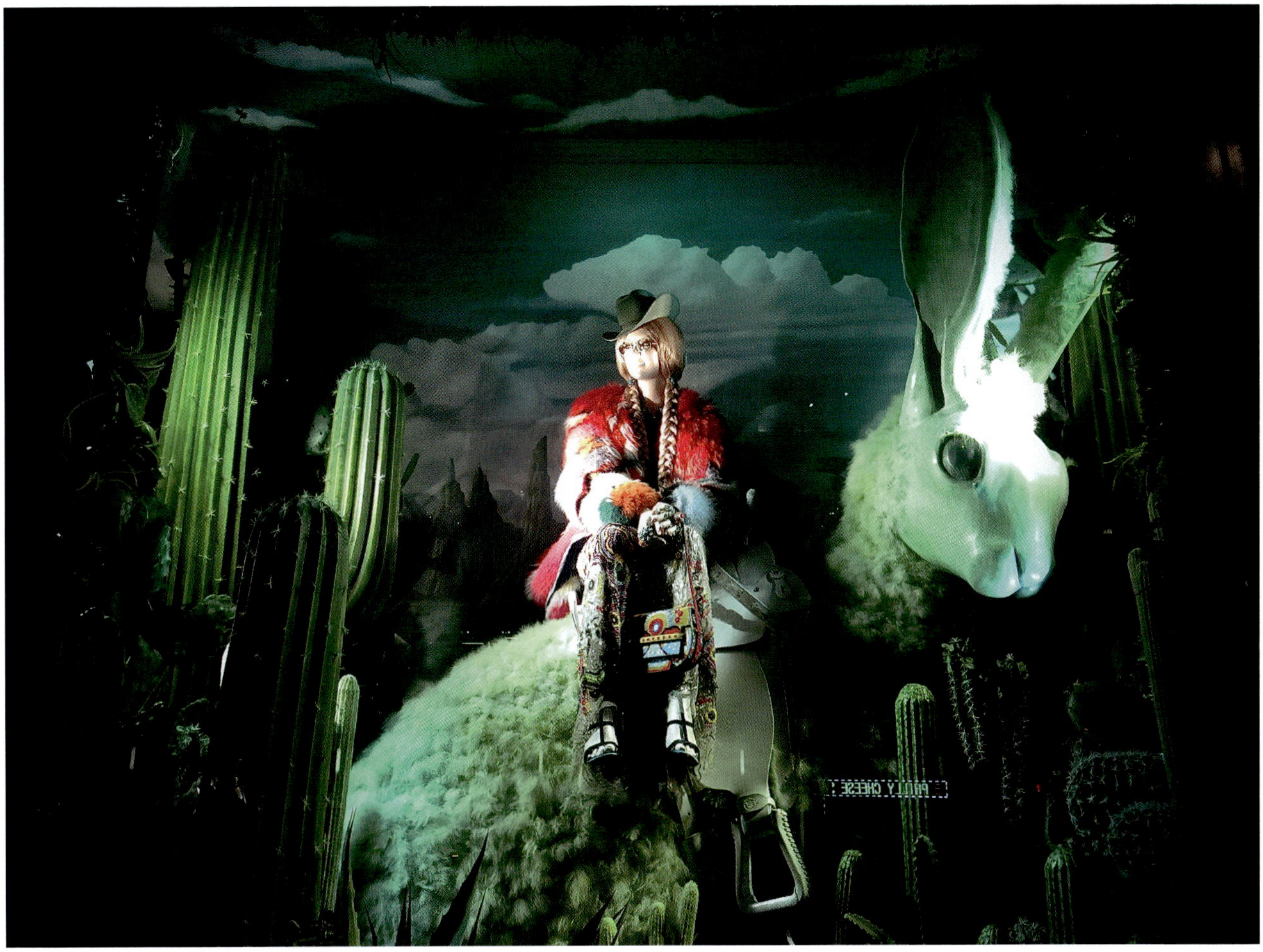

FIGURE 7.28. Bergdorf-Goodman New York holiday window, saguaro forest, white rabbit. 2016. Photograph by James Stave.

The saguaro's far-flung imagery has salable uses much as it always has. The 2016 holiday windows of New York's Bergdorf Goodman featured a dreamy interpretation of the arboreal desert (Figure 7.28). Store display director David Hoey set the stage. Hoey explained, "The windows are like delirious remakes of classic giant dioramas seen in a natural history museum" (Minton 2016; Silver 2016). The saguaro had its moment in the store's lead window. Entitled "The Hitch Hiker," the window featured a stylish mannikin sidesaddle on a Lewis Carroll rabbit brought up short in a dreamy saguaro forest.

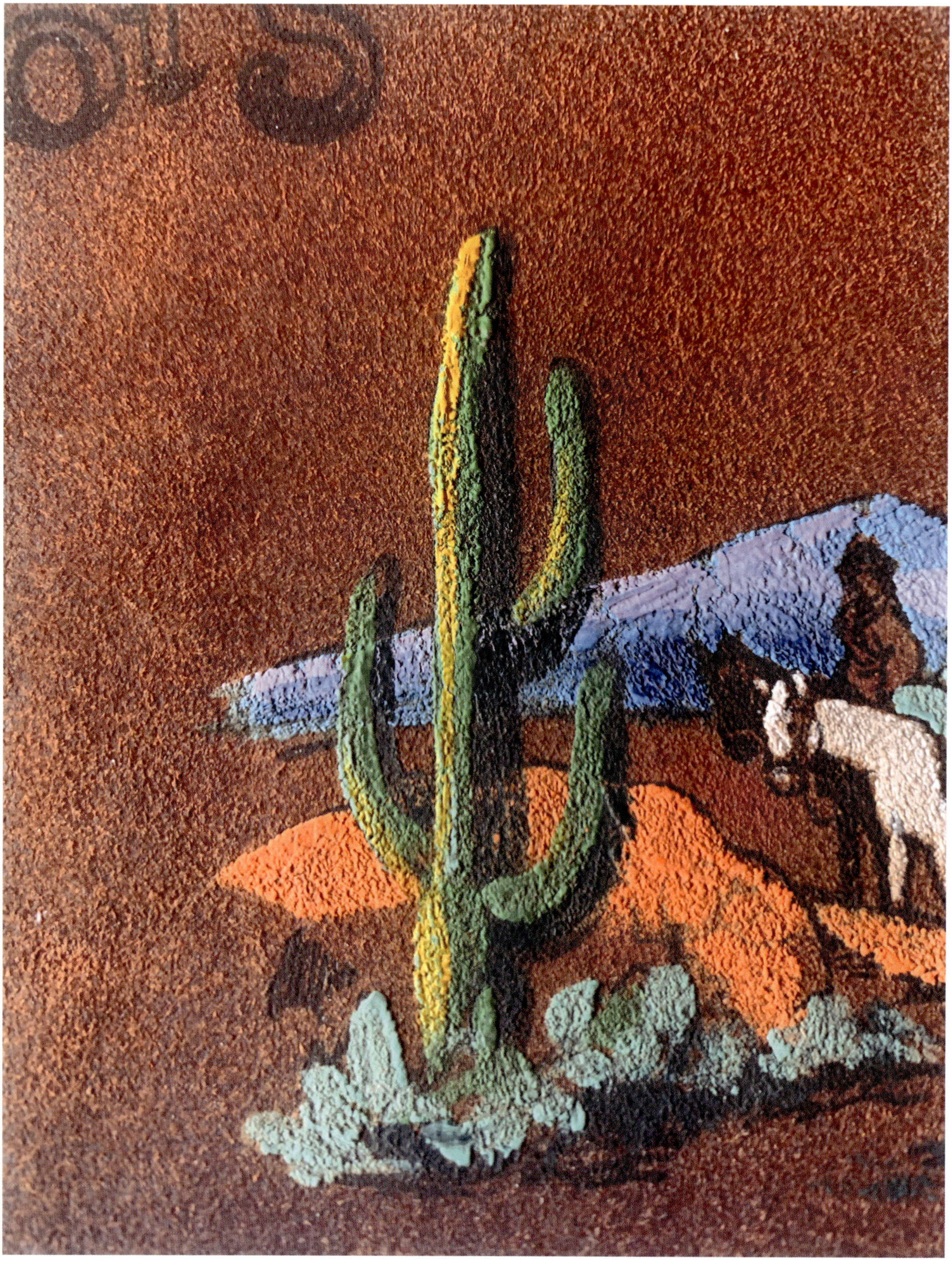

AFTERWORD
THE SAGUARO MOMENT

After the first renderings of the Sonoran Desert in woodcuts and engravings, the saguaro's appearance in unexpected places from the Pacific coast to the nation's world's fairs shaped ideas about the plant as a western symbol. Championed by some and derided by others, the saguaro became an uncertain symbol even in its own habitat. Growing an imaginative and far-flung range with a robust tourist trade, its welcoming arms took on an iconic significance by framing the desert for social engagement and leisurely consumption.

Saguaro iconography is nothing if not an expression of a human relationship with a plant that has historically sustained one's curious nature. Looking back upon people posing with the plant in snapshots, pictures and postcards, we can see ourselves.

We measure ourselves.

We express ourselves.

We take stock and care.

In the moment, we touch.

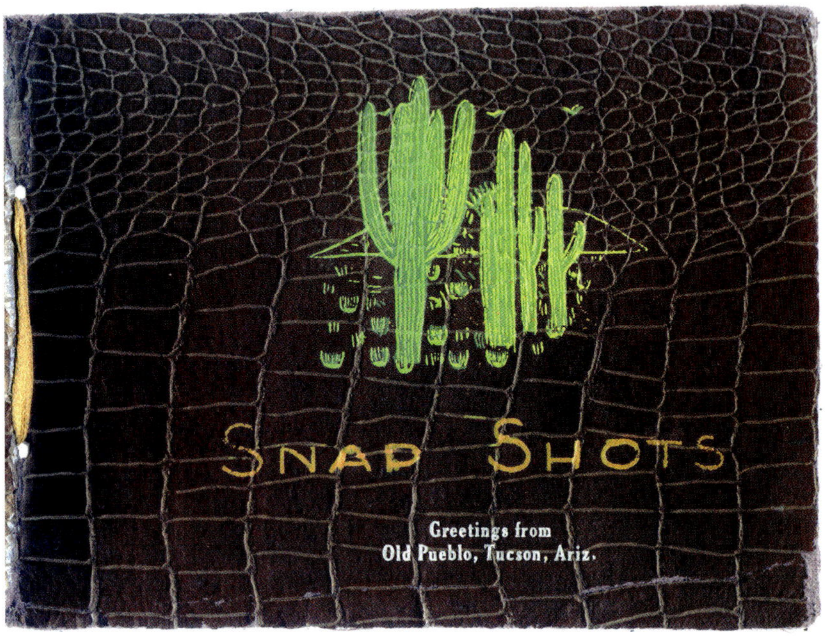

FIGURE 8.1. Plastic embossed photo album cover, Greetings from Old Pueblo, Tucson, Arizona, ca. 1935.

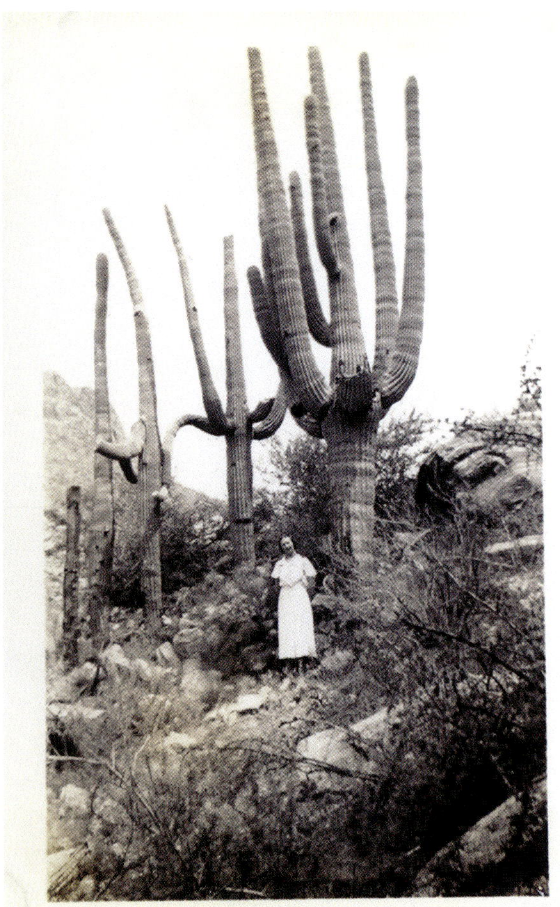

FIGURE 8.2 Photograph of Carrie Stouch of Gary, Indiana posing with a saguaro in Sabino Canyon, 1937.

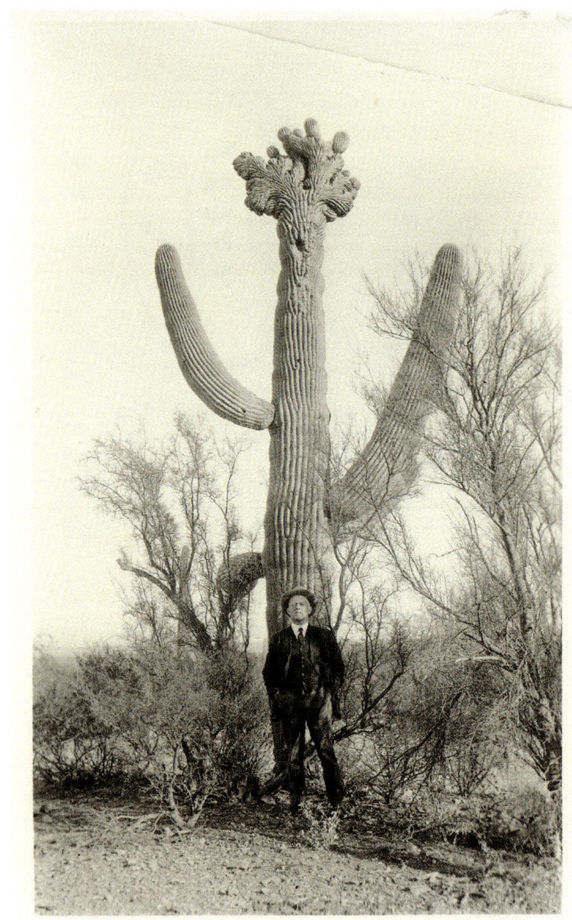

FIGURE 8.3. Photograph of man posing with a cristate saguaro, ca. 1937.

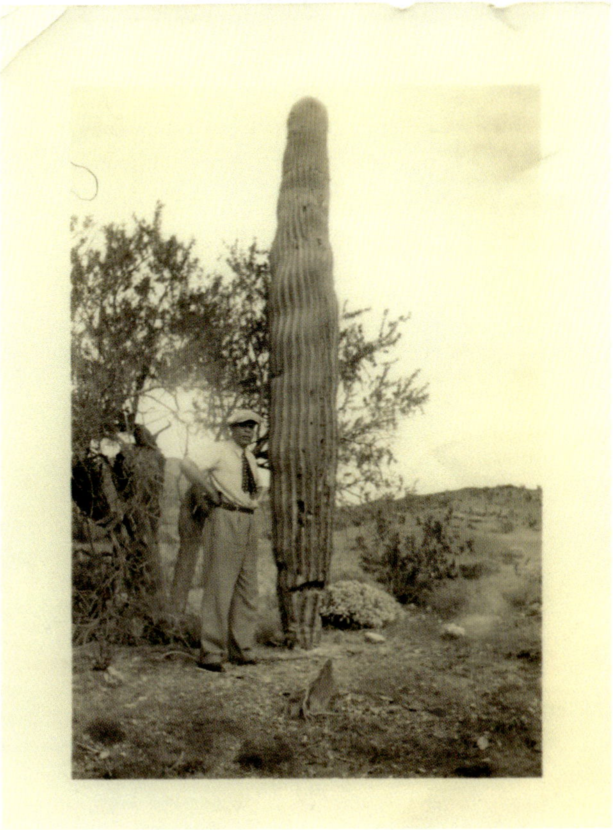

FIGURE 8.4. Photograph of man posing with a saguaro. February 9, 1940.

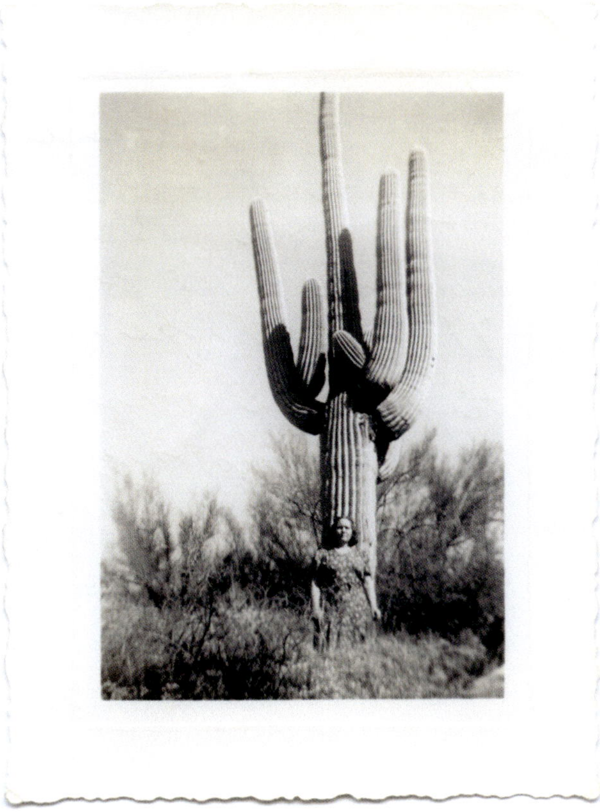

FIGURE 8.5. Photograph of woman posing with a saguaro. Written on the reverse: "These cactus are thick all over these hills and this is as pretty as I've seen. Every one has a different shape. I'll get some more cactus pictures. They are nice scenery." Ca. 1942.

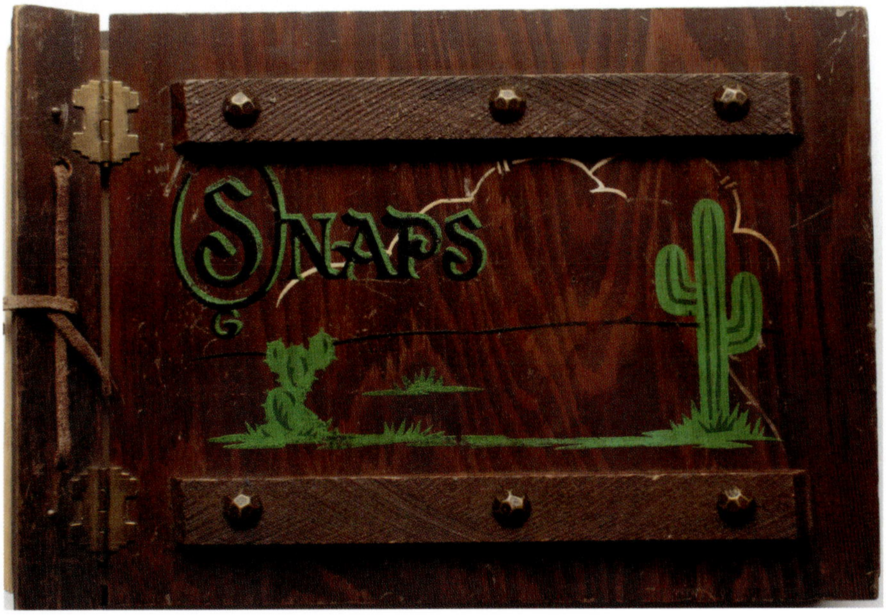

FIGURE 8.6. Wooden painted photo album cover, Snaps, ca. 1954.

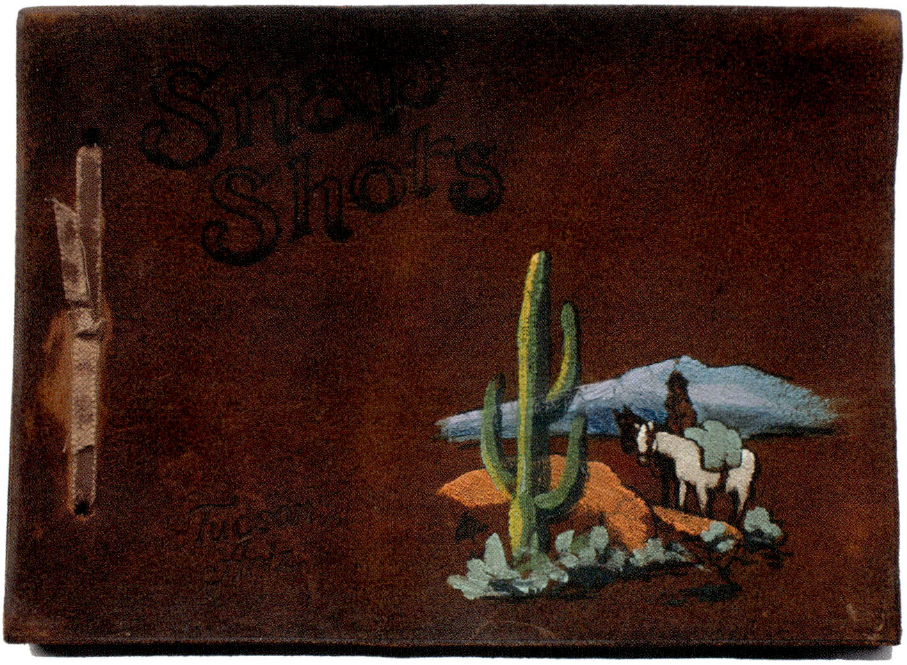

FIGURE 8.7. Leather painted photo album cover, Tucson, Arizona, ca. 1960.

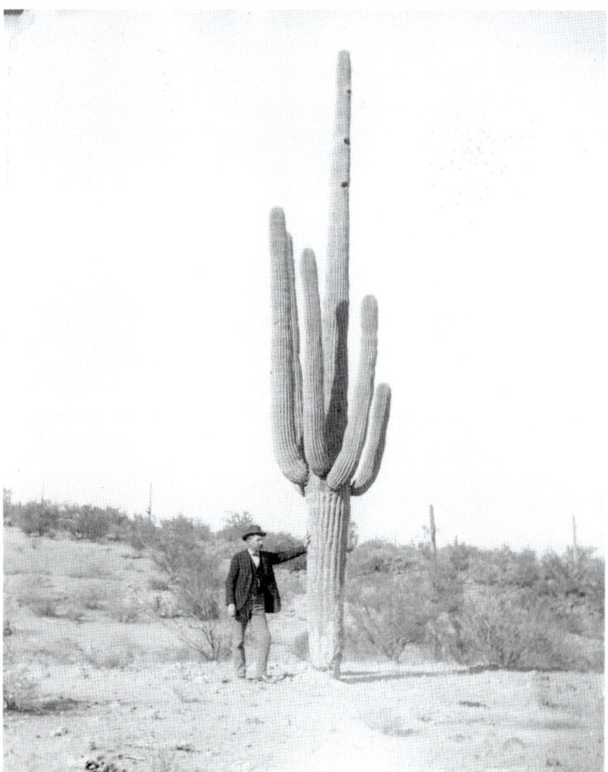

FIGURE 8.8. Photograph of man in jacket touching a saguaro, ca. 1919.

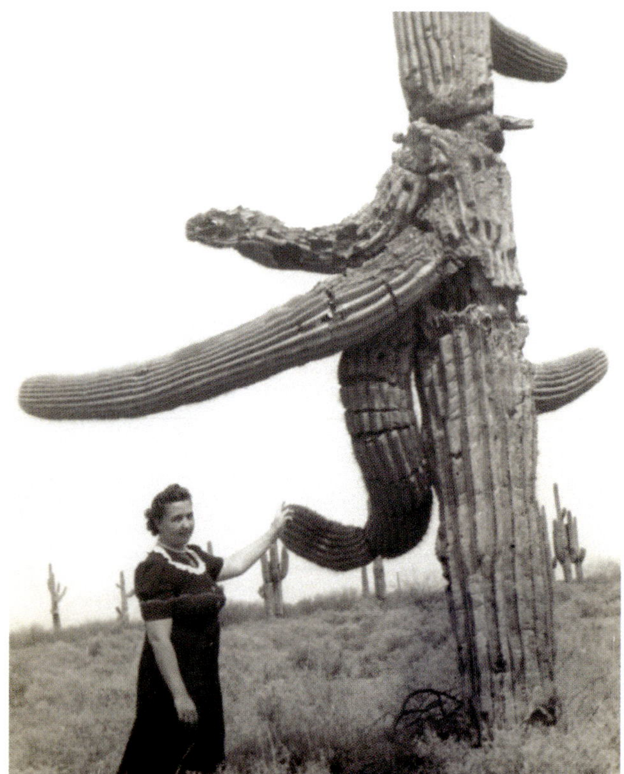

FIGURE 8.9. Photograph of woman touching the tip of a scarred saguaro arm, 1945.

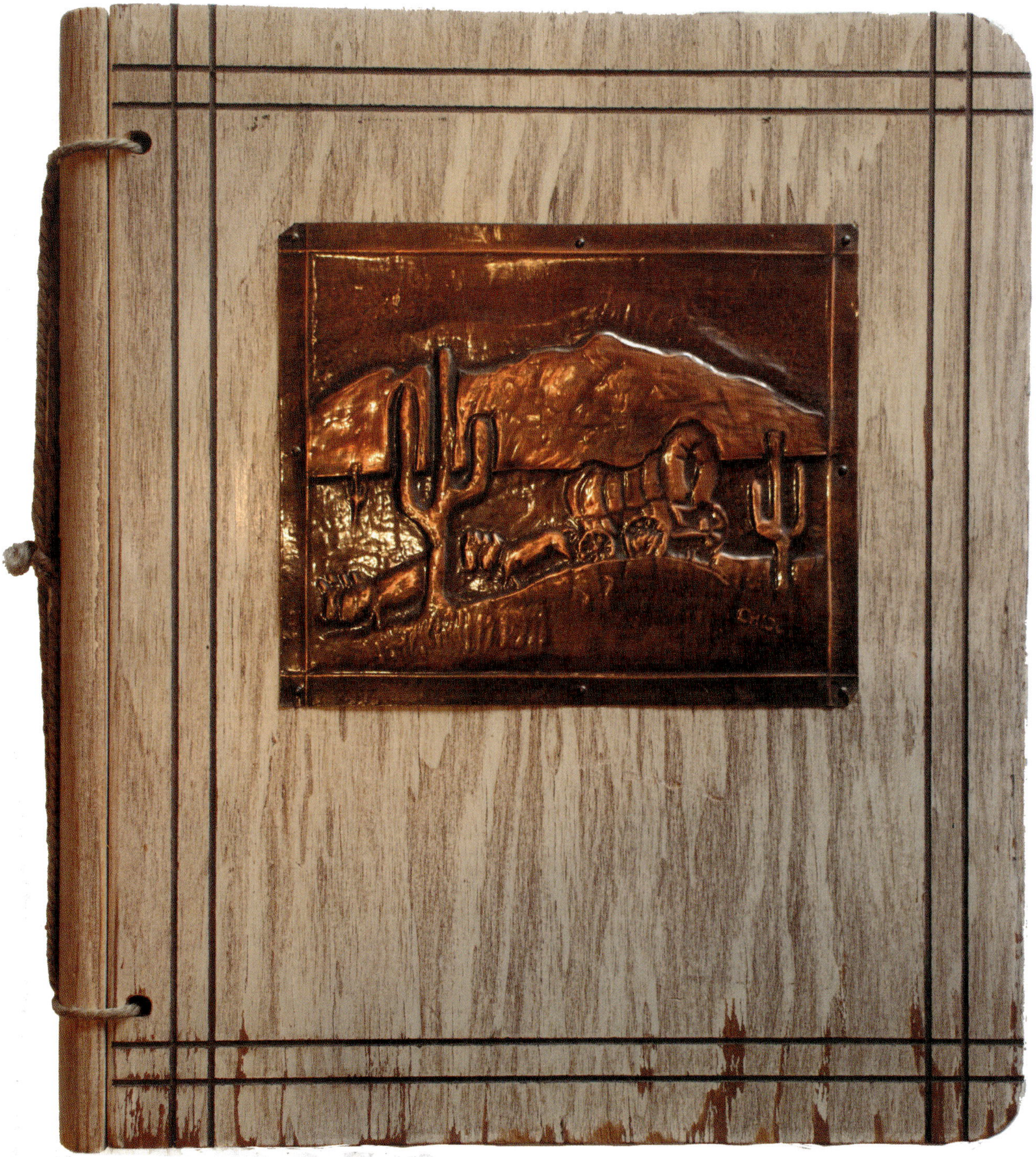

FIGURE 8.10. Copper embossed scrapbook cover, ca. 1970.

ACKNOWLEDGMENTS

I wish to thank the following individuals whose special knowledge and collections informed my study of saguaro iconography. At the Smithsonian Institution Libraries, Jim Roan, Katrina Brown, Lilla Vekerdy, Erin Rushing, Robin Everly, and Alexandra Alvis entertained my many requests for scans of their books and prints. At the National Museum of American History, Helena Wright, Joan Boudreau, Shari Stout, and Sarah Oakman generously made materials available, as did Kenneth Wurdack and Alice Tangerini, Department of Botany, National Museum of Natural History, who pointed me to the saguaro specimen sheets of the National Herbarium.

Specialists in the visual record of botanical exploration, gardens and related world's fair displays shared photographs including Stephen Sinon and Esther Jackson, New York Botanical Garden; Ann Morgan Dodge, John Hay Library, Brown University; Adam Haas, Phipps Conservancy; Andrew Colligan, Missouri Botanical Garden; Michelle McCoy, Chicago Public Library Special Collections and Preservation Division; Graham Greer, Newberry Library; and Nathaniel Parks, Ryerson and Burnham Archives Research Center, The Art Institute of Chicago.

Collections specialists and curators of Arizona and the Southwest contributed their kind advice and primary source material for study including Diane Dittemore, Arizona State Museum; Bob Clancy and Shannon Stone, Tubac State Historic Park and Museum; Erika Castano, University of Arizona Library Special Collections; Matthew Messbarger, Arizona State University Library; Jerry Freund, Tucson Medical Center Archives; Tom Schmidt, Sharlot Hall Museum; Rachel Black, Carlos Lozano, Perri Pyle and Rebekah Tabah, Arizona Historical Society Library; Jennifer Friedman, UMass Amherst Libraries; Hope Ketcham Geeting, David M. Rubenstein Rare Book and Manuscript Library, Duke University; Mark Seigwarth and Shelbi Lindesmith, Boyce Thompson Arboretum; Becky Senf, Center for Creative Photography; George Ferguson, University of Arizona Herbarium; Khaleel Saba, Western Archaeological and Conservation Center, National Park Service; and Stephanie Mahan, Arizona Capitol Museum.

One of the pleasures of a project that spans past and present are the personal reminiscences shared along the way, especially cued by primary source material. Russell True of the White Stallion Ranch shared his parents' colorful collection of travel and tourism literature, while Carol Moore described riding the horses pictured in several examples within. Jan Flickenger Whittemore shared her reminiscence about her young career as a Tucson Sunshine Climate Club model. Judi and Bud Busche opened their home to inspect the picture window framed by the photographers Charles and Lucile Herbert. Carolyn Manley Robinson shared her father's early Tucson photographs and stories. Bill Peachey and Bill Thornton acquainted me firsthand with the pleasures of the Ironwood Forest National Monument. Loreli Panico described her research on the history of the cactus sun-suit. Christine Murphey invited me to see the saguaro photographs at her parents' real estate office. Cecilia Hunter of St. Philips in the Hills generously shared church historical records about the framing of its foothills window. Douglas C. Towne shared his pioneering interpretive study of the traveling saguaro. Carl H. Abbott graciously allowed me to publish his father's photograph of Edith Hamlin and Maynard Dixon. Dean Jeffrey and James Stave generously allowed me to publish their respective saguaro roadside and street photographs. Mady Berry kindly shared photos of her contemporary fiber art. Wyatt McSpaden, Imp Kid, Claire Hoffman, and D'ana Nunez shared their contemporary photographs further documenting the saguaro moment. I am deeply indebted to Ron and Marcia Spark for their thoughtful reminiscences and cool hunter eyes.

Early in the project I benefited from a conversation with Peggy Larson of Tucson's Arizona Sonora Desert Museum about the history of the museum's founding and the shaping of its public displays in the 1950s. As if a bookend near my project's completion, I benefited from a conversation with performance artist Kimi Eisele, whose Standing with Saguaros is a compelling model of public art, at one with the spirit of the early cultural and botanical champions of the Sonoran Desert.

Demion Clinco and Andie Zelnio of the Tucson Historic Preservation Foundation helped shape and publish an illustrated chapter-length portion of the early manuscript. Jim Cook of the Western National Parks Association kindly took an interest in the manuscript and helped steer it to publication. At the Desert Laboratory, University of Arizona, Benjamin T. Wilder and Anna Seiferle-Valencia helped refine the narrative for the *Proceedings* generous full-page graphic format, rendered by designer Amanda Gonzalez Moreno. Artist Paul Mirocha contributed a saguaro habitat map. I am further indebted to the critical comment of my readers Professor Jennifer L. Jenkins and Professor Emeritus Roger L. Nichols, University of Arizona. I also wish to thank the anonymous readers of the Desert Laboratory who graciously offered critical advice and comments. It goes without saying that any errors or interpretive shortcomings are mine.

WLB
Tucson, Arizona

FIGURE CREDITS

FRONTISPIECE

Magazine cover, *Pictorial America—Arizona Edition*, 1950. Photograph by Ray Manley

PREFACE: THE SOCIAL SAGUARO

Figure I. Photoengraving, Joseph Wood Krutch, *Voice of the Desert*, 1954. Photograph by Ray Manley with the permission of Carolyn Manley Robinson.

Figure II. Copy photograph, Maynard Dixon and Edith Hamlin, Tucson, Arizona, 1942. Photograph by Charles Abbott with the permission of Carl H. Abbott.

Figure III. Photograph, Cactus ballet. Photograph by Charles Herbert. Charles and Lucile Herbert: Western Ways Features Manuscript and Photograph Collection, MS 1255, box 2 Photo Negatives Group B, MS 125 and "The Cactus Ballet," MS 1255, box 10, f 90, Arizona Historical Society, Tucson.

Figure IV. Woodcut, cover of *Harper's New Monthly Magazine* 29 (October 1864), from the author's collection.

Figure V. Edward S. Curtis, photograph, Maricopa women, 1907, Library of Congress, accessed November 8, 2020. https://www.loc.gov/item/90710181/

Figure VI. Edward S. Curtis, photograph, Qahatika women, 1907, Library of Congress, accessed November 8, 2020. https://www.loc.gov/item/96502068/

Figure VII. Copy photograph, Charles W. Herbert, 1949, courtesy of Carolyn Manley Robinson.

Figure VIII. Advertisement, Tucson Sunshine Climate Club. *National Geographic*, 1952, from the author's collection.

CHAPTER ONE: CACTUS WORDS AND PICTURES

Figure 1.1. Range of the saguaro. Map by Paul Mirocha. Saguaro distribution based on Yetman et al. (2020).

Figure 1.2. Woodcut, "'Monster Cactus' at the Royal Botanic Gardens, Kew," in *Illustrated London News* (October 17, 1846), p. 246, from the author's collection.

Figure 1.3. Woodcut, "Kew Cactus from Real Del Monte—(Cereus Senilis)," in *Illustrated London News* (July 17, 1847), p. 48, from the author's collection.

Figure 1.4. Letter page, William H. Emory to George Engelmann, January 14, 1848, George Engelmann Papers, collection 1 RG 4/1/5/1, Missouri Botanical Garden, St. Louis.

Figure 1.5. Lithograph, "*Cereus giganteus*," in W.H. Emory, *Notes* (1848) pp. 72–73, Smithsonian Institution Libraries, Washington, D.C.

Figure 1.6. Lithograph, "Chain of Natural Spires on the Gila," in W.H. Emory, *Notes* (1848) pp. 78–79, Smithsonian Institution Libraries, Washington, D.C.

Figure 1.7. Lithograph, "Saguaro Trees," in J. Froebel, *Seven Years' Travel in Central America, Northern Mexico, and the Far West of the United States* (R. Bentley, London, 1859), pp. 508–509, Smithsonian Institution Libraries, Washington, D.C.

Figure 1.8. Engraving, "View Along the Gila—*Cereus Giganteus*," in W.H. Emory, *Report* Vol. II G. Engelmann, *Cactaceae of the Boundary* (1859), frontispiece, Smithsonian Institution Libraries, Washington, D.C.

Figure 1.9. John Russell Bartlett, *Tucson, Arizona, from the eastern flanks of Sentinel Peak in the Tucson Mountains. Looking northeast across the Santa Cruz River Valley*, July 17–18, 1852, pencil, sepia and wash, 12.7 × 18.3 inches, 1852.07.17, John Carter Brown Library, Providence.

CHAPTER TWO: CACTUS DISPLAYS

Figure 2.1. Woodcut, "Arizona Garden – Cacti 13 ft. high," in *Harper's Weekly* 31 (April 16, 1887), p. 277, from the author's collection.

Figure 2.2. Woodcut, advertisement, A. Blanc & Co., "Rare Cacti," in *Century* (March 1889, unpaginated), from the author's collection.

Figure 2.3. Photoengraving, "Cactus-House, Holly Point, Heaton Mersey," in *Gardeners' Chronicle* (March 9, 1901), after p. 164, UMass Amherst Library, Amherst.

Figure 2.4. Woodcut, cover of *Baltimore Cactus Journal* 1, no. 1, (July 1894, cover), David M. Rubenstein Rare Book and Manuscript Library, Duke University, Durham.

Figure 2.5. Woodcut, "The Cactus Club of Baltimore," in *Harper's Bazaar* (February 24, 1894) p. 148), HathiTrust, Harvard University, Cambridge.

Figure 2.6. Photoengraving, Mexican cactus display mounted by the Secretary of Development, Horticulture Hall, World's Columbian Exposition, Chicago, 1893, from the author's collection.

Figure 2.7. Photoengraving, "Cactus bed at Carroll Park, Baltimore, 1894," in *Baltimore Cactus Journal* 1, no. 2 (August 1894), p. 15, from the author's collection.

Figure 2.8. Stereoview, "The cactus garden," Prospect Park greenhouses, Brooklyn, NY, U.S.A.,1907. Digital file number 1s06876u, Library of Congress, Washington, D.C. https://www.loc.gov/item/2017649023/

Figure 2.9. Photograph, Saguaro with arms, Horticultural Hall, World's Columbian Exposition, Chicago, 1893, WCE/CDA-V/59, Special Collections and Preservation Division, Chicago Public Library, Chicago.

Figure 2.10. Cyanotype, Cactus display outside Horticultural Hall, World's Columbian Exposition, Chicago, 1893, from the author's collection.

Figure 2.11. Cyanotype, Columnar saguaro, World's Columbian Exposition, Chicago, 1893, from the author's collection.

Figure 2.12. Stereoview, "The great cactus, Horticultural Building, World's Fair, Chicago,"1893, from the author's collection.

Figure 2.13. C.D. Arnold, photograph, World's Columbian Exposition, Territorial Building, World's Columbian Exposition Photographs by C.D. Arnold, digital file # 198902.02_043-141.2.14, Ryerson and Burnham Archives, The Art Institute of Chicago, Chicago.

Figure 2.14. Postcard, "Hole in Rock, near Phoenix, Arizona," published by Central News Co., Phoenix, Arizona, 1922, from the author's collection.

Figure 2.15. Postcard, "Hole in the Rock, Papago, near Phoenix, Arizona [with water]," Curt Teich & Co., Chicago, Herz Post Cards, San Diego, Calif., 1936, from the author's collection.

Figure 2.16. Realphoto postcard, "Giant Cactus S.P. Depot L.A.," 1893, National Herbarium, National Museum of Natural History, Smithsonian Institution, Washington, D.C.

Figure 2.17. Postcard, "Greetings from San Antonio, Tex. Giant cactus, Southern Pacific Depot," Joske Bros., San Antonio, Texas–Germany, postmarked 1907, from the author's collection.

Figure 2.18. Realphoto postcard, "California Building, Panama Pacific International Exposition," San Francisco, 1915, from the author's collection.

Figure 2.19. Postcard, "Things Grow Large in California," Detroit Publishing Co., 1905–1906, from the author's collection.

Figure 2.20. Postcard, "The Arizona Cactus Garden, Hotel Del Monte, Monterey Co., California," 1901–1907, from the author's collection.

Figure 2.21. Postcard, "Arizona Cactus Garden, Hotel del Monte, Cal.," Albertype Co., Brooklyn, N.Y., published by Palace Drug Co., Monterey, Ca., 1898–1906, from the author's collection.

CHAPTER THREE: THE ARBOREAL DESERT

Figure 3.1. Photograph, Cactus House, Phipps Conservatory, Pittsburgh, April 14, 1903, RU 000221, J. Horace McFarland Company Collection, Archives of American Gardens, Smithsonian Institution, Washington, D.C.

Figure 3.2. Photograph, Shaw's Garden main gate with columnar saguaros, 1897, PHO 1983-0384, Missouri Botanical Garden, St. Louis.

Figure 3.3. Postcard, "Shaw's Garden, St. Louis, Mo.," Dresden-Leipsig–Berlin made in Germany, published by the St. Louis News Company, St. Louis, Mo., postmarked 1908, from the author's collection.

Figure 3.4. Photograph, Shaw's garden saguaro in open crate with creosote bush, 1914, GPN 1982-0520, Missouri Botanical Garden, St. Louis.

Figure 3.5. Photograph, *Cereus giganteus*, about 25 years old, 13 ½ feet high, near Tucson, Arizona, in *American Gardening* (May 17, 1902), p. 319.

Figure 3.6. Photograph, load of boxed cacti, Torres, Sonora, scrapbook, Daniel Trembly MacDougal collection, MS 452, box 45, #FZ, Arizona Historical Society, Tucson.

Figure 3.7. Copy photograph, Santa Catalina mountains from east of Desert Laboratory, April 13, 1914, folder Tumamoc Hill Views, Desert Laboratory Collection, Desert Laboratory on Tumamoc Hill, Tucson.

Figure 3.8. Postcard, "Desert Botanical Laboratory, Tucson, Arizona," published by Benham Co., Los Angeles, ca. 1912, from the author's collection.

Figure 3.9. Copy photograph on scrapbook page, fruit harvest pole used as a corner boundary marker, Desert Laboratory, MS 452, box 45, #M, Arizona Historical Society, Tucson.

Figure 3.10. Photograph, saguaro next to Desert Laboratory building, MS 452, B34, F474, #A1-31, Arizona Historical Society, Tucson.

Figure 3.11. Photograph, Mrs. E.S. Spalding measuring a saguaro, 1909, MS 452, box 34, folder 458, img 8875, Arizona Historical Society, Tucson.

Figure 3.12. Photoengraving, Dendrograph recording growth and changes in volume of young tree-cactus, July 1925, in MacDougal, "Laboratory for Plant Physiology," *Annual Report* (Carnegie Institution of Washington, Washington, D.C.), p. 19. Desert Laboratory Collection, Desert Laboratory on Tumamoc Hill, Tucson.

Figure 3.13. Photoengraving, Desert Laboratory exhibit, "Plants of the Sand," Shreeve, January 3, 1937, in *News Service Bulletin Staff Edition* 10 (Carnegie Institution of Washington, Office of Publications, Washington, D.C.) p. 90, Desert Laboratory Collection, Desert Laboratory on Tumamoc Hill, Tucson.

Figure 3.14. Photoengraving, Boyce Thompson Arboretum stone greenhouse with saguaro, in *Purpose-History-Dedication, Superior, AZ, July 1930* (Boyce Thompson Arboretum), p. 9. Smithsonian Institution Libraries, Washington, D.C.

Figure 3.15. Postcard, "Administration Building, Desert Botanical Garden in Papago Park, Tempe, Arizona," Curteich-Chicago distributed by Lollesgard Specialty Co., Tucson and Phoenix, Az, 1950, from the author's collection.

Figure 3.16. Realphoto postcard, "Capitol Grounds, Phoenix, Ariz.," ca. 1925, from the author's collection.

Figure 3.17. Realphoto postcard, Capitol Grounds, Phoenix, Arizona, 1912, from the author's collection.

Figure 3.18. Postcard, "Main Building-University of Arizona Tucson, Ariz.," annotated, "This is the cactus garden in front," R. Rasmesson, hand-colored, postmarked 1909, from the author's collection.

Figure 3.19. Photoengraving, "The suwarro (*Cereus giganteus*) in bloom in the conservatories of the New York Botanical Garden," in *Journal of the New York Botanical Garden* 6 (August 1905), p. 131, Smithsonian Institution Libraries, Washington, D.C.

Figure 3.20. Mary Emily Eaton, pen and ink watercolor, *Carnegiea gigantea*, 1912. National Museum of Natural History, Smithsonian Institution, Washington, D.C.

CHAPTER FOUR: ON THE MARGINS

Figure 4.1. Postcard, "Giant Cactus," annotated, "Here is a strange plant that grows on the desert. There is [*sic*] a lot of them to be seen from the train," PC 98 B14 cactus 69900, Arizona Historical Society, Tucson.

Figure 4.2. Postcard, "Giant Cactus, Salt River Valley, Phoenix, Ariz.," Albertype Co., Brooklyn, N.Y., R.L. Balke, Indian Trader, Phoenix, Ariz., ca. 1908, from the author's collection.

Figure 4.3. Postcard, "Giant Cactus, Kingman, Ariz.," Albertype Co., Brooklyn, N.Y., published for Hotel Beale, Kingman, Ariz., ca. 1907, from the author's collection.

Figure 4.4. Postcard, "Giant Cacti of the Plains," Thayer Publishing Co., Denver, Colo., ca. 1910, from the author's collection.

Figure 4.5. Poster, "Jan'y 31, 1884, The Beginning of the Society of Arizona Pioneers, by Sam Hughes," pen and ink watercolor, Arizona Historical Society, Tucson.

Figure 4.6. Booklet photogravure, "Trees and lawns greet you on arrival—Southern Pacific Station," "Tucson—Arizona Man-Building in the Sunshine Climate," Tucson Sunshine Climate Club, 1922–1924, from the author's collection.

Figure 4.7. Postcard, "S.P. Grounds and Depot, Tucson, Ariz.," published by Curt Teich Co., annotated, 1915, from the author's collection.

Figure 4.8. Stereoview, Buehman cactus still life in studio, "Arizona Bouquet Cactus and Spanish Bayonet," from the author's collection.

Figure 4.9. Copy photograph, Saguaros, Buehman Studios BN 204958, Henry and Albert Buehman Memorial Collection, Buehman-Subjects-Cactus–Saguaro, Arizona Historical Society, Tucson.

Figure 4.10. Stereoview, Carleton Watkins, "*Cereus giganteus* in blossom. Arizona," ca. 1875. History Room, California State Library, Sacramento. https://csl.primo.exlibris group.com/permalink/01CSL_INST/hbrh2d/alma990015613060205115

Figure 4.11. Glass plate positive lantern slide, "The Effects of Irrigation on Arizona Desert Giant, Salt River Valley," Keystone View Company #209, from the author's collection.

Figure 4.12. Stereoview, "Giant cacti of the desert in alfalfa field made production by irrigation, Arizona," Keystone View Company, from the author's collection.

Figure 4.13. Brochure cover, "Health Wealth Golden Opportunities," Tucson Chamber of Commerce, ca. 1906, from the author's collection.

Figure 4.14. Advertisement, Tucson Chamber of Commerce, "Tucson is the Metropolis of Arizona!" *Plant World*, June 1911. Desert Laboratory Collection, Desert Laboratory on Tumamoc Hill, Tucson.

Figure 4.15. Postcard, "Sabino Canyon, Catalina Mountains, Tucson, Arizona," reverse, "For Tucson Literature address Chamber of Commerce, Tucson, Arizona," Albertype Co., Brooklyn, N.Y., published by R. Rasmessenn, from the author's collection.

Figure 4.16. Booklet, "Tucson, Arizona, in the Residence District," ca. 1911, from the author's collection.

Figure 4.17. Postcard, "A Navajo Blanket Weaver," Detroit Publishing Co., 1901–1902, hand stamped, "Rasmessen's Curio Store, Tucson, Arizona," from the author's collection.

Figure 4.18. Postcard, "On the Road to San Xavier Mission, Tucson, Ariz.," ca. 1909, published by R. Rasmessen, from the author's collection.

Figure 4.19. Territorial seal with saguaro and deer, Tubac State Historic Park and Museum, Tubac.

Figure 4.20. Great Seal of the State of Arizona, 1912, Arizona Capitol Museum, Phoenix.

Figure 4.21. Brochure with envelope, "Tucson Arizona," Tucson Chamber of Commerce, 1930, from the author's collection.

Figure 4.22. Photoengraving, "Scenes Around Tucson," Tucson Chamber of Commerce, 1930, from the author's collection.

Figure 4.23. Advertisement, Sunshine Climate Club, "Must Hearts Break," in *Literary Digest* (December 22, 1922), p. 42, from the author's collection.

Figure 4.24. Booklet cover, "Tucson—Arizona Man-Building in the Sunshine Climate," Tucson Sunshine Climate Club, 1922–1924, from the author's collection.

Figure 4.25. Booklet foreword, "Tucson—Arizona Man-Building in the Sunshine Climate," Tucson Sunshine Climate Club, 1922–1924, from the author's collection.

Figure 4.26. Booklet cover, "Phoenix Arizona Where Winter Never Comes," 1924, from the author's collection.

CHAPTER FIVE: FRAMING UP

Figure 5.1. Copy photograph, Harold Bell Wright home landscaped courtyard, ca. 1922, Tucson Medical Center Archive, Tucson.

Figure 5.2. Copy photograph, grounds crew posing with wrapped saguaro, 1927, Tucson Medical Center Archive, Tucson.

Figure 5.3. Copy photograph, grounds crew preparing to load wrapped saguaro, 1927, Tucson Medical Center Archive, Tucson.

Figure 5.4. Copy photograph, grounds crew with loaded saguaro, 1927, Tucson Medical Center Archive, Tucson.

Figure 5.5. Realphoto postcard, Zuni Dining Room exterior, Desert Sanatorium, Tucson, 1930, from the author's collection.

Figure 5.6. Photoengraving, Zuni Dining Room, Desert Sanatorium, Tucson, 1928, Tucson Medical Center Archive, Tucson.

Figure 5.7. Copy photograph, Rutger Porter and Harold Bell Wright, 1927, Tucson Medical Center Archive, Tucson.

Figure 5.8. Copy photograph, entrance to Desert Sanatorium, Tucson Medical Center Archive, Tucson.

Figure 5.9. Postcard, "Main Entrance, The Desert Sanatorium, Tucson, Arizona," ca. 1928, Tucson Medical Center Archive, Tucson.

Figure 5.10. R.B. Streets, photograph, Homer Shantz, February 22, 1930, neg. AA-6-1930, MS 481, box 50, Homer Shantz Negative Collection, University of Arizona Special Collections, accessed November 8, 2020, http://uair.library.arizona.edu/item/285817

Figure 5.11. Homer Shantz, photograph, chandelier-type saguaro, February 22, 1931, B-7-1931, MS 481, box 51, Homer Shantz Negative Collection. University of Arizona

Special Collections, accessed November 8, 2020, http://uair.library.arizona.edu/item/285990

Figure 5.12. Homer Shantz, photograph, saguaro stand, February 20, 1933, A-2-1933, MS 481, box 53, Homer Shantz Negative Collection. University of Arizona Special Collections, accessed November 8, 2020, http://uair.library.arizona.edu/item/286478

Figure 5.13. Homer Shantz, photograph, watermelon saguaro, waiting car, September 22, 1935, T-1-1935, MS 481, box 55, Homer Shantz Negative Collection, University of Arizona Special Collections, Tucson.

Figure 5.14. Homer Shantz, photograph, Queen of the Night in bloom, cactus garden west of Main building, June 15, 1932, X-12-1932, MS 481, box 53, Homer Shantz Negative Collection, University of Arizona Special Collections, accessed November 8, 2020, http://uair.library.arizona.edu/item/286348

Figure 5.15. Postcard, "Aerial View, University of Arizona, Tucson, Ariz.," photograph by Ben Gross, Tucson, Arizona, Eagle Post Card View Co., Inc., New York, postmarked 1943, from the author's collection.

Figure 5.16. R.B. Streets, photograph, workers in cactus garden arroyo east of Main building, ca. 1929, folder 2: gardens - cactus gardens photographs, N-10,039, University of Arizona Special Collections, Tucson.

Figure 5.17. Homer Shantz, photograph, cactus garden arroyo east of Main building, April 1931, I-10-1931, Homer Shantz Negative Collection, MS 481, box 51, University of Arizona Special Collections, Tucson.

Figure 5.18. Homer Shantz, photograph, Boojum trees, Mexican garden, azu_shantz_19320707_z_3_m Homer Shantz Negative Collection, MS 481 Box 53, University of Arizona Special Collections, accessed November 8, 2020, http://uair.library.arizona.edu/item/286364

Figure 5.19. Copy photograph, Charles Herbert with Western Ways staff, MS 1255, box 3, folder 31A Activity–Offices, Arizona Historical Society, Tucson.

Figure 5.20. Copy photograph, Lucile Herbert with magazines, MS 1255, box 3, folder 37 Activity–Publications, Arizona Historical Society, Tucson.

Figure 5.21. Color transparency by Charles Herbert, Tucson Home with Picture Window, ca. 1947, MS 1255, Neg. Group C, B3, Arizona Historical Society, Tucson..

Figure 5.22. Newspaper sheet, "Dream House Located with Camera Eye," *Detroit News* (August 3, 1947), MS 1056, Charles and Lucile Herbert: Western Ways Features Manuscript and Photograph Collection, box 15, folder 215 Picture Windows, Arizona Historical Society, Tucson.

Figure 5.23. Photograph, moving giant cactus, final digging, 1939, MS 1255, Negative Group B, B2, Arizona Historical Society, Tucson.

Figure 5.24. Photograph, moving giant cactus, putting mattress, 1939, MS 1255, Negative Group B, B2, Arizona Historical Society, Tucson.

Figure 5.25. Photograph, moving giant cactus, arches view, 1939, MS 1255, Negative Group B, B2, Arizona Historical Society, Tucson.

Figure 5.26. Photograph by Maynard Parker, living room mural installation, Eleven Arches, 1939, Maynard Parker Collection, Huntington Library, accessed November 8, 2020, http://hdl.huntington.org/cdm/compoundobject/collection/p15150coll5/id/16384/rec/8

Figure 5.27. Photo offset negative, Cabat-Gill Advertising Company, logo, Cabot-Gill Collection, Tucson Historic Preservation Foundation, Tucson.

Figure 5.28. Brochure, Lazy K Bar, living room b/w, with picture window framing a saguaro view, Cabot-Gill Collection, Tucson Historic Preservation Foundation, Tucson.

Figure 5.29. Brochure, Lazy K Bar, living room white walls, with picture window framing a saguaro view, courtesy of Russell True.

Figure 5.30. Brochure, Lazy K Bar, living room pink walls, with picture window framing a saguaro view, courtesy of Russell True.

Figure 5.31. Photograph, Mural Room dining room, El Presidio Hotel, B3F28, box 3, MS 1255, *Western Ways*. Arizona Historical Society, Tucson.

Figure 5.32. Menu cover, El Presidio Hotel Mural Room, Cabot-Gill Collection, Historic Preservation Foundation, Tucson.

Figure 5.33. Postcard, "Cactus Bar, Salome Van's, Salome, Arizona," Frasher Fotos, ca. 1939, from the author's collection.

Figure 5.34. Realphoto postcard, Herb Wood Cacticraft Cactus Furniture, Tucson, Arizona, ca. 1947, from the author's collection.

Figure 5.35. Photograph mounted in saguaro rib frame. Photograph by U.S. Forest Service, ca. 1930. Frame by William Bucher, 1932. NMAH AG 124450.06 cat. 50, 19, National Museum of American History, Smithsonian Institution, Washington, D.C.

Figure 5.36. Color transparency by Charles Herbert, desert dioramas, MS 1255, Photo negatives, group C, B3, Arizona Historical Society, Tucson.

Figure 5.37. Completed Knauff saguaro diorama under glass, from the author's collection.

Figure 5.38. Copy photograph, campus cactus garden arroyo east of Main building, ca. 1937, folder 2: gardens-cactus gardens photographs, N-7385, University of Arizona Special Collections, Tucson.

Figure 5.39. Copy photograph, man inspecting vandalized saguaros, campus cactus garden, 1952, folder 2: gardens-cactus gardens photographs, 4264-6, University of Arizona Special Collections.

CHAPTER SIX: LIBERTIES TAKEN

Figure 6.1. Brochure cover, "Mesa, Arizona, in Arizona's Valley of the Sun," ca. 1940, from the author's collection.

Figure 6.2. Brochure cover, "Phoenix Arizona in the Valley of the Sun," ca. 1954, from the author's collection.

Figure 6.3. Envelope covers with guest ranch cactus graphics, 1936-1958, from the authors collection.

Figure 6.4. Realphoto postcard, "Saguaro Lake Ranch, Mesa, Arizona," postmarked 1945, from the author's collection.

Figure 6.5. Brochure cover, "Arizona and the Valley of the Sun," Santa Fe Railroad, 1950, from the author's collection.

Figure 6.6. Pamphlet cover, "Apache Trail of Arizona Southern Pacific," by William Howard Bull, 1924, from the author's collection.

Figure 6.7. Brochure cover, "Scenic Grandeur of the West," Southern Pacific Railroad, 1943, from the author's collection.

Figure 6.8. Photoengraving, Giant Cactus, Arizona, from portfolio Scenic Grandeur of the West, Southern Pacific Railroad, 1943, from the author's collection.

Figure 6.9. Brochure cover, stylized saguaro print, Hotels and Guest Ranches in Arizona, Southern Pacific Railroad, November 1933, from the author's collection.

Figure 6.10. Russell Lee, photograph, Cactus light standard in front of hotel in Phoenix, Arizona, May 1940, Farm Security Administration–Office of War Information Photograph Collection, Library of Congress, Washington, D.C. https://www.loc.gov/item/2017786304/

Figure 6.11. Booklet title page, "Water for Thirteen Cities," Metropolitan Water District of Southern California, Los Angeles, 1935, from the author's collection.

Figure 6.12. Brochure cover, "Accommodations in Arizona's Valley of the Sun," Phoenix Chamber of Commerce, Phoenix, 1952, from the author's collection.

Figure 6.13. Photoengraving, model on horse, ca. 1952, from the author's collection.

Figure 6.14. Advertisement (detail), Sunshine Climate Club, "More sunshine in Tucson than any other resort city," in *Holiday* 8 (October 1950), p. 115, from the author's collection.

Figure 6.15. Advertisement (detail), "Western Ways Guest Ranch Promotion Shots by Ray Manley," in *The Magazine Tucson's Annual Desert Pictorial*, 1951, p. 81, from the author's collection.

Figure 6.16. Postcard, generic Southern Arizona guest ranch [Double U Guest Ranch], Kodachrome original by Ray Manley, published and distributed by Bob Petley, Phoenix, Arizona, postmarked 1953, from the author's collection.

Figure 6.17. Postcard, "Saddle and Surrey Guest Ranch Moonlight Ride," published by Ray Manley Commercial Photography, Tucson, Arizona, ca. 1958, from the author's collection.

Figure 6.18. Harry Sinclair Drago, novel cover, *Stagecoach Kingdom* (Novel Selections, New York, 1943), from the author's collection.

Figure 6.19. Paint by number "Painted Desert," Craft Master New Artist Series NA-443, 1964, 18 × 24 inches, from the author's collection.

Figure 6.20. Postcard, "Hello! From Arizona," distributed free by the Arizona Development Board, Phoenix, Arizona, Micro Color by Tyler printing, Phoenix, Arizona, 1962, from the author's collection.

Figure 6.21. Map, Arizona Road Map 1962, Golden Anniversary Statehood of Arizona, from the author's collection.

Figure 6.22. Photograph, John Margolies, Hilltop Steak House sign, Route 1, Saugus, Massachusetts, Library of Congress, accessed November 8, 2020. http://loc.gov/pictures/resource/mrg.01572/

Figure 6.23. Postcard, "Traveler's Motel," Yuma, Arizona, color photo by Bob Von Luchene, published by Petley Studios, Phoenix, Arizona, ca. 1960, from the author's collection.

Figure 6.24. Postcard, "Cactus Motel," Denver, Colorado, distributed by Clark A. Brandenburg, Denver, Colo., ca. 1962. From the author's collection.

Figure 6.25. Postcard, "El Rancho Motel," High Springs, Florida, Dexter Press, West Nyack, N.Y., published by Jack Taylor, ca. 1958, from the author's collection.

Figure 6.26. Postcard, "Las Vegas Motel," Eatontown, New Jersey, published by N.P.C. Studios, Avon-by-the-Sea, N.J., ca. 1962, from the author's collection.

Figure 6.27. Photograph, Westland Shopping Center saguaro sign, Richmond, Virginia, 2017, courtesy of Dean Jeffrey. https://www.flickr.com/photos/29276830@N02/4963803699/in/photolist-x7Tk4Z-3p7ijE-i365vS-8yCPot-8yCMUp-7F7dLs

CHAPTER SEVEN: THE VIRAL RANGE

Figure 7.1. Postcard, annotated "Saguaro National Monument Tucson Arizona," published by Western Ways, Tucson, Arizona, postmarked 1947, from the author's collection.

Figure 7.2. Photograph, Mary Thorbecke and Roy Drachman, 1941. Photograph by Charles Herbert. MS 1255 Negative Group B, B2, Cactus Gags, Arizona Historical Society, Tucson.

Figure 7.3. Photograph, Mary Thorbecke and Roy Drachman, 1941. Photograph by Charles Herbert. MS 1255 Negative Group B, B2, Cactus Gags, Proof A., Arizona Historical Society, Tucson.

Figure 7.4. Photograph, Merri Ciocetti sitting on barrel cactus, 1940. Photograph by Charles Herbert. MS 1255, negative group B, B2, Cactus Gags, frame 37, Arizona Historical Society, Tucson.

Figure 7.5. Envelope, "Tops in photography," Photocenter saguaro logo, 1947, Cabot-Gill Collection, Tucson Historic Preservation Foundation, Tucson.

Figure 7.6. Copy photograph, Donald O'Connor, *Curtain Call at Cactus Creek*, Universal, 1950, from the author's collection.

Figure 7.7. Copy photograph, model seated in golden barrel cactus garden, All Year 'Round Club of Southern California, 1957, from the author's collection.

Figure 7.8. Copy photograph, Cactus disease control, pulling down a saguaro, Saguaro National Monument, 1941, PC 151, F8, #56349, Arizona Historical Society, Tucson.

Figure 7.9. Copy photograph, Cactus disease control, workers burying cut sections of saguaros, Saguaro National Monument, 1941, PC 151, F8, #56353, Arizona Historical Society, Tucson.

Figure 7.10. Photograph, Dr. Alice Boyle treating saguaro with bleach solution, 1956, MS 1255, negative group B, B2, Cactus Disease, Arizona Historical Society, Tucson.

Figure 7.11. Ray Manley, brochure cover, "Tucson," ca. 1952, from the author's collection.

Figure 7.12. Copy photograph, Seasons Greetings, Sunshine Climate Club, 1955, from the author's collection.

Figure 7.13. Newsletter cover, *Sunshine News* All-Publicity Issue, April 1954, Arizona Historical Society, Tucson.

Figure 7.14. Copy photograph, swimsuit model, ca. 1958, MS 1120 Riddell, box 2, folder 5, 5307, Arizona Historical Society, Tucson.

Figure 7.15. Copy photograph, swimsuit model, ca. 1958, MS 1120 Riddell box 2, folder 25, 5304, Arizona Historical Society, Tucson.

Figure 7.16. Copy photograph, western wear model, ca. 1958, MS 1120 Riddell box 2, folder 17, 5389, Arizona Historical Society, Tucson.

Figure 7.17. Copy photograph, Ruth Eggemeyer, Saguaro National Monument, 1941, Western Archaeological and Conservation Center, Tucson.

Figure 7.18. Brochure cover, "Cactus Forest Drive," Saguaro National Monument, February 1954, from the author's collection.

Figure 7.19. Fabric silkscreen print, "Saguaro Land," by Harwood Steiger, photograph by Jude Ignacio and Gerardine Vargas, Tucson Historic Preservation Foundation, Tucson.

Figure 7.20. Magazine cover, 29th Annual Rodeo Madison Square Garden, 1954, from the author's collection.

Figure 7.21. Photograph, Rockmount Western shirt sleeve detail, from the author's collection.

Figure 7.22. Photograph, Woman with jeans jacket and cactus pattern skirt before Prada fashion show, Milan Fashion Week street style on September 22, 2015, A. Astes/ Alamy Stock Photo.

Figure 7.23. Photograph, sport shirt with saguaro print fabric, ca. 2016, Native Youth.

Figure 7.24. Instagram screen shot, saguaro costume by Mady Berry, 2016, photograph by and courtesy of @imp_kid.

Figure 7.25. Photograph, Siren by Mady Berry, 2016, courtesy of Wyatt McSpaden.

Figure 7.26. Photograph, saguaro installation by artist Office Kovacs, Coachella Valley Music and Arts Festival, 2019, Claire Hoffman/Bizbash https://www.bizbash.com/style-decor/event-design/media-gallery/21065429/event-design-ideas-from-coachella-parties-and-activations

Figure 7.27. Photograph, Desert Chill party, Instagram, Palm Springs, 2019 Manifold and D'ana Nunez https://www.wearemanifold.com/work/instagram-at-coachella-desert-chill/

Figure 7.28. Photograph, Wild West Fantastical, 2016, courtesy of James Stave https://www.flickr.com/photos/56923898@N02/31496178781/in/faves-42458236@N03/

AFTERWORD: THE SAGUARO MOMENT

Figure 8.1. Photo album cover, Snap Shots, ca. 1935, Tucson Medical Center Archive, Tucson.

Figure 8.2. Photograph, Carrie Stouch of Gary, Indiana posing with a saguaro, Sabino Canyon, May 24, 1937, Tucson Medical Center Archive, Tucson.

Figure 8.3. Photograph, man posing with crestate saguaro, ca. 1937, PC 1000–Saguaros–47628019, Arizona Historical Society, Tucson.

Figure 8.4. Photograph, man with tie, February 9, 1940, from the author's collection.

Figure 8.5. Photograph, woman posing with saguaro, ca. 1942, from the author's collection.

Figure 8.6. Photo album cover, painted wood, Snaps, ca. 1954, from the author's collection.

Figure 8.7. Photo album cover, painted leather, Snap Shots Tucson, Ariz, ca. 1960, from the author's collection.

Figure 8.8. Photograph, man in jacket touching a saguaro, ca. 1919, PC 1000 Brown, FW 92687, Arizona Historical Society, Tucson.

Figure 8.9. Photograph, woman touching the tip of a scarred saguaro arm, 1945, from the author's collection.

Figure 8.10. Photo album cover, copper embossed, ca. 1970, from the author's collection.

ENDPAPERS

Stamps, Phoenix, Arizona, Cabat-Gill Collection, Tucson Historic Preservation Foundation, Tucson.

Stamps, Tucson, Arizona, MS 1400, Cabat-Gill Collection, Arizona Historical Society, Tucson.

LITERATURE CITED

A. Blanc and Company. 1886. *Catalogue and Hints on Cacti.* A. Blanc and Company, Philadelphia.

A. Blanc and Company. 1889. "Rare Cacti." *Century* 37:unpaginated.

A. Blanc and Company. 1893. *Cacti, Rare Plants and Bulbs Offered by A. Blanc and Co.* A. Blanc and Company, Philadelphia.

Agricultural Experiment Station. 1908. "Planting of native economic cacti continued." Nineteenth Annual Report. University of Arizona, Tucson, pp. 353–354.

Ahnmark, E.B. and D.E. Swann. 2008. "A history of saguaro cactus monitoring in Saguaro National Park, 1939–2007." Natural Resource Report NPS/SODN/NPR 2009/093. National Park Service, Fort Collins. https://irma.nps.gov/DataStore/DownloadFile/15317

Allnutt, H. 1877. *The Cactus and Other Tropical Succulents.* Estates Gazette Office, London.

American Gardening. 1900. "Current news and notes: The New York Botanical Garden." *American Gardening*, July 7, 1900, p. 451.

American Gardening. 1902. "The giant cacti of Arizona." *American Gardening*, May 17, 1902, p. 319.

Angell, B. 2003. "Mary Emily Eaton and her Cactaceae paintings." *Botanical Artist* 15:28. https://www.asba-art.org/article/botanica-collected-mary-emily-eaton

Arizona Silver Belt. 1893a. "One of the most characteristic and attractive displays that Arizona could possibly offer." *Arizona Silver Belt* [Globe, Arizona], May 13, 1893, p. 1.

Arizona Silver Belt. 1893b. "Arizona at the fair." *Arizona Silver Belt* [Globe, Arizona], September 9, 1893, p. 1.

Arizona Daily Star. 1880. Advertisement. "Henry A. Buehman, landscape and general photographer." *Arizona Daily Star*, October 3, 1880, p. 2.

Arizona Daily Star. 1901. "The Old Fox." *Arizona Daily Star*, December 6, 1901, p. 8.

Arizona Daily Star. 1905. "To boost Tucson." *Arizona Daily Star*, May 6, 1905, p. 12.

Arizona Daily Star. 1906. "J.M. Ormsby." *Arizona Daily Star*, March 14, 1906, p. 6.

Arizona Daily Star. 1908. "Vegetation of Old Pueblo 30 years ago." *Arizona Daily Star*, October 25, 1908, p. 9.

Arizona Daily Star. 1912a. "Pictures of first train were taken." *Arizona Daily Star*, November 22, 1912, p. 2.

Arizona Daily Star. 1912b. "Want ad pictures will boost Tucson." *Arizona Daily Star*, November 27, 1912, p. 8.

Arizona Daily Star. 1914a. "Ad Committee will use *Post* columns." *Arizona Daily Star*, February 7, 1914, p. 5.

Arizona Daily Star. 1914b. "Desert work results will be exhibited." *Arizona Daily Star*, April 1, 1914, p. 8.

Arizona Daily Star. 1915a. "Medical papers to tell about Tucson." *Arizona Daily Star*, July 2, 1915, p. 4.

Arizona Daily Star. 1915b. "Plan 'ad' campaign." *Arizona Daily Star*, July 23, 1915, p. 2.

Arizona Daily Star. 1918a. "Publicity for Tucson." *Arizona Daily Star*, July 3, 1918, p. 2.

Arizona Daily Star. 1918b. "50,000 postcards being distributed as publicity stunt." *Arizona Daily Star*, July 7, 1918, p. 2.

Arizona Daily Star. 1918c. "100,000 postcards to be given away as plan to advertise Tucson." *Arizona Daily Star*, August 31, 1918, p. 2.

Arizona Daily Star. 1918d. "Canteen puts Tucson on map while giving pleasure to soldiers." *Arizona Daily Star*, September 17, 1918, p. 6.

Arizona Daily Star. 1921. "30,000 postcards distributed by C of C." *Arizona Daily Star*, May 27, 1921, p. 3.

Arizona Daily Star. 1929a. "Arboretum to open April 6." *Arizona Daily Star*, March 25, 1929, p. 4.

Arizona Daily Star. 1929b. "Investors to Sunshine Climate Club." *Arizona Daily Star*, September 19, 1929, p. 6.

Arizona Daily Star. 1933. "An undesirable consolidation." *Arizona Daily Star*, April 3, 1933, p. 8.

Arizona Daily Star. 1936. "Social science will be topic." *Arizona Daily Star*, April 16, 1936, p. 3.

Arizona Daily Star. 1940a. "Chuck Abbott's chuck wagon publicizes Tucson in the East." *Arizona Daily Star*, April 30, 1940, p. 2.

Arizona Daily Star. 1940b. "Sunshine Club plans drive for $25,000 annual budget." *Arizona Daily Star*, August 30, 1940, p. 2.

Arizona Daily Star. 1941. "Abbott-Henderson wedding is told." *Arizona Daily Star*, July 4, 1941, p. 5.

Arizona Daily Star. 1945. "Tucson Sunshine Club photo used in German paper." *Arizona Daily Star*, February 27, 1945, p. 2.

Arizona Daily Star. 1946. "The Sunshine Climate Club." October 4, 1946 p. 22.

Arizona Daily Star. 1947. "Film features Knauff Family's dioramas." May 27, 1947, p. 16.

Arizona Daily Star. 1948. "Britons react to picture story of moving 350-year-old giant saguaro cactus." *Arizona Daily Star*, February 8, 1948, p. 14-A.

Arizona Daily Star. 1950. "Tucson's national ad campaign of Sunshine Club getting started." *Arizona Daily Star*, September 14, 1950, p. 8

Arizona Daily Star. 1952. "Booster club's fete arranged." *Arizona Daily Star*, March 30, 1952, p. 21.

Arizona Daily Star. 1956. "110-year-old item tells of grim death sentence." *Arizona Daily Star*, February 23, 1956, p. 6.

Arizona Daily Star. 1957. "Harrison rites set Saturday." *Arizona Daily Star*, August 29, 1957, p. B1.

Arizona Daily Star. 1962. "New telephone books to go out Saturday." *Arizona Daily Star*, June 14, 1962, p. 13.

Arizona Daily Star. 1968. "Columnist impressed by TMC." *Arizona Daily Star*, February 2, 1968, p. 21.

Arizona Daily Sun. 1962. "Statehood fete fund bill dies in House unit." *Arizona Daily Sun* [Flagstaff], March 6, 1962, p. 8.

Arizona Highways. 1946. "Cactus." *Arizona Highways* 22, February 1946, pp. 45–48.

Arizona Republic. 1890. "The Capitol site." *Arizona Republic*, September 15, 1890, p. A1.

Arizona Republic. 1899. "A corner of cacti." *Arizona Republic*, May 22, 1899, p. 4.

Arizona Republic. 1903. "A cactus 'orchard.'" *Arizona Republic*, September 19, 1903, p. 6.

Arizona Republic. 1904. "Scientific expedition." *Arizona Republic*, July 13, 1904, p. 6.

Arizona Republic. 1909. "New superintendent for Capitol grounds." *Arizona Republic*, May 8, 1909, p. 4.

Arizona Republic. 1919. "Former legislator from Maine visits state he assisted." *Arizona Republic*, April 13, 1919, p. 2.

Arizona Republic. 1924a. "Three-day campaign for $50,000." *Arizona Republic*, September 28, 1924, p. 30.

Arizona Republic. 1924b. "Clubs to hear of advertising campaign plans." *Arizona Republic*, September 30, 1924, p. 5.

Arizona Republic. 1924c. "Kiwanians hear coast man talk on advertising." *Arizona Republic*, October 1, 1924, p. 8.

Arizona Republic. 1934a. "Vessel to get cacti display." *Arizona Republic*, April 10, 1934, p. 12.

Arizona Republic. 1934b. "Cactus garden to enhance Arizona exhibit at Chicago." Arizona Republic, April 25, 1934, p. 3.

Arizona Republic. 1937. "Cactus Society adopts slogan." *Arizona Republic*, February 16, 1937, p. 8.

Arizona Republic. 1940. "Cactus light to be moved." *Arizona Republic*, January 8, 1940, p. 5.

Arizona Republic. 1948. "Only cactus street light moved to Jaycee building." *Arizona Republic*, July 23, 1948, p. 3.

Arizona Republic. 1961. "Our cover story." *Arizona Republic*, September 3, 1961, p. 114.

Arizona Weekly Citizen. 1893a. "The officials of the Southern Pacific." *Arizona Weekly Citizen*, January 7, 1893, p. 3.

Arizona Weekly Citizen. 1893b. "Pima County exhibit." *Arizona Weekly Citizen*, May 20, 1893, p. 3.

Arizona Weekly Citizen. 1893c. "It is admitted by the world 's fair management." *Arizona Weekly Citizen*, May 26, 1893, p. 2.

Arizona Weekly Citizen. 1893d. "Arizona at the fair: Its exhibit is not the very most creditable." *Arizona Weekly Citizen*, July 8, 1893, p. 1.

Arizona Weekly Citizen. 1893e. "Of the cactus exhibit for Arizona at the world's fair." *Arizona Weekly Citizen*, August 5, 1893, p. 4.

Arrington, L.J. 1969. "From Apache hunting to hosting America." *Arizona Review* 17, August–September, pp. 3–5.

Ascarza, W. 2010. *Tucson Mountains*. Arcadia Publishing, Charleston.

Atkinson, A. 1939. "Recalling our several conversations about the cactus garden." Memo to Mr. Bray, January 30, 1939. Alfred Atkinson, box 18, folder 9, AZ 426. University of Arizona Special Collections, Tucson.

Atkinson, A. 1945. Letter from Atkinson to Starkweather, December 8, 1945. Alfred Atkinson, box 18, folder 9, AZ 426. University of Arizona Special Collections, Tucson.

Atlanta Constitution. 1940. "Four Indian tribes end use of swastika." *Atlanta Constitution*, February 26, 1940, p. 2.

Avery, B. 1962. "Amended measure may pass." *Arizona Republic*, February 3, 1962, p. 5.

Bailey, L.H. 1893. "The Columbian Exposition cacti." *Garden and Forest*, October 11, 1893, p. 429.

Balestrero, P. 1972. "The wonderful world-wide Western Ways of Charles and Lucille Herbert." *Arizona Highways* 48:4–11, 34–39.

Ball, P. 1987. *A Photographic History of the University of Arizona 1885–1983*. University of Arizona Foundation, Tucson.

Baltimore Cactus Journal. 1894a. "Our cactus bed at Carroll Park, Baltimore." *Baltimore Cactus Journal* 1:1, 13.

Baltimore Cactus Journal. 1894b. "Cacti in bloom." *Baltimore Cactus Journal* 1:11.

Baltimore Sun. 1889. "To cultivate the cactus." *Baltimore Sun*, December 14, 1889, p. 6.

Baltimore Sun. 1890. "Large consignment of cactus plants." *Baltimore Sun*, February 8, 1890, p. 6.

Barnhart, J.H. 1918. "Historical sketch of the Torrey Botanical Club." *Memoirs of the Torrey Botanical Club* 17:12–21.

Bartlett, J.R. 1854. *Explorations and Incidents in Texas, New Mexico, California, Sonora, and Chihuahua, Connected with the United States and Mexican Boundary Commission, During the Years 1850, '51, '52, and '53*. 2 vols. D. Appleton and Company, New York.

Bean, W.J. 1908. *The Royal Botanical Gardens, Kew: Historical and Descriptive*. Cassell and Company, London.

Black, J. (translator). 1822. *Political Essay on the Kingdom of New Spain by Alexander de Humboldt*. Vol. 2. Longman, Hurst, Rees, Orman, and Brown, London.

Black, J.A. 1890. "Commissioner of Immigration." *Arizona: The Land of Sunshine and Silver, Health and Prosperity, the Place for Ideal Homes*. Republican Book and Job Print, Phoenix.

Blake, M.E. 1883. *On the Wing: Rambling Notes of a Trip to the Pacific*. Lee and Shepard, Boston.

Blake, W.P. 1870. *The Route for a Railroad to the Pacific Ocean Near the Thirty-Fifth Parallel of Latitude: Its Resources and Advantages*. Atlantic and Pacific Railroad, Mo-nos, Boston.

Blanc, T.A. 2000. *Oasis in the City: The History of the Desert Botanical Garden*. Heritage Publishers, Phoenix.

Boom, B.M. 1996. "Botanical expeditions of the New York Botanical Garden." *Journal of the New York Botanical Garden* 48:297–307.

Boston Globe. 1940. "Dissension arises among Indians over discarding swastika." *Boston Globe*, February 28, 1940, p. 9.

Boston Globe. 1964. "Hilltop Steak House: 'Man's meal—steak, potatoes, salad.'" *Boston Globe*, September 27, 1964, p. D4.

Bowers, J. 1990. "The Naming of the saguaro." *Carnegie Institution of Washington Newsletter*, November, p. 4.

Boyce Thompson Southwestern Arboretum. 1930. *Boyce Thompson Southwestern Arboretum: Purpose–History–Dedication*. Boyce Thompson Southwestern Arboretum, Superior.

Boyle, A.M. 1948. *Further Studies of the Bacterial Necrosis of the Giant Cactus.* PhD dissertation, University of Arizona, Tucson.

Bradford, P.V. 1992. *Ota Benga: The Pygmy in the Zoo*. Dell Publishing, New York.

Brissman, L. 2019. "Outdoor curiosities: Westland Shopping Center cactus sign." *Richmond Times-Dispatch*, August 11, 2019. https://www.richmond.com/discover-richmond/outdoor-curiosities-westland-shopping-center-cactus-sign/article_858b4cce-5df4-58c9-9cab-cf0c3d760fa5.html

British Cactus and Succulent Society, Manchester Branch. 2011. "Manchester Branch's involvement at Wythenshawe since 2011." http://manchester.bcss.org.uk/home/manchester-branch-and-wythenshaw-park/

Britton, N.L. 1896. "Botanical gardens." *Bulletin of the Torrey Botanical Club*, September 25, 1896, pp. 331–345.

Britton, N.L. 1899. Letter to Daniel T. MacDougal, March 18, 1899. MacDougal papers, box 2, folder 14, MS 0452. Arizona Historical Society, Tucson.

Britton, N.L. 1908a. Letter to Andrew Carnegie, November 2, 1908. Carnegie papers, v. 157, Manuscript Division, Library of Congress, Washington, D.C.

Britton, N.L. 1908b. Letter to Daniel T. MacDougal, November 6, 1908. MacDougal papers, box 9, folder 119, MS 0452. Arizona Historical Society, Tucson.

Britton, N.L. and J.N. Rose. 1908. "A new genus of Cactaceae." *Journal of the New York Botanical Garden* 9:185–188.

Britton, N.L. and J.N. Rose. 1919–1920. *The Cactaceae: Descriptions and Illustrations of Plants of the Cactus Family*. 2 vols. Carnegie Institution of Washington, Washington, D.C.

Brown, M. 1975. "Botanical Gardens finds a home." *Tucson Daily Citizen*, March 31, 1975, p. 12.

Brown, T.A. 1998. *19th Century Horticulturalists and Plant Raisers*. Thomas A. Brown, Petaluma.

Broyles, B. 1987. "Adventure in the Pinacate." *Journal of Arizona History* 28:155–188.

Bryant, Edwin. 1849. *What I Saw in California: Being the Journal of a Tour; by the Immigrant Route and South Pass of the Rocky Mountains, across the Continent of North America, the Great Desert Basin, and through California, in the Years 1846, 1847*. D. Appleton and Company, New York.

Bucher, W.F. n.d. [1932?]. Typescript, "Our trees and their woods." Accession files 115767 and 124450, 1933, National Museum of American History, Smithsonian Institution, Washington, D.C.

Burtner, M. 2011. *Crowning the Queen of the Sonoran Desert: Tucson and Saguaro National Park. An Administrative History*. University of Arizona, Tucson. http://npshistory.com /publications/sagu/adhi.pdf

Byrkit, J.W. 1992. "Land, sky, and people: The Southwest defined." *Journal of the Southwest* 34:256–387.

Cain, J. 2004. "Rudolph Ulrich's Arizona Gardens." *Pacific Horticulture* 65:3–11. http:// www.pacifichorticulture.org/articles/rudolph-ulrichs-arizona-gardens/

Cain, J. 2005. "Rudolph Ulrich's Arizona Gardens." *Cactus and Succulent Journal* 77:6–13.

Cain, J. and M. Graham. 2003. "Rudolph Ulrich on the San Francisco Peninsula." *Eden, Journal of the California Garden and Landscape Historical Society* 6:1–8.

Caldwell, W.N. 1962. *Arizona Symbols: Anniversary Edition*. Smyre Company, Phoenix.

Campbell, J.B. (editor). 1892. "Giant cactus from Arizona." *The World's Columbian Exposition Illustrated* 2:236.

Cannon, W.A. 1903. Letter to D.T. MacDougal. November 12, 1903. MacDougal papers, box 3, folder 23, MS 0452, Arizona Historical Society, Tucson.

Carman, M.D. 2001. *Under the Copper Dome: The Arizona State Capitol, 1898-1974*. Museum Division of the Arizona State Library, Archives and Public Records. Arizona Capitol Museum, Phoenix.

Carnegie Institution. 1937. "The vegetation of the desert. Exhibition Representing Results of Research Activities of the Carnegie Institution December 14, 15, and 16, 1935." Carnegie Institution, Washington, D.C., pp. 32–34.

Carroll, L. 1876. *The Hunting of the Snark—An Agony, in Eight Fits*. Macmillan and Co., London

Casa Grande Dispatch. 1934. "Arizona exhibit draws attention." *Casa Grande Dispatch*, June 8, 1934, p. 5.

Casaday, L.W. 1950. *The Role of the Tourist Industry in Arizona's Expanding Economy*. Bureau of Business Research, Tucson.

Castetter, E.F. and W.H. Bell. 1937. "The aboriginal utilization of the tall cacti in the American Southwest. Ethnobiological Studies in the American Southwest IV." *University of New Mexico Bulletin*. University of New Mexico Press, Albuquerque.

Castle, L. 1884. *Cactaceous Plants: Their History and Culture. With Numerous Illustrations.* Horticultural Press, London.

Chambers, G.W. n.d. [1955?] Papers. Typescript. "Sunshine Modeling Club." George W. Chambers papers, box 2, folder 97 Sunshine Club publications, MS 1079, Arizona Historical Society, Tucson.

Chicago Daily Tribune. 1940. "Touch me not." *Chicago Daily Tribune*, March 23, 1940, p. 2.

Clarke, A.B. 1852. *Travels in Mexico and California: Comprising a Journal of a Tour from Brazos Santiago, Through Central Mexico, by Way of Monterey, Chihuahua, the Country of the Apaches, and the River Gila, to the Mining Districts of California*. Wright and Hasty, Boston.

Clarke, D.L. 1966. "Soldiers under Stephen Watts Kearny." *California Historical Quarterly* 45:132–138.

Cocozza, P. 2017. "Looking sharp! How the cactus became the world's most wanted plant." *The Guardian* May 31, 2017. https://www.theguardian.com/lifeandstyle/2017/may/31/cactus-cool-cartier-designer-plants

Cohn, J. 2003. "A new view of saguaros." *BioScience* 53:123–216.

Cooper, E.S. 1989. "The Buehmans of Tucson: A famly tradition in Arizona photography." *Journal of Arizona History* 30:251–278.

Cooper, E.S. 1995. *The Buehman Studio Tucson in Focus*. Arizona Historical Society, Tucson.

Cooper, T.C. 1973. "Arizona Highways: From engineering pamphlet to prestige magazine." Master's thesis, University of Arizona, Tucson.

Courtney, D. 2018. "The Texanist: There ain't no saguaro cactus in Texas. Got it?" *Texas Monthly*, May 6, 2018. https://www.texasmonthly.com/being-texan/the-texanist-no-saguaro-in-texas/

Coville, F.V. 1903. Letter to R.H. Forbes, June 18, 1903. MacDougal Papers, box 3, folder 24, MS 0452, Arizona Historical Society, Tucson.

Coville, F.V. and D.T. MacDougal. 1903. *Desert Botanical Laboratory of the Carnegie Institution*. Carnegie Institution, Washington, D.C.

Craig, P. 2005. *Centennial History of the Carnegie Institution of Washington Vol. IV: The Department of Plant Biology*. Cambridge University Press, New York.

Crosswhite, F.S. 1980. "The annual saguaro harvest and crop cycle of the Papago, with reference to ecology and symbolism." *Desert Plants* 2:2–61. https://repository.arizona.edu/bitstream/handle/10150/550735/dp_02_01-002-061.pdf?sequence=1&isAllowed=y

d'Emilio, S. and S. Campbell. 1991. *Visions and Visionaries: The Art and Arts of the Santa Fe Railway*. Peregrine Smith Books, Salt Lake City.

Daniels, D. 1968. "Photography's wet-plate interlude in Arizona Territory: 1864–1880." *Journal of Arizona History* 9:171–194.

de Candolle, P.A. 1829. *Revue de la famille des cactees*. Chez A. Berlin, Paris.

Dehler, G.H. 2013. *The Most Defiant Devil: William Temple Hornaday and His Controversial Crusade to Save American Wildlife*. University of Virginia Press, Charlottesville.

Department of State Office of the Secretary of State. 2019. "History of the Great Seal of Arizona. Department of State Office of the Secretary of State, Phoenix." Accessed October 06, 2020. https://azsos.gov/sites/default/files/state_seal_history.pdf

Desert Magazine. 1960. "Moving a saguaro cactus." *Desert Magazine* November 23, 1960, pp. 5–7.

Desert Sun. 1936. "Chuck Abbott broils 3000 steaks in 28 states this season." [Palm Springs] *Desert Sun*, October 16, 1936, p. 8.

Detroit News Pictorial. 1947. "Dream house located with camera eye." *Detroit News Pictorial*, August 3, 1947, p. 2.

DeVillemarette, C. 2018. "Fabric of the West: The iconic textile designs of Harwood Steiger." In: D. Clinco (Ed.), *Tucson Modernism Week 2018*. Tucson Historic Preservation Foundation, Tucson, pp. 22–29.

Dimmit, M.A. 2000. "Cactaceae (cactus family)." In: S.J. Phillips and P.W. Comus (Eds.). *A Natural History of the Sonoran Desert*. Arizona-Sonora Desert Museum Press, Tucson, pp. 184–193.

Dimmock, T. 1890. "Henry Shaw: A biographical sketch." *Missouri Garden Annual Report* 1890:7–25.

Dittemore, D. 2013. "Mrs. Schrader's House of Baskets: A forgotten Tucson treasure." *Journal of Arizona History* 54:397–418.

Downum, C.E., P.R. Fish, and S.K. Fish. 2009. "Refining the role of cerros de trincheras in Southern Arizona settlement." *Kiva* 75:261–287.

Drachman, R. 1980. *This Is Not A Book. Just Memories*. R.P. Drachman, Tucson.

Drezner, T.D. 2014. "The keystone saguaro (Carnegiea gigantean, Cactaceae): A review of its ecology, associations, reproduction, limits, and demographics." *Plant Ecology* 215:581–595. http://dx.doi.org/10.1007/s11258-014-0326-y

Duncan, F. 1929. "From a nuisance to a cult; or, the rise of the cactus." *Los Angeles Times*, August 18, 1929, p. G2.

Dunkak, H.M. 2007. *Knowledge, Truth and Service: The New York Botanical Garden, 1891 to 1980*. University Press of America, Lanham.

Durbin, C.B. 1965. *History of Tucson Medical Center*. Tucson Medical Center, Tucson.

Emory, W.H. 1847. Letter to John Torrey, June 17, 1847. William H. Emory and John Torrey correspondence. Esther T. Merz Library, New York Botanical Garden, Bronx.

Emory, W.H. 1848a. Letter to George Engelmann, January 14, 1848. George Engelmann Papers, collection 1 RG4/1/5/1. Missouri Botanical Garden Archives, St. Louis.

Emory, W.H. 1848b. Letter to George Engelmann, September 13, 1848. George Engelmann Papers, collection 1 RG4/1/5/1. Missouri Botanical Garden Archives, St. Louis.

Emory, W.H. 1848c. *Notes on a Military Reconnoissance, from Fort Leavenworth, in Missouri, to San Diego, in California, Including Part of the Arkansas, Del Norte, and Gila Rivers*. Wendell and Van Benthuysen, Washington, D.C.

Emory, W.H. 1852. Letter to George Engelmann, April 20, 1852. George Engelmann Papers, collection 1 RG4/1/5/1. Missouri Botanical Garden Archives, St. Louis.

Emory, W.H. 1853. Letter to John Torrey, November 5, 1853. William H. Emory and John Torrey correspondence. Esther T. Merz Library, New York Botanical Garden, Bronx.

Emory, W.H. 1857. *Report on the United States and Mexican Boundary Survey*. Vol. 1. A.O.P. Nicholson, Washington, D.C.

Emory, W.H. 1859. *Report on the United States and Mexican Boundary Survey.* Vol. 2. A.O.P. Nicholson, Washington, D.C.

Engelmann, G. 1848. Letter to John Torrey, April 3, 1848. George Engelmann and John Torrey correspondence. Esther T. Merz Library, New York Botanical Garden, Bronx.

Engelmann, G. 1849. Letter to John Torrey, October 10, 1849. George Engelmann and John Torrey correspondence. Esther T. Merz Library, New York Botanical Garden, Bronx.

Engelmann, G. 1852. "Notes on the *Cereus giganteus* of South Eastern California and Some Other Californian Cactaceae." *American Journal of Science and Arts* 14:335–339.

Engelmann, G. 1854. "Further notes on *Cereus giganteus* of Southeastern California, with a short account of another allied species in Sonora." *American Journal of Science and Arts* 17:231–235.

Engelmann, G. 1859. "Cactaceae of the boundary." In: William H. Emory, *Report on the United States and Mexican Boundary Survey.* Vol. 2. A.O.P. Nicholson, Washington, D.C.

Engelmann, G. 1986. "Instructions for the collection and preservation of botanical specimens." *Annals of the Missouri Botanical Garden* 73:504–507.

Evans, S. 1981. "Henry Buehman, Tucson photographer 1874–1912." *History of Photography* 5:59–81.

Evening Star. 1893. "The cactus." [Washington] *Evening Star*, December 15, 1893, p. 15.

Evening Star. 1924. "Advertising is held best community aid." [Washington] *Evening Star*, May 5, 1924, p. 28.

Family Circle. 1947[?]. "Cactus cabinet maker." *Family Circle*, August 1947[?], Western Ways Features Manuscript and Photograph Collection, box 3, folder 34 Activity—Publications, MS 1255. Arizona Historical Society, Tucson.

Farish, T.E. 1889a. *Central and Southwestern Arizona: The Garden of America.* Arizona Daily Gazette, Phoenix.

Farish, T.E. 1889b. *Southeastern Arizona, Its Varied Climate and Wonderful Resources, by T.E. Farish, Commissioner of Immigration. Inducements to Immigrants.* Arizona Daily Gazette, Phoenix.

Farmer, J. 2013. *Trees in Paradise: A California History*. W.W. Norton, New York.

Felger, R.S. and M.B. Moser. 1974. "Columnar cacti in Seri Indian culture." *Kiva* 39:257–275.

Ferg, A. 1979. "The petraglyphs of Tumamoc Hill." *Kiva* 45:95–118.

Flick, A.J. 1998. "Mystery model." *Tucson Daily Citizen* May 29, 1998, p. 17.

Forbes, R.H. 1904. Letter to Daniel T. MacDougal, May 16, 1904, MacDougal Papers, box 4, folder 34, MS 0452. Arizona Historical Society, Tucson.

Francaviglia, R. 1994. "Elusive lands: Changing geographic images of the Southwest." In R.V. Francaviglia and D.E. Narrett (Eds.) *Essays on the Changing Images of the Southwest*. Texas A and M University Press, College Station, pp. 8–39.

Froebel, J. 1859. *Seven Years' Travel in Central America, Northern Mexico and the Far West of the United States*. R. Bentley, London.

Frontz, K. and D. Tackenburg. 2014. "Promoting Tucson: Rudolf Rasmessen's legacy of photographic postcards, 1905–1920." *Journal of Arizona History* 55:187–204.

Fuller, H. 1960. "Iron cactus JC's prize possession." *Arizona Republic*, December 18, 1960, p. 23-E.

Gardeners' Chronicle. 1901. "A cactus-house." *Gardeners' Chronicle,* March 9, 1901, p. 156.

Geographical Review. 1959. "Homer Leroy Shantz." *Geographical Review* 49:278–280.

Gray, A. 1854. *Plantae Novae Thurberianae: The Characters of Some New Genera and Species of Plants in a Collection Made by George Thurber, Esq., of the Late Mexican Boundary Commission Chiefly in New Mexico and Sonora*. Metcalf and Company, Cambridge.

Green, T. 2018. *Carlton Watkins: Making the West American*. University of California Press, Oakland.

Griffiths, D. and C.H. Thompson. 1929. "Cacti." *United States Department of Agriculture Circular* No. 66. Government Printing Office, Washington, D.C.

Grove, C. 2005. *Henry Shaw's Victorian Landscapes: The Missouri Botanical Garden and Tower Grove Park*. University of Massachusetts Press, Amherst.

Grubb, R.D. 1984. *Portrait of Progress: A Story of Tucson Medical Center*. Raim, Tucson.

Hall, G.H. n.d. [1940?] Typescript. "Picture window." Folder 215 Picture Windows. Western Ways Features Manuscript and Photograph Collection, box 15, MS 1056. Arizona Historical Society, Tucson.

Hardy, R.W.H. 1829. *Travels in the Interior of Mexico in 1825, 1826, 1827, and 1828*. Henry Colburn and Richard Bentley, London.

Harper's Weekly. 1887. "Arizona Garden—Cacti 13 ft. high." *Harper's Weekly* 31, April 16, 1887, p. 277.

Harper's Bazaar. 1894. "The Cactus Club of Baltimore." *Harper's Bazaar*, February 24, 1894, p. 147.

Harper's Weekly 1902. "Cactus-hunting in the Sonoran Desert." *Harper's Weekly* 46, May 10, 1902, p. 590.

Hartford Courant. 1940. "Indians ban swastika resentful of Nazis." *Hartford Courant*, February 26, 1940, p. 8.

Hartmann, G.H. and W.K. Hartmann. 1979. "Prehistoric trail systems and related features on the slopes of Tumamoc Hill." *Kiva* 45:39–69.

Hassrick, P.H. and M.N. Besaw (editors). 2015. *Painted Journeys: The Art of John Mix Stanley*. University of Oklahoma Press, Norman.

Hastings, J.R. and R.M. Turner. 1965. *The Changing Mile: An Ecological Study of Vegetation Change with Time in the Lower Mile of an Arid and Semiarid Region*. University of Arizona Press, Tucson.

Henderson, E. 1987. "Looking backward." *Journal of Arizona History* 28:409–430.

Henry, B. 1993. "Photographer captured the West." *Arizona Daily Star*, November 29, 1993, p. 43.

Herbert, C.W. n.d. [1940?]. Letter to Roy Drachman. Western Ways Features Manuscript and Photograph Collection, box 83, folder Tucson–Sunshine Club correspondence (8), MS 1056. Arizona Historical Society, Tucson.

Herbert, C.W. n.d. [1941?]. Typescript. "A view with a room." Western Ways Features Manuscript and Photograph Collection, box 15, folder 215 Picture Windows, MS 1056. Arizona Historical Society, Tucson.

Herbert, L.W. 1960. "Antibiotics for sick saguaros." *Westways* 52:22–23.

Hertrich, W. 1949. *The Huntington Botanical Gardens, 1905–1949: Personal Recollections of William Hertrich, Curator Emeritus*. Huntington Library, San Marino.

Hilpert, B. 1983. "Arizona goes to the fair: The World's Columbian Exposition of 1893." *Arizona and the West* 25:261–276.

Hine, R.V. 1968. *Bartlett's West: Drawing the Mexican Boundary*. Yale University Press, New Haven.

Hinton, R.J. 1878. *The Hand-Book to Arizona*. Upham and Company, San Francisco.

Hodge, H.C. 1877. *Arizona As It Is; or, the Coming Country*. Hurd and Houghton, New York.

Hoffman, C. 2019. "Coachella 2019: 6 design trends to steal for your next event." *BizBash* April 25, 2019. https://www.bizbash.com/style-decor/event-design/media-gallery/21065429/event-design-ideas-from-coachella-parties-and-activations

Holiday. 1947a. "Cactus products." *Holiday*, December 2, 1947, p. 55.

Holiday. 1947b. "Synthetic West." *Holiday*, December 2, 1947, pp. 58–59.

Holly, G. 2001. *How the West Was Worn*. Harry N. Abrams, New York.

Hondagneu-Sotelo, P. 2014. *Paradise Transplanted: Migration and the Making of California Gardens*. University of California Press, Oakland.

Hooker, W.J., 1847. *Kew Gardens, or a Popular Guide to the Royal Botanic Gardens of Kew*. Longman, Brown, Green, and Longmans, London.

Hooker, W.J. (editor). 1849. *Hooker's Journal of Botany and Kew Garden Miscellany*. Vol. 1. Rowe, Benham, and Reeve, London.

Hornaday, W.T. 1908. *Camp-Fires on Desert and Lava*. Charles Scribner's Sons, New York.

Howard, K.L. and D.F. Pardue. 1996. *Inventing the Southwest: The Fred Harvey Company and Native American Art*. Northland Publishing, Flagstaff.

Howard, K.L. and D.F. Pardue. 2016. *Over the Edge: Fred Harvey at the Grand Canyon and in the Great Southwest*. Rio Nuevo Publishers, Tucson.

Hughes, J. and R. Mottram. 2006. "The Darrah Collection." *Cactus World* 24: 65–69.

Humphries, R.R. 1974. *The Boojum and Its Home*. University of Arizona Press, Tucson.

Hungerford, E. 1923. "A Study in consistent railroad advertising." *Santa Fe Magazine* 17:43–49. Reprinted from *Printers' Ink*.

Hyde, A.F. 1990. *An American Vision: Far Western Landscape and National Culture, 1820–1920*. New York University Press, New York.

Illustrated London News. 1845. "The monster cactus (*Echinocactus Stainesii*)." *Illustrated London News*, January 4, 1845, p. 12.

Illustrated London News. 1846. "'The monster cactus' at the Royal Botanic Gardens, Kew." *Illustrated London News*, October 17, 1846, p. 246.

Illustrated London News. 1847. "New Cactus from Real Del Monte–(*Cereus Senilis*)." *Illustrated London News*, July 17, 1847, p. 48.

Ingram, K.C. 1927. "Salesmanship in print" for S.P. service. *Southern Pacific Bulletin* 15:15–16.

Ingram, K.C. 1929. "How advertising sells railroad 'goods.'" *Southern Pacific Bulletin* 17:5–7.

James, H.L. 1969. "History of the United States-Mexican boundary survey—1858–1955." In Cordoba, D.A. (Ed.), *The Border Region (Chihuahua, Mexico & USA)*. New Mexican Geological Society 20th Annual Fall Field Conference Guidebook. New Mexico Geological Society, Socorro. https://nmgs.nmt.edu/publications/guidebooks/downloads/20/20_p0040_p0055.pdf

Jaynes, A.B. 1903. Letter to Coville and MacDougal. February 17, 1903. MacDougal Papers, box 4, folder 28, MS 0452. Arizona Historical Society, Tucson.

Jenkins, J.L. 2016. *Celluloid Pueblo: Western Ways Films and the Invention of the Postwar Southwest*. University of Arizona Press, Tucson.

Journal of Horticulture and Home Farmer. 1907. "The Manchester cacti." *Journal of Horticulture and Home Farmer*, May 2, 1907, p. 357.

Kearney, T.H., R.H. Peebles, and Collaborators. 1942. *Flowering Plants and Ferns of Arizona*. United States Government Printing Office, Washington, D.C.

Keatley, V.B. 1949. Typescript. "Photographs must accompany articles." November 10, 1949. Western Ways Features Manuscript and Photograph Collection, box 12, folder Ray Manley, Manuscript 1056. Arizona Historical Society, Tucson.

Kelsey, Robin. 2007. *Archive Style: Photographs and Illustrations for U.S. Surveys, 1850–1890*. University of California Press, Berkeley.

Kimmelman, A.J. 1987. "Luring the tourist to Tucson: Civic promotion during the 1920s." *Journal of Arizona History* 28:135–154.

Klein, K.L. 1993. "Frontier products: Tourism, consumerism, and southwestern public lands, 1890–1990." *Pacific Historical Review* 62:39–71.

Koppelkamm, S. 1981. *Glasshouses and Winter Gardens of the Nineteenth Century*. Rizzoli, New York.

Kunze, R.E. 1909. *Price-List of Cactacea Collected and Cultivated by Dr. Richard Kunze*. H.H. McNeil, Phoenix.

Langdon, T.C. 1975. "Harold Bell Wright: Citizen of Tucson." *Journal of Arizona History* 16:77–98.

Lee, S.H. assisted by J. Petrie. 2019. *An Abundant Arboretum*. S.H., Superior.

Leininger, P. 1923. "The case for railroad publicity." *University Journal of Business* 1:43–440.

Life. 1941. "Strange are the uses of desert blooms." *Life* 10, April 7, 1941, p. 57.

Lightle, P.T. 1947. "Field observations on bacterial necrosis of the giant cactus." Master's thesis, University of Arizona, Tucson.

Lemmon, K. 1962. *The Covered Garden*. Museum Press, London.

Lockwood, F.C. 1932. *Pioneer Days in Arizona: From the Spanish Occupation to Statehood.* McMillan, New York.

Logan, M.F. 1995. *Fighting Sprawl and City Hall: Resistance to Urban Growth in the Southwest*. University of Arizona Press, Tucson.

Logan, M.F. 2006. *Desert Cities: The Environmental History of Phoenix and Tucson*. University of Pittsburgh Press, Pittsburgh.

Los Angeles Times. 1893. "Railroad affairs. The Southern Pacific importing mammoth Arizona cacti." *Los Angeles Times*, March 29, 1893, p. 8.

Los Angeles Times. 1931. "Water District data compiled." March 16, 1931, *Los Angeles Times*, p. 26.

Los Angeles Times. 1935. "Fair water show opens." *Los Angeles Times*, June 17, 1935, p. 5.

Los Angeles Times. 1940. "Indians bar swastika design as protest against Nazis." *Los Angeles Times*, February 26, 1940, p. 1.

Los Angeles Times. 2013. "Postcards from the West: Standing before the Lone Cypress." *Los Angeles Times*, May 19, 2013.

Luckingham, B. 1982. *The Urban Southwest: A Profile History of Albuquerque–El Paso–Phoenix–Tucson*. Texas Western Press, El Paso.

Luckingham, B. 1993. "The promotion of Phoenix." In G. Wesley Johnson (Ed.), *Phoenix in the Twentieth Century: Essays in Community History*. University of Oklahoma Press, Norman, pp. 84–91.

Lynch, D. 2014. "From test plots to large lots: The gardens of San Marino, California as natural and social laboratories." *UCLA Historical Journal* 25. https://escholarship.org/uc/item/7022f109

Lyons, G. 1969. "Huntington Desert Garden." Reprinted from the *Cactus and Succulent Journal* 41.

Lyons, G. 2007. *Desert Plants: A Curator's Introduction to the Huntington Desert Garden*. Huntington Library, San Marino.

MacDougal, D.T. 1902. "Report of Dr. D.T. MacDougal, First Assistant, on an expedition to Arizona and Sonora." *Journal of the New York Botanical Garden* 3–4:89–99.

MacDougal, D.T. 1903a. Letter to F.V. Coville. March 18, 1903. MacDougal papers, box 3, folder 24, MS 0452. Arizona Historical Society, Tucson.

MacDougal, D.T. 1903b. Letter to F.V. Coville. March 28, 1903. MacDougal papers, box 3, folder 24, MS 0452. Arizona Historical Society, Tucson.

MacDougal, D.T. 1903c. Letter to W.A. Cannon. September 25, 1903. MacDougal papers, box 3, folder 23, MS 0452. Arizona Historical Society, Tucson.

MacDougal, D.T. 1904. Letter to R.L. Forbes. May 10, 1904. MacDougal papers, box 4, folder 34, MS 0452. Arizona Historical Society, Tucson.

MacDougal, D.T. 1905a. "The suwarro." *Journal of the New York Botanical Garden* 6:131.

MacDougal, D.T. 1905b. "Suwarro or saguaro." *Journal of the New York Botanical Garden* 6:149–150.

MacDougal, D.T. 1905c. "The Suwarro, or tree cactus." *Journal of the New York Botanical Garden* 6:129–133.

MacDougal, D.T. 1925. "Laboratory for Plant Physiology." *Annual Report*. Carnegie Institution of Washington, Washington, D.C., pp. 17–22.

Maloney, C.J. 2012. *World's Fair Gardens: Shaping American Landscapes*. University of Virginia Press, Charlottesville.

Manifold. 2019. "Instagram at Coachella: Desert Chill." https://www.wearemanifold.com/work/instagram-at-coachella-desert-chill/

Manning, R. 1939. Cartoon detail. "Reg Manning's Big Parade. If Phoenix adopts cactus style lamp posts." *Arizona Republic*, January 29, 1939, p. 36.

Martin, D.D. 1960. *The Lamp in the Desert: The Story of the University of Arizona*. Tucson: University of Arizona Press.

Masse, W.B. 1979. "An intensive survey of prehistoric dry farming systems near Tumamoc Hill in Tucson, Arizona." *Kiva* 45:141–186.

Maulsby, F.R. and Tucson Arizona Chamber of Commerce. 1911[?]. *Tucson, Arizona*. Sunset Magazine Homeseekers' Bureau, San Francisco.

McAuliffe, J.R. 1996. "Saguaro cactus dynamics." In: W.L. Halvorson and G.E. Davis (Eds.), *Science and Ecosystem Management in the National Parks*. University of Arizona Press, Tucson, pp. 96–131.

McCormick, R.C. 1865. *Arizona: Its Resources and Prospects. A Letter to the Editor of the New York Tribune*. D. Van Nostrand, New York.

McGee, W.J. 1896. "Expedition to Seriland." *Science* 3:493–505.

McGinnies, W.G. 1981. *Discovering the Desert: Legacy of the Carnegie Desert Botanical Laboratory*. University of Arizona Press, Tucson.

McKelvey, S.D. 1955. *Botanical Exploration of the Trans-Mississippi West, 1790–1850*. Arnold Arboretum of Harvard University, Jamaica Plain.

McVaugh, R. 1956. *Edward Palmer: Plant Explorer of the American West*. University of Oklahoma Press, Norman.

Meikle, J.L. 2015. *Postcard America: Curt Teich and the Imagining of a Nation, 1931–1950*. University of Texas Press, Austin.

Metropolitan Museum of Art. 2020. Lithograph. Metropolitan Museum of Art, New York. https://www.metmuseum.org/about-the-met/curatorial-departments/drawings-and-prints/materials-and-techniques/printmaking

Mickulas, P. 2007. *Britton's Botanical Empire: The New York Botanical Garden and American Botany, 1888–1929*. New York Botanical Garden Press, New York.

Miller, B. 1988. "Oh, to dine in Saugus, Mass." April 6, 1988, *New York Times*, p. C1.

Miller, P. 1768. *The Gardners Dictionary*. Printed for the author, London.

Middleton, P.H. 1914. "A cactus farm." *American Homes and Gardens*, November 1, 1914, p. 11.

Minton, M. 2016. "Bergdorf Goodman unveils 2016 holiday windows." *Architectural Digest*, November 16, 2016. https://www.architecturaldigest.com/story/bergdorf-goodman-unveils-2016-holiday-windows

Mitich, Larry W. 1996. "A.A. Nichol: Student of Arizona Flora." *Haseltonia* 4:34–38. http://cactusandsucculentsociety.org/cssaarchives/Andrew%20Alexander%20Nichol2.pdf

Moran, J. 2016. "Trends with benefits: Why are cacti everywhere in fashion right now?" *Papermag*, March 17, 2016. https://www.papermag.com/cactus-1668820843.html

Mowry, S. 1863. *The Geography and Resources of Arizona and Sonora: An Address Before the American Geographical and Statistical Society. New York, February 3, 1859*. A. Roman and Company, San Francisco and New York.

Mowry, S. 1864. *Arizona and Sonora: The Geography, History and Resources of the Silver Region of North America*. Harper and Brothers, New York.

Munk, J.A. 1912. "A visit to Doctor Kunze." *California Eclectic Medical Journal* 5:35–37.

National Geographic. 1953. "Six bales of hay and 43 mattresses swaddle a fragile 12-ton saguaro on moving day." *National Geographic* 104:375.

National Park Service. 2016. "Saguaro National Forest." Accessed October 6, 2020. https://www.nps.gov/sagu/learn/historyculture/upload/Saguaro-Cactus-State-Forest-Brief.pdf

Negri, S. 2010. *Movers and Shakers and the Creation of Tucson Mountain Park.* Pima County Natural Resources, Parks and Recreation, Tucson.

New York Daily News. 1954. "Louise N. Grace." *New York Daily News,* February 11, 1954, p. 83.

New York Evangelist. 1873. "A magnificent flower garden." *New York Evangelist*, November 27, 1873, pp. 44, 48.

New York Times. 1921. "Bronx cacti end summer vacation." *New York Times*, September 25, 1921, p. 36.

New York Times. 1937. "11 Murals shown by Louise N. Grace." 1937. *New York Times*, December 22, 1937, p. 31.

New York Tribune. 1893. "Rare plants at Pittsburgh." *New York Tribune*, December 7, 1893, p. 5.

New York Tribune. 1894. "Personal." *New York Tribune*, January 15, 1894, p. 6.

New York Tribune. 1900. "The Prospect Park cactus collection." *New York Tribune*, September 23, 1900, p. 26.

Newton, L.E. 1961. "The first American book on cacti." *National Cactus and Succulent Journal* 16:85.

Newton, L.E. 1964. "An overlooked booklet on cacti." *National Cactus and Succulent Journal* 19:13.

Nudie, J.L. 2004. *Nudie: The Rodeo Tailor*. Gibbs-Smith, Salt Lake City.

Olmstead, F.L. 1891. Letter to Rudolph Ulrich, March 24, 1891. Olmsted Associates Records, A8: 210. In: D. Schuyler and G. Kaliss (Eds.), *The Papers of Frederick Law Olmstead. Vol. 9. The Last Great Projects, 1890–1895.* Johns Hopkins University Press, Baltimore, 2015, pp. 329–331.

Orsi, R.J. 2005. *Sunset Limited: The Southern Pacific Railroad and the Development of the American West, 1850–1930.* University of California Press, Berkeley.

Otero, L.R. 2010. *La Calle: Spatial Conflicts and Urban Renewal in a Southwest City.* University of Arizona Press, Tucson.

Palmquist, P.E. 1983. *Carleton Watkins: Photographer of the American West*. University of New Mexico Press for the Amon Carter Museum, Albuquerque.

Palmquist, P.E. 1987. "'It is hot as h---': Carleton E. Watkins' photographic excursion through Southern Arizona." *Journal of Arizona History* 28:353–390.

Parker, D. 1989. Mr. F.A. Walton. *British Cactus and Succulent Journal* 7:34–37.

Pattie, J.O. 1962 [1831]. *The Personal Narrative of James O. Pattie: The 1831 Edition, Unabridged. Introduction by William H. Goetzman.* J.B. Lippincott Company, Philadelphia and New York.

Perkins, D.W. 1885. *Practical Common Sense Guide Book through the World's Industrial and Cotton Centennial Exposition at New Orleans.* Lane S. Hart, Harrisburg.

Pfeiffer, L. 1843. *Abbildung and Beschreiburg Bluhender Cacteen.* Verlag von Theodor Fischer, Cassel.

Phoenix-Arizona Club. 1924a. Advertisement. "Play where winter never comes – Phoenix Arizona." *Outlook*, October 29, 1924, p. 343.

Phoenix-Arizona Club. 1924b. Advertisement. "See the Apache Trail America's most scenic highway – Phoenix Arizona." *Outlook*, November 24, 1924, p. 423.

Pittsburgh Post-Gazette. 1902. "Phipps offers a cactus house." *Pittsburgh Post-Gazette*, January 8, 1902, p. 3.

Pomeroy, E.S. 1957. *In Search of the Golden West: The Tourist in Western America.* Knopf, New York, pp. 19–20.

Pool, R.J. 1916. "On the behavior of an excised branch of the sahuaro." *The Plant World* 19:17–22.

Pope, M. 2020. "Survey expeditions into the American West." Sonoran Desert Florilegium Program. Tucson Cactus and Succulent Society, Tucson. https://art-botanical.org/historical_surveys.html

Pyle, E. 1942. "Photographer or cook, Esther Henderson finally married that 'dude ranch guy.'" [St. Petersburg, Florida] *Tampa Bay Times*, April 9, 1942, p. 26.

Rasmessen, R. n.d. [1910?] *Picturesque Tucson and Vicinity: Handcolored Views of the Land of Sunshine.* Rudolph Rasmessen, Tucson.

Raymond, M.W. 1915. *The Architecture and Landscape Gardening of the Exposition.* Paul Elder and Company, San Francisco.

Reveal, J.L. 2010. "Asa Gray and the botanical exploration of the American West." *Harvard Papers in Botany* 15:309–316.

Richards, E. and I. Bell (editors). 2005. Esther Henderson interview. Digital Library Collections, University of California Santa Cruz. http://digitalcollections.ucsc.edu/cdm/compoundobject/collection/p265101coll13/id/3338/show/3336

Riddell, R.M. n.d. [1960?]. Papers. Typescript with copy photographs. "Some helpful data on how to become a good Sunshine Club model." Box 1, folder 5, MS 1120. Arizona Historical Society, Tucson.

Robbins, L.H. 1935. "The odd cactus: Other lands want to put America's useful plant in their arid regions." *New York Times*, June 23, 1935, p. SM11.

Rogers, W.A. 1899. "Conquering a desert." *Harper's Weekly* 43:594, 609.

Rose, J.N. 1911–1916. "List of cacti which have died at greenhouse since January 1, 1911." Joseph Nelson Rose field notes, Department of Botany, Smithsonian Institution Archives, SIA Acc. 12-052, RU000221. Smithsonian Institution, Washington, D.C.

Rowe, J. 1997. *Photographers in Arizona 1850–1920: A History and Directory.* Carl Muntz Publishing, Nevada City.

Rowe, J. 2006. *Arizona Real Photo Postcards History and Portfolio.* Carl Muntz Publishing, Nevada City.

Rowe, J. 2010. "Have a camera, will travel: Arizona roadside images by Burton Frasher." *Journal of Arizona History* 51:337–366.

Rowley, G.D. 1966. "In retrospect." *National Cactus and Succulent Journal* 21:34–35.

Rowley G.D. 1997. *A History of Succulent Plants.* Strawberry Press, Mill Valley.

Rowley, G.D. 1995. "National Cactus Society centenary—Cactus stirrings in London 1895." *British Cactus and Succulent Journal* 13:42–47.

Rowley, R. 1984. "UA's cacti have survived despite 'progress.'" *Arizona Daily Star*, May 12, 1984, p. 35.

Runte, A. 1991. "Promoting the golden west: Advertising and the railroad." *California History* 70:62–75.

Ryan, R. 1991. "East-side union halls: Where craft workers met, 1887–1917." In: E. Fee, L. Shopes and L. Zeidman (Eds.), *The Baltimore Book: New Views of Local History.* Temple University Press, Temple, pp. 112–113.

Rydell, M. n.d. [1951?]. Typescript. "A cactus by any other name." January 1951, p. 8. Western Ways Features Manuscript and Photograph Collection, box 4, folder Cactus, MS 1056. Arizona Historical Society, Tucson.

San Bernadino County Sun. 1931. "Metropolitan Water District exhibit interests crowds attending Twenty-First National Classic." *San Bernadino County Sun*, February 27, 1931, p. 6.

San Bernardino Sun. 1940. "What next!" *San Bernardino Sun*, March 27, 1940, p. 2.

San Francisco Examiner. 1915. "S.P. to dedicate fair building Thursday." *San Francisco Examiner*, March 8, 1915, p. 9.

Sandweiss, M.A. 2006. "A stereoscopic view of the American West." *Princeton University Library Chronicle* 67:271–289.

Seavey, C.A. 1990. "Government graphics: The development of illustration in U.S. federal publications, 1817–1861." *Government Publications Review* 17:121–142.

Seelye, K.Q. 2013. "The sun sets on a symbol of western-themed dining." *New York Times*, October 20, 2013, p. 14.

Shantz, H.L. 1937. "The saguaro forest." *National Geographic* 71:515–532.

Shaw, E.A. 1982. "Augustus Fendler's collection list: New Mexico, 1846–1847." *Contributions from the Gray Herbarium of Harvard University* 212:1–70.

Shaw, E.A. 1986. "Changing botany in North America: 1835–1869." The role of George Engelmann. *Annals of the Missouri Botanical Garden* 73:508–519.

Shepstone, H.J. 1901. "The romance of the cactus." *London Magazine*, September 1901, pp. 22–23.

Sheridan, T.E. 1995. *Arizona: A History*. University of Arizona Press, Tucson.

Shirer, J. 1954. "Arizona's tourist trade in 1953." *Arizona Business and Economic Review* 3:1–6.

Shreeve, F. 1937. "Plants of the sand." *Carnegie Institution of Washington News Service Bulletin Staff Edition* 10:90.

Silver, D. 2016. "Meet the creative minds behind NYC's elaborate holiday windows." *Observer*, November 23, 2016. https://observer.com/2016/11/meet-the-creative-minds-behind-nycs-elaborate-holiday-windows/

Sisson, P. 2018. "Why cactuses and succulents are the perfect plants for this cultural moment." *Curbed*, April 4, 2018. https://www.curbed.com/2018/4/4/17199044/cactus-succulents-trend-interior-design-landcaping

Smith, S.L. 1995. "History of the Hotel Del Monte's Arizona Garden." Calhoun, Institutional Archive of the Naval Postgraduate School. Dudley Knox Library, Monterey.

Snead, J.E. 2002. "Lessons of the ages: Archaeology and the construction of cultural identity in the American Southwest." *Journal of the Southwest* 44:17–34.

Sonnichsen, C.L. 1987. *Tucson: The Life and Times of an American City*. University of Oklahoma Press, Norman.

Soule, O.H. 1970. "Dr. George Engelmann: The first man of cacti and a complete scientist." *Annals of the Missouri Botanical Garden* 57:135–144.

Southern Pacific Company Passenger Department. 1900. *The New Arizona: A Brief Review of Its Resources, Development, Industries, Soil, Climate and Especially Its Advantages for Homemaking*. Southern Pacific Company, San Francisco.

Southern Pacific Lines. 1928. Advertisement. "Into the land of vanished peoples." *Southern Pacific Bulletin* 16:back cover.

Spude, R. 1989. "Shadow catchers: A portrait of Arizona's pioneer photographers, 1863–1893." *Journal of Arizona History* 30:233–250.

St. Louis Post–Dispatch. 1914. "11-foot, 1,000-lb. cactus now at Shaw's Garden." *St. Louis Post–Dispatch*, July 26, 1914, p. 3B.

Starkweather, L.E. 1939. Letter to Atkinson, February 18, 1939. Alfred Atkinson, box 18, folder 9, AZ 426. University of Arizona Special Collections, Tucson.

Starkweather, L.E. 1945. Letter to Gov. Sidney P. Osborn, December 13, 1945. Alfred Atkinson, box 18, folder 9, AZ 426. University of Arizona Special Collections, Tucson.

Starkweather, L.E. 1946. Letter to President [Atkinson] and Board of Regents, February 4, 1946. Alfred Atkinson, box 18, folder 9, AZ 426. University of Arizona Special Collections, Tucson.

Starr, K. 1993. "Watering the Land." *Southern California Quarterly* 75:303–332.

Steenbergh, W.F. 1969. "Critical factors during the first years of life of the saguaro cactus (*Cereus giganteus*) at Saguaro National Monument, Arizona." *Ecology* 50:825–834.

Steenbergh, W.F. and C.H. Lowe. 1977. *Ecology of the Saguaro. Vol. 2.* National Park Service Monograph Series, No. 8. Department of the Interior, Washington, D.C. https://www.nps.gov/parkhistory/online_books/science/8/index.htm

Stern, J. and M. Stern. 1993. *Way Out West.* Harper Collins, New York.

Stieber, M.T. and C. Lange. 1986. "Augustus Fendler (1813–1883): Professional plant collector: selected correspondence with George Engelmann." *Annals of the Missouri Botanical Garden* 73:520–531.

Streets, R.B. and A.L. Finch. 1940. "Plan for development of cactus garden area." Memo from Campus Committee to Atkinson, October 3, 1940. Alfred Atkinson, box 18, folder 9, AZ 426. University of Arizona Special Collections, Tucson.

Sturgis, R. n.d. [1954?]. "Tucson's model high school." *Men: Your Goodyear Dealer's Magazine.* Western Ways Features Manuscript and Photograph Collection, box 19, folder 276, Tucson – Data, History, MS 1056. Arizona Historical Society, Tucson.

Styles, T.E. 2012. *Missouri Botanical Garden.* Arcadia Publishing, Charleston.

Sunshine Climate Club, Tucson, Arizona. 1922a. Advertisement. "Unlock the best that's in you." *New York Times*, November 22, 1922, p. 34.

Sunshine Climate Club, Tucson, Arizona. 1922b. Advertisement. "Spring days now! Come to Tucson." *Outlook*, November 22, 1922, p. 542.

Sunshine Climate Club, Tucson, Arizona. 1922c. Advertisement. "Are you paying the price of health—and not getting it?" *New York Times*, November 29, 1922, p. 34.

Sunshine Climate Club, Tucson, Arizona. 1922d. Advertisement. "Golf on New Year's Day." *St. Louis Post-Dispatch*, December 3, 1922, p. A30B.

Sunshine Climate Club, Tucson, Arizona. 1922e. Advertisement. "Must hearts break for want of sunshine?" *Literary Digest* 75, December 30, 1922, p. 44.

Sykes, G. 1944. *A Westerly Trend*. Arizona Pioneers Historical Society, Tucson.

Thomas, D. 1978. *The Southwestern Indian Detours: The Story of the Fred Harvey/Santa Fe Railway Experiment in "Detourism."* Hunter Publishing Company, Phoenix.

Thompson, G.A. 1815. *The Geographical and Historical Dictionary of America and the West Indies Containing an Entire Translation of the Spanish Work of Colonel Don Antonio de Alcedo*. Vol. 5. Carpenter and Son; Longman, Hurst, Rees, Orme, and Brown; White, Cochrane and Company; Murray, and Deighton, London, Oxford, and Cambridge.

Thornber, J.J. 1906. "The Toumey cactus garden." *The Plant World* 9:273–277.

Torre, D. 2017. *Cactus*. Reaktion Books, London.

Towne, D.C. 2000. "The mysteries of the wandering cactus unearthed: A monograph on the commercial use of the saguaro symbol." Accessed October 6, 2020. http://www.neonmotel.com/saguaro.html

Trelease, W. and A. Gray (editors). 1887. *The Botanical Works of the Late George Engelmann Collected for Henry Shaw*. John Wilson and Son, Cambridge.

Trendbible. 2019. "Why the cactus is the new pineapple." Accessed March 27, 2019. https://www.trendbible.com/why-the-cactus-is-the-new-pineapple/

Trennert, R.A. 1988. "A different perspective: Victorian travelers in Arizona, 1860–1900." *Journal of Arizona History* 29:349–370.

Truettner, W.H. (editor). 1991. *The West as America: Reinterpreting Images of the Frontier, 1820–1920*. Smithsonian Institution Press, Washington, D.C.

Tucson Botanical Gardens. 2020. "History." Accessed October 6, 2020. https://tucsonbotanical.org/history/

Tucson Chamber of Commerce. 1904. *Health Wealth Golden Opportunities*. Norman Pierce, San Francisco.

Tucson Chamber of Commerce. 1908. Advertisement. "Keep your eye on Tucson." *Plant World* 11:unpaginated.

Tucson Chamber of Commerce. 1911. "Tucson is the metropolis of Arizona!" *Plant World* 14:unpaginated.

Tucson Daily Citizen. 1911. "Praises Tucson's descriptive booklets." *Tucson Daily Citizen*, April 3, 1911, p. 4.

Tucson Daily Citizen. 1922. "Noted author." *Tucson Daily Citizen*, March 26, 1922. Harold Bell Wright biographical file, Arizona Historical Society, Tucson.

Tucson Daily Citizen. 1929. "Activity began with the purchase of land in that section by Harold Bell Wright, and his building of a palatial home." *Tucson Daily Citizen*, April 28, 1929, p. 2.

Tucson Daily Citizen. 1938. "May see last of Old Main." *Tucson Daily Citizen*, November 18, 1938, p. 1.

Tucson Daily Citizen. 1944. "Navy restores UA 'Old Main' to usefulness." *Tucson Daily Citizen*, December 25, 1944, p. 5.

Tucson Daily Citizen. 1947a. "Desert wood artist dies." *Tucson Daily Citizen*, April 21, 1947, p. 2.

Tucson Daily Citizen. 1947b. "25-year-old club starts campaign." *Tucson Daily Citizen*, September 19, 1947, p. 2.

Tucson Daily Citizen. 1947c. "Photocenter. Cowboy photographer Chuck Abbott, portrait artist Esther Henderson start elaborate Tucson photo studio." *Tucson Daily Citizen*, November 16, 1947, p. 14.

Tucson Daily Citizen. 1952. "Three 100-Year-Old U.A. saguaros destroyed." *Tucson Daily Citizen*, November 22, 1952, p. 1.

Tucson Daily Citizen. 1960. "C of C, Sunshine Club seek to function harmoniously." *Tucson Daily Citizen*, May 3, 1960, p. 22.

Tucson Historic Preservation Foundation. 2012. *Tucson Health Seekers: Design, Planning, and Architecture in Tucson for the Treatment of Tuberculosis. National Register of Historic Places Multiple Properties Documentation Form.* Tucson Historic Preservation Foundation, Tucson. https://preservetucson.org/stories/tucson-health-seekers-context-study/

Tucson Historic Preservation Foundation. 2020. *Desert Admen: The Forgotten Legacy of the Cabat-Gill Advertising Agency*. Tucson Historic Foundation, Tucson. https://preservetucson.org/stories/desert-ad-men/

Tullsen, H. 1902. "Is the giant cactus doomed?" *Plant World* 5:53.

Turner, R.M., R.H. Webb, J.E. Bowers, J.R. Hastings. 2003. *The Changing Mile Revisited: An Ecological Study of Vegetation Change with Time in the Lower Mile of an Arid and Semiarid Region*. Arizona University Press, Tucson.

Tyler, Ron. 1999. "Illustrated government publications related to the American West, 1843–1863." In: Edward C. Carter II (Ed.), *Surveying the Record: North American Scientific Exploration to 1930*. American Philosophical Society, Philadelphia, pp. 147–172.

Unger, H.F. 1964. "Traveling cactus." *Arizona Days and Ways Magazine. Arizona Republic*, November 22, 1964, p. 27.

United States Centennial Commission. 1876. *International Exhibition, 1876, Official Catalogue Complete in One Volume*. John R. Nagle and Company, Philadelphia.

United States National Museum. 1933. *Annual Report of the United States National Museum for the Year 1932*. Government Printing Office, Washington, D.C.

United States Senate. 1850. "Report of the Secretary of the Senate, showing the expenditures from the contingent fund of the Senate, during the year ending December 2, 1849."

University Campus Committee. 1939. "Cactus garden." Memo to President Atkinson, February 10, 1939. Alfred Atkinson, box 18, folder 9, AZ 426. University of Arizona Special Collections, Tucson.

University of Arizona. September 2008. "Tumamoc Hill Cultural Resources Policy and Management Plan." https://pdc.arizona.edu/file/TumamocManagementPlan_final.pdf

University of Arizona Library Special Collections. 1942. Typescript. Homer LeRoy Shantz. March 12, 1942.

University of Arizona Library Special Collections. 2020a. "Homer LeRoy Shantz." Accessed April 15, 2020. https://president.arizona.edu/person/homer-leroy-shantz

University of Arizona Library Special Collections. 2020b. "Atkinson, Alfred." Accessed April 15, 2020. https://president.arizona.edu/person/alfred-atkinson

VanderMeer, P.R. 2010. *Desert Visions and the Making of Phoenix, 1860–2009*. University of New Mexico Press, Albuquerque.

Vasey, G. 1884. "A catalogue of the forest trees of the United States which usually attain a height of sixteen feet or more. United States Centennial Commission." *Report of the Board on Behalf of the United States Executive Departments at the international Exhibition Hall at Philadelphia, Pa., 1876*. Vol. 2. Government Printing Office, Washington, D.C.

Venegas, M. 1759. *A Natural and Civil History of California*. 2 vols. James Rivington and James Fletcher, London.

Walton, F.A. 1899. "Adventures in search of wild cacti." *Wide World* 4:288–296.

Washington Post. 1909. "Name a cactus for Carnegie." *Washington Post*, January 31, 1909, p. SM4.

Washington Post. 1940. "Swastika purged on Indian blankets." *Washington Post*, February 27, 1940, p. X5.

Watkins, W.N. 1911. Memorandum to William Ravenel July 11, 1931. Accession files 115767 and 124450. National Museum of American History, Smithsonian Institution, Washington, D.C.

Watson, W. 1889. *Cactus Culture for Amateurs*. L. Lippincott Gill, London.

Webb, R.H., D.E. Boyer, R.M. Turner, S.H. Bullock. 2007. "The Desert Laboratory Repeat Photography Collection—An invaluable archive documenting landscape change." U.S. Geological Survey, Tucson. https://pubs.usgs.gov/fs/2007/3046/fs2007-3046.pdf

Weber, N.F. 1998. *Cleve Gray*. Harry N. Abrams, New York.

Weigle, M. 1992. "Exposition and mediation: Mary Colter, Erna Fergusson, and the Santa Fe/Fred Harvey popularization of the native Southwest, 1902–1940." *Frontiers: A Journal of Women Studies* 12:116–150.

Weinberg, L. 2019. "Instagram will bring its first branded house to Coachella." *Hollywood Reporter*, April 10, 2019. https://www.hollywoodreporter.com/news/instagram-will-bring-first-branded-house-coachella-1200961

Werne, J.R. 2007. *The Imaginary Line: A History of the United States and Mexican Boundary Survey, 1848–1857*. Christian University Press, Fort Worth.

Western Ways. n.d. [1939?] Typescript. "Useless cactus finds a place in milady's wardrobe." Western Ways Features Manuscript and Photograph Collection, box 4, folder Cactus Candy–Cactus Ballet–Cactus Craft, MS 1056. Arizona Historical Society, Tucson.

Western Ways. n.d. [1941?]. Typescript. "The people of the state of Arizona are faced with a problem." Western Ways Features Manuscript and Photograph Collection, box 4, folder Cactus Disease, MS 1056. Arizona Historical Society, Tucson.

Western Ways. n.d. [1942?]. Typescript. "Only in Arizona." Western Ways Features Manuscript and Photograph Collection, box 4, folder 58 Cactus Disease, MS 1056. Arizona Historical Society, Tucson.

Western Ways. 1949. Typescript. "So. Arizona guest ranches 1949 list of folders at Western Ways." Western Ways Features Manuscript and Photograph Collection, box 8, folder Dude Ranches, Miscellaneous Data, MS 1056. Arizona Historical Society, Tucson

Western Ways. n.d. [1950?]. Typescript. "Cactus gags." Western Ways Features Manuscript and Photograph Collection, box 4, folder Cactus Candy–Cactus Ballet–Cactus Craft, MS 1056. Arizona Historical Society, Tucson.

Western Ways. 1954. Clipping. "THS Girls, Sizes 10 to 14, Chosen as Sunshine Models." October 21, 1954. Western Ways Features Manuscript and Photograph Collection, box 19, folder 276 Tucson– Data–History, MS 1056. Arizona Historical Society, Tucson.

Western Ways. 1956. "Sickly saguaros." June 17, 1956. Western Ways Features Manuscript and Photograph Collection, box 4, folder Cactus Diseases, MS 1056. Arizona Historical Society, Tucson.

White, C.A. 1896. *Memoir of George Engelmann. 1809–1884. Read before the National Academy, April 1896.* Judd and Detweiler, Washington, D.C.

Wilcox, D.R. and S.M. Larson. 1979. "Introduction to the Tumamoc Hill survey." *Kiva* 45:1–14.

Wilder, B.T. and C. O'Meara. 2015. "The continuum of desert research." *Journal of the Southwest* 57:163–186.

Wilder, J.C. 1967. "The years of a desert laboratory." *Journal of Arizona History* 8:179–199.

Wilson, B. 1913[?], *Tucson, Arizona.* Sunset Magazine Homeseekers' Bureau, San Francisco.

Winship, G.P. 1896. *The Coronado Expedition, 1540–1542.* Government Printing Office, Washington, D.C.

Winsor, R.H. n.d. [1962?] Typescript. "A real picture window." Western Ways Features Manuscript and Photograph Collection, box 15, folder 215 Picture Windows, MS 1056. Arizona Historical Society, Tucson.

Wislizenus, G. 1848. *A Memoir of a Tour to Northern Mexico Connected with Col. Doniphan's Expedition in 1846 and 1847.* Tippin and Streeper, Washington, D.C.

Wright, H.B. 1924. "Why I did not die." *American Magazine* 97:13–15, 82, 84, 86, 88, 90.

Wright, H.B. 1930. Letter to John E. Harrison, Jr., December 22, 1930. Correspondence and Documents Relating to the Establishment of Saguaro National Monument, 1930–1937, AZ 170. University of Arizona Special Collections, Tucson.

Yeager, G. 1977. "Porfirian commercial propaganda: Mexico in the world industrial expositions." *The Americas* 34:230–243.

Year 'Round Club of Southern California. 1957. "Handle with care." *Boston Globe*, April 7, 1957, p. B7.

Yetman, D. 2007. *The Great Cacti: Ethnobotany and Biogeography.* University of Arizona Press, Tucson.

Yetman, D., A. Búrquez, K. Hultine, M. Sanderson. 2020. *The Saguaro Cactus: A Natural History.* University of Arizona Press, Tucson.

Young, J.V. 1968. "Saga of the saguaro, the giant of the desert." *Cactus and Succulent Journal of Great Britain* 30:63–64.

Youth's Companion. 1902. "Dying out." *Youth's Companion*, March 27, 1902, p. 161.

Zenga, M.E. 2001. "Advertising the Southwest." *Journal of the Southwest* 43:281–315.

Zimmerman, T. 1985. "Paradise promoted: Boosterism and the Los Angeles Chamber of Commerce." *California History* 64:22–33.

Library of Congress Cataloging-in-Publication Data
Names: Bird, William L., Jr., author.
Title: In the arms of saguaros : iconography of the giant cactus / William L. Bird Jr.
Description: Tucson : University of Arizona Press, 2023. | Includes bibliographical references.
Identifiers: LCCN 2023016714 (print) | LCCN 2023016715 (ebook) | ISBN 9780816552832 (paperback) | ISBN 9780816552849 (ebook)
Subjects: LCSH: Plants—Symbolic aspects. | Saguaro. | Plants in art. | Sonoran Desert—Symbolic representation.
Classification: LCC QH104.5.S58 B57 2023 (print) | LCC QH104.5.S58 (ebook) | DDC 508—dc23/eng/20230503
LC record available at https://lccn.loc.gov/2023016714
LC ebook record available at https://lccn.loc.gov/2023016715